Pradip Debnath, Hari Mohan Srivastava, Yeol Je Cho (Eds.)
Fixed Point Theory and Functional Analysis

Also of Interest

Nonlinear Functional Analysis and Applications
Jesús Garcia-Falset, Khalid Latrach, 2023
ISBN 978-3-11-103096-8, e-ISBN (PDF) 978-3-11-103181-1
in De Gruyter Series in Nonlinear Analysis and Applications
ISSN 0941-813X

Differential Equations, Fourier Series, and Hilbert Spaces
Lecture Notes at the University of Siena
Raffaele Chiappinelli, 2023
ISBN 978-3-11-129485-8, e-ISBN (PDF) 978-3-11-130252-2

General Topology and Analysis
Fixed Point Theorem and Continuous Linear Mappings
Yamina Miloudi, 2025
ISBN 978-3-11-914741-5, e-ISBN (PDF) 978-3-11-220663-8

Applied Nonlinear Functional Analysis
An Introduction
Nikolaos S. Papageorgiou, Patrick Winkert, 2024
ISBN 978-3-11-128421-7, e-ISBN (PDF) 978-3-11-128695-2

Algebraic Topology
Constructions, Retractions, and Fixed Point Theory
Smail Djebali, 2025
ISBN 978-3-11-151736-0, e-ISBN (PDF) 978-3-11-151738-4

Fixed Point Theory and Functional Analysis

Metric Spaces, Banach Spaces, Hilbert Spaces

Edited by
Pradip Debnath, Hari Mohan Srivastava, Yeol Je Cho

DE GRUYTER

Mathematics Subject Classification 2020
Primary: 47H10, 54H25; Secondary: 54E50

Editors
Pradip Debnath
Tezpur University
Department of Mathematical Sciences
Tezpur-784028
India

Prof. Dr. Hari Mohan Srivastava
University of Victoria
Dep. of Mathematics and Statistics
3800 Finnerty Road (Ring Road)
Victoria V8W 3R4
BC Canada
harimsri@math.uvic.ca

Yeol Je Cho
Gyeongsang National University
Department of Mathematics Education
Jinju-daero
Jinju-Si 52828
Republic of Korea

ISBN 978-3-11-914503-9
e-ISBN (PDF) 978-3-11-221576-0
e-ISBN (EPUB) 978-3-11-221593-7

Library of Congress Control Number: 2025943582

Bibliographic information published by the Deutsche Nationalbibliothek
The Deutsche Nationalbibliothek lists this publication in the Deutsche Nationalbibliografie; detailed bibliographic data are available on the Internet at http://dnb.dnb.de.

Cover image: wragg / E+ / Getty Images
Typesetting: VTeX UAB, Lithuania
Printing and binding: CPI books GmbH, Leck

www.degruyter.com
Questions about General Product Safety Regulation:
productsafety@degruyterbrill.com

Preface

Fixed-point theory and functional analysis are trending research topics in mathematical and physical sciences. The main sources of information on these topics are scattered among a variety of journals and proceedings. The book provides a simple interface for an amateur as well as an advanced learner in fixed-point theory and functional analysis. This book showcases the latest developments and indicates future research directions.

This book seeks to highlight the most recent developments in fixed-point theory and functional analysis by presenting insights from renowned scientists, physicists and mathematicians globally.

The book offers a comprehensive overview of the latest advancements in the field, featuring original contributions and surveys. Readers will find a wealth of useful tools and techniques to deepen their understanding of recent progress in mathematical and functional analysis and its applications in physics and engineering. Each chapter emphasizes new research avenues, making this book ideal for graduate students, faculty, and researchers looking to broaden their knowledge of fixed-point theory and functional analysis and their practical uses. The computational aspects in Banach and Hilbert spaces are well explored. Only basic knowledge in analysis, topology, basic computation and functional analysis is needed to fully appreciate the material.

This book consists of 17 chapters.

The first chapter of the book presents an interesting survey on the history of contraction mappings.

The 2nd chapter presents a study of fixed points of Hardy–Rogers and Ciric–Reich–Rus-type interpolative cyclic contractions.

The 3rd chapter discusses hybrid-type fixed-point results, whereas Chapter 4 presents refinements and reverses of Jensen tensorial inequality.

In the 5th chapter, some new summation formulas as modular relations are explained.

Chapter 6 provides some new geometric properties on Banach lattices.

In Chapter 7, applications of some nonlinear principles and fixed-point theorems for non-self contractions are given.

In Chapter 8, we present a new study on the order of convergence of a Traub-type method.

Chapter 9 focuses on some fixed-point results for expansive mappings.

Extended Kantorovich's results on Newton's method for solving generalized equations on Hilbert space are elaborated in Chapter 10.

Some new integral-type Berezin radius inequalities are discussed in Chapter 11.

Local convergence analysis of a hybrid Gauss–Newton method under a majorant condition is studied in Chapter 12.

Chapter 13 explores convergence region for a family of King–Traum method for solving nonlinear equations.

https://doi.org/10.1515/9783112215760-202

Chapter 14 presents an interesting study on new integral inequalities for Atangana–Baleanu fractional integrals.

Chapter 15 provides an analysis of fixed-point theory in the setting of non-Archimedean fuzzy b-metric spaces.

Fixed points of set-valued generalized rational graphic contractions are discussed in Chapter 16.

Finally, Chapter 17 explores a new generalization of Kannan's fixed-point theorem via simple F-contraction.

Pradip Debnath (India)
Hari Mohan Srivastava (Canada)
Yeol Je Cho (South Korea)

Contents

About the Editors

Pradip Debnath

Pradip Debnath is an Associate Professor in the Department of Mathematical Sciences at Tezpur University, India. He featured in the *World's Top 2 % Scientists* List prepared by Stanford University and published by Elsevier for the consecutive years 2023 and 2024. He is also served as a remunerative Associate Editor of the journal "Heliyon" (Mathematics section), published by Elsevier. He serves as Editorial Board member of several international journals such as Scientific Reports, Research in Mathematics, PLOS One, TWMS Journal of Applied and Engineering Mathematics, etc. He was previously an Assistant Professor (in Mathematics) at the Department of Applied Science and Humanities, Assam University, Silchar (a central university), India. Prior to that, he was working as an Assistant Professor in the Department of Mathematics at the North Eastern Regional Institute of Science and Technology (NERIST), India. He earned his Ph.D. in Mathematics from the National Institute of Technology Silchar, India. His research interests include fixed-point theory, functional analysis, soft computing and mathematical statistics. He has published more than 70 papers in various journals of international repute and is an active reviewer for more than 65 international journals. He has reviewed more than 450 papers for various journals. He is also a reviewer for *Mathematical Reviews* published by the American Mathematical Society. He has published 9 books as the lead editor with publishers like Springer, CRC Press and World Scientific. Dr. Debnath has worked in collaboration with leading mathematicians worldwide such as from countries like Canada, Serbia, Portugal, South Korea, Thailand, Saudi Arabia, Iran, Spain, Tunisia, etc. He is a topical advisory panel member of the journals *Axioms* and *Fractal and Fractional* and guest editor of several special issues for different journals such as Symmetry, Axioms and Contemporary Mathematics. He has successfully guided Ph.D. students in the areas of nonlinear analysis, soft computing and fixed-point theory. He has completed a major Basic Science Research Project in fixed-point theory funded by the UGC, the Government of India. Having been an academic gold medalist during his post-graduation studies from Assam University, Silchar, Dr. Debnath has qualified several national-level examinations in Mathematics in India.

Hari M. Srivastava

Hari M. Srivastava is Professor Emeritus at the Department of Mathematics and Statistics, University of Victoria, Canada. He has been a Clarivate Analytics (Thomson-Reuters) Highly Cited Researcher for the years 2015, 2017, 2018, and 2020. He has also been listed and ranked in sixth place in General Mathematics

https://doi.org/10.1515/9783112215760-204

among the Top 2 % Scientists in the World in 2021, 2022, 2023, 2024 (published by Stanford University in the US). In fact, it is not possible to list all his tremendous academic achievements in just a few sentences. The details of the same may be found on his regularly updated website www.math.uvic.ca/harimsri/.

Yeol Je Cho

Yeol Je Cho is Emeritus Professor at the Department of Mathematics Education, Gyeongsang National University, Jinju, Korea, and Distinguished Professor at the School of Mathematical Sciences, the University of Electronic Science and Technology of China, Chengdu, Sichuan, China. In 1984, he received his Ph.D. in Mathematics from Pusan National University, Pusan, Korea. He has been a fellow of the Korean Academy of Science and Technology, Seoul, Korea, since 2006, and a member of several mathematical societies. He has organized international conferences on nonlinear functional analysis and applications, fixed point theory and applications and workshops and symposiums on nonlinear analysis and applications. He has published over 400 papers, 20 monographs and 12 books with renowned publishers from around the world. His research areas are nonlinear analysis and applications, especially fixed-point theory and applications, some kinds of nonlinear problems, that is, equilibrium problems, variational inequality problems, saddle point problems, optimization problems, inequality theory and applications, stability of functional equations and applications. He has delivered several invited talks at international conferences on nonlinear analysis and applications and is on the editorial boards of 10 international journals of mathematics.

Sehie Park

1 From Banach to Rus–Hicks–Rhoades: A history of contraction nappings

In memories of Art Kirk, Kaz Goebel, and Wataru Takahashi

Abstract: There have appeared thousands of works about generalizations of the Banach contraction on metric spaces and their extended artificial spaces. In this survey article, we show that many known such results on metric spaces are consequences of our generalized forms of the Banach contraction principle for weak contractions or the Rus–Hicks–Rhoades (RHR) maps on quasi-metric spaces. We list a large number of examples of metric fixed point theorems which follow from our principles. Moreover, we add critical comments on many related papers in order to improve them.

Keywords: Banach contraction, Rus–Hicks–Rhoades contraction, Suzuki type maps, fixed point, quasi-metric

MSC 2020: 06A75, 47H10, 54E35, 54H25, 58E30, 65K10

1.1 Prologue

Let (X, q) be a quasi-metric space. A selfmap $f : X \to X$ is called a Banach contraction with a constant $\alpha \in [0, 1)$ if

$$q(fx, fy) \leq \alpha\, q(x, y) \quad \forall x, y \in X.$$

A selfmap $f : X \to X$ is called a weak contraction or a Rus–Hicks–Rhoades contraction (or simply an RHR map) with $\alpha \in [0, 1)$ whenever

$$q(fx, f^2x) \leq \alpha\, q(x, fx) \quad \forall x \in X.$$

Since 2023, we have been studying RHR maps and related topics in [47–52]. We obtained the generalized Banach contraction principle and the RHR contraction principle with their applications to scores of examples in the literature.

The present survey is to introduce the history and contents of such researches and to give examples of applications of the RHR contraction principle derived by a large number of authors. Consequently, this will enhance the reader's understanding of metric fixed-point theory.

Sehie Park, The National Academy of Sciences, Seoul 06579, Republic of Korea; and Department of Mathematical Sciences, Seoul National University, Seoul 08826, Republic of Korea, e-mail: park35@snu.ac.kr

https://doi.org/10.1515/9783112215760-001

This survey is organized as follows: Section 1.2 is to introduce Jachymski's 2023 theorem, which motivated our present research. In Section 1.3, we recall forerunners of the RHR maps. Section 1.4 concerns logical equivalents of the RHR principle based on our 2023 Metatheorem. In Section 1.5, we introduce the RHR contraction principle and its direct consequence, the generalized Banach contraction principle.

In Sections 1.6–1.8, we introduce examples of various forms of the RHR maps. Section 1.6 is for relatively earlier ones, Section 1.7 for maps derived by Suzuki and his colleagues, and Section 1.8 for relatively new ones. Finally, Section 1.9 is for the epilogue.

This survey would be a precious supplement of the history of metric fixed pint theory.

For the terminology related quasi-metric spaces, see [1, 49].

1.2 Jachymski's 2003 theorem

Let $\mathrm{Max}(\preceq)$ (resp. $\mathrm{Min}(\preceq)$) be the set of maximal (resp. minimal) elements of an ordered set $(X, \preceq)$, and $\mathrm{Fix}(f)$ (resp. $\mathrm{Per}(f)$) denotes the set of all fixed (resp. periodic) points of a map $f : X \to X$ for a nonempty set X.

In our previous articles, we introduced many examples of maps $f : X \to X$ satisfying $\mathrm{Per}(f) = \mathrm{Fix}(f) \neq \emptyset$. Such sets X can have more rich properties by applying the following main theorem of Jachymski [24, Theorem 2]:

Jachymski's 2003 theorem. *Let X be a nonempty abstract set and $T : X \to X$. The following statements are equivalent:*

(a) $\mathrm{Per}(T) = \mathrm{Fix}(T) \neq \emptyset$.
(b) (Zermelo) *There exists a partial ordering $\preccurlyeq$ such that every chain in $(X, \preccurlyeq)$ has a supremum and T is progressive with respect to $\preccurlyeq$ (that is, $x \preccurlyeq Tx$ for all $x \in X$).*
(c) (Caristi) *There exists a complete metric d and a lower semicontinuous function $\varphi : X \to \mathbb{R}^+$ such that T satisfies the Caristi condition $d(x, Tx) \leq \varphi(x) - \varphi(Tx)$ for all $x \in X$.*
(d) *There exists a complete metric d and a d-Lipschitzian function $\varphi : X \to \mathbb{R}^+$ such that T satisfies the Caristi condition and T is nonexpansive with respect to d; i. e.*

$$d(Tx, Ty) \leq d(x, y) \quad \textit{for all } x, y \in X.$$

(e) (Hicks–Rhoades) *For each $\alpha \in (0, 1)$, there exists a complete metric d such that T is nonexpansive with respect to d and*

$$d(Tx, T^2x) \leq \alpha d(x, Tx) \quad \textit{for all } x \in X.$$

(f) *There exists a complete metric d such that T is continuous with respect to d and for each $x \in X$, the sequence $(T^n x)_{n=1}^{\infty}$ is convergent (the limit may depend on x).*

(g) *There exists a partition of* X, $X = \bigcup_{\gamma\in\Gamma} X_\gamma$, *such that all the sets* X_γ *are nonempty, T-invariant and pairwise disjoint, and for all* $\gamma \in \Gamma$, $T|_{X_\gamma}$ *has a unique periodic point.*

(h) *For each* $a \in (0,1)$, *there exists a partition of* X, $X = \bigcup_{\gamma\in\Gamma} X_\gamma$, *and complete metrics* d_γ *on* X_γ *such that all the sets* X_γ *are nonempty; T-invariant and pairwise disjoint; and*

$$d_\gamma(Tx, Ty) \le a d_\gamma(x,y) \quad \textit{for all } x, y \in X.$$

Remark 1.1. Jachymski's 2003 theorem is a rich source of relatives of the Rus–Hicks–Rhoades theorem, and each of (b)–(h) seems to be order theoretic fixed-point theorems. For them, we state our own comments in [46] as follows:

(a) This could be $\mathrm{Max}(\preccurlyeq) \subset \mathrm{Fix}(T) = \mathrm{Per}(T) \ne \emptyset$ by defining $\preccurlyeq$ on X.

(b) Zermelo's theorem is improved in [46] and has many equivalents there. Note that its conclusion should be as above in (a).

(c) Caristi's theorem has been improved in [46] and its conclusion should be as in (a). Moreover, the lower semicontinuity there can be replaced by the one *from above*.

(d) This is a variant of Caristi's theorem and its conclusion should be as in (a).

(e) This is the origin of our study in the present article and (e) shows the importance of RHR maps in mathematics.

1.3 Forerunners of the RHR contractions

In 1977, Billy E. Rhodes in his historical article [58] listed the basic 25 contractive conditions. All of them imply the RHR maps except

$$(6),\ (10),\ (13),\ (22),\ (24),\ \text{and } (25).$$

Each of these implies as follows:
(6), (10), (22): $d(fx, f^2x) < d(x, fx)$;
(13): $d(fx, f^2x) < d(x, f^2x)$;
(24): Ćirić [13];
(25): $d(fx, f^2x) < \max\{d(x, fx), d(x, f^2x)\}, x \ne fx$.

Therefore, the other 19 conditions are applicable to our Theorem P for f-orbitally complete quasi-metric spaces. Most of such results appears in the present article.

Berinde [6] in 2003 mentioned the so called Banach orbital condition $d(Tx, T^2x) \le a\, d(x, Tx)$, for all $x \in X$, studied by various authors in the context of fixed point theorems, see for example Kasahara, Hicks and Rhoades, Ivanov, Rus and Taskovic.

Berinde–Pacurar [8] in 2022 defined a *graphic contraction (orbital contraction)* and give examples:

Banach contraction, Kannan mapping, Ćirić-Reich-Rus contraction, Bianchini mapping, Chatterjea mapping, Zamfirescu mapping, Ćirić quasi-contraction, Hardy and Rogers contraction, Berinde's almost contraction,

In 2023, Berinde, Petrusęl and I. A. Rus [9] listed previous names of the RHR maps as graphic contraction, iterative contraction, weakly contraction, Banach mapping, ...

When the present author studied metric fixed-point theory in 2023, he encountered the theorems of Rus [60] in 1973 and Hicks–Rhoades [22] in 1979 (from Jachymski [24, Section 2]). That is why he named the Rus–Hicks–Rhoades map in [47].

1.4 Equivalents of the Rus–Hicks–Rhoades theorem

In our previous work [47] in 2023, the Rus–Hicks–Rhoades theorem was shown to hold for metric spaces. Later, we found that it holds for quasi-metric spaces in [48]. In this section, we show its equivalent formulations.

Let (X,δ) be a quasi-metric space and $\mathrm{Cl}(X)$ denote the family of all nonempty closed subsets of X (not necessarily bounded). For $A, B \in \mathrm{Cl}(X)$, set

$$H(A,B) = \max\{\sup\{\delta(a,B) : a \in A\},\ \sup\{\delta(b,A) : b \in B\}\},$$

where $\delta(a,B) = \inf\{\delta(a,b) : b \in B\}$. Then H is called a generalized Hausdorff quasi-metric since it may have infinite values.

Based on our 2023 Metatheorem [45] and the RHR theorem, we obtained the following [49, 50]:

Theorem H. *Let (X,q) be a quasi-metric space, and $0 < \alpha < 1$. All maps $f : X \to X$ and multimaps $T : X \to \mathrm{Cl}(X)$ are supposed to be continuous.*

Then the following equivalent statements hold:

(0) *(X,q) is complete.*

(α) *For a multimap $T : X \to \mathrm{Cl}(X)$, there exists an element $v \in X$ such that $H(T(v),T(w)) > \alpha\,\delta(v,w)$ for any $w \in X \backslash \{v\}$.*

(β) *If $\mathfrak{F}$ is a family of maps $f : X \to X$ such that, for any $x \in X\backslash\{f(x)\}$, there exists a $y \in X\backslash\{x\}$ satisfying $q(f(x),f(y)) \le \alpha\, q(x,y)$, then $\mathfrak{F}$ has a common fixed element $v \in X$, that is, $v = f(v)$ for all $f \in \mathfrak{F}$.*

(γ) *If $\mathfrak{F}$ is a family of maps $f : X \to X$ satisfying $q(f(x),f^2(x)) \le \alpha\, q(x,f(x))$ for all $x \in X\backslash\{f(x)\}$, then $\mathfrak{F}$ has a common fixed element $v \in X$, that is, $v = f(v)$ for all $f \in \mathfrak{F}$.*

(δ) *Let $\mathfrak{F}$ be a family of multimaps $T : X \to \mathrm{Cl}(X)$ such that, for any $x \in X\backslash T(x)$, there exists $y \in X\backslash\{x\}$ satisfying $H(T(x),T(y)) \le \alpha\, q(x,y)$. Then $\mathfrak{F}$ has a common fixed element $v \in X$, that is, $v \in T(v)$ for all $T \in \mathfrak{F}$.*

(ϵ) *If $\mathfrak{F}$ is a family of multimaps $T : X \to \mathrm{Cl}(X)$ satisfying $H(T(x),T(y)) \le \alpha\, q(x,y)$ for all $x \in X$ and any $y \in T(x)\backslash\{x\}$, then $\mathfrak{F}$ has a common stationary element $v \in X$, that is, $\{v\} = T(v)$ for all $T \in \mathfrak{F}$.*

(η) *If Y is a subset of X such that for each $x \in X \setminus Y$ there exists a $z \in X \setminus \{x\}$ satisfying $H(T(x), T(z)) \leq \alpha\, q(x, z)$ for a $T : X \to \mathrm{Cl}(X)$, then there exists a $v \in X \cap Y = Y$.*

Remark 1.2. (1) Note that there are many characterizations of the metric completeness; see [51]. Theorem H covers some of them. It is well known that the Banach contraction does not characterize the metric completeness. However, the extended RHR principle does by Theorem H(γ).

(2) Note that Theorem H(β) properly extends the Banach principle, (γ) the RHR principle, and (δ), (ϵ) the Nadler and Covitz–Nadler theorems. Moreover, Theorem H gives unified short-cut proofs of such extensions.

1.5 Basic principles in ordered fixed-point theory

In this section, we introduce some of our results in ordered fixed-point theory. The following in Park [47–49, 52] is called the RHR contraction principle:

Theorem P. *Let (X, q) be a quasi-metric space and let $T : X \to X$ be an RHR map; that is,*

$$q(T(x), T^2(x)) \leq \alpha\, q(x, T(x)) \quad \textit{for every } x \in X,$$

where $0 < \alpha < 1$.

(i) *If X is T-orbitally complete, then, for each $x \in X$, there exists a point $x_0 \in X$ such that*

$$\lim_{n\to\infty} T^n(x) = x_0$$

and

$$q(T^n(x), x_0) \leq \frac{\alpha^n}{1-\alpha} q(x, T(x)), \quad n = 1, 2, \ldots,$$
$$q(T^n(x), x_0) \leq \frac{\alpha}{1-\alpha} q(T^{n-1}(x), T^n(x)), \quad n = 1, 2, \ldots.$$

(ii) *x_0 is a fixed point of T, and, equivalently,*

(iii) *$T : X \to X$ is orbitally continuous at $x_0 \in X$.*

Example 1.1. For any quasi-metric space (X, q), let $T = 1_X$ be the identity map. Then, Theorem P holds for 1_X.

Example 1.2. Let

$$X := \{-1\} \cup \{0\} \cup \left\{\frac{1}{n} : n = 1, 2, \ldots, 100\right\}.$$

Let $q : X \times X \to \mathbb{R}$ be the ordinary metric except $q(-1,0) = 1$ and $q(0,-1) = 0$. Let $T : X \to X$ be a map such that

$$T(-1) = 0, T(0) = 0, \quad \text{and} \quad T\left(\frac{1}{n}\right) = \frac{1}{n+1}.$$

Then we can check

$$q(Tx, T^2x) \leq \alpha\, q(x, Tx) \quad \text{with } \alpha = \frac{100}{102}.$$

Therefore, Theorem P works.

The following form of the RHR theorem is a consequence of Theorems P and H and useful in practice, see [49, 50].

Theorem H(γ1). *Let (X,q) be a quasi-metric space, and $0 < \alpha < 1$. If $f : X \to X$ is a continuous map satisfying*

$$q(f(x), f^2(x)) \leq \alpha\, q(x, f(x)) \quad \textit{for all } x \in X \backslash \{f(x)\},$$

and X is f-orbitally complete, then f has a fixed element $v \in X$, that is, $v = f(v)$.

Moreover, the following consequence of Theorem P in Park [47–49, 52] is the generalized Banach contraction principle:

Theorem Q. *Let (X,q) be a quasi-metric space and let $T : X \to X$ be a generalized Banach contraction, that is, for each $x \in X$, there exists a $y \in X\backslash\{x\}$ such that*

$$q(T(x), T(y)) \leq \alpha\, q(x,y) \quad \textit{where } 0 < \alpha < 1.$$

(i) *If X is T-orbitally complete, then, for each $x \in X$, there exists a point $x_0 \in X$ such that*

$$\lim_{n\to\infty} T^n(x) = x_0$$

and

$$q(T^n(x), x_0) \leq \frac{\alpha^n}{1-\alpha} q(x, T(x)), \quad n = 1, 2, \ldots,$$
$$q(T^n(x), x_0) \leq \frac{\alpha}{1-\alpha} q(T^{n-1}(x), T^n(x)), \quad n = 1, 2, \ldots.$$

(ii) *x_0 is the unique fixed point of T (equivalently, $T : X \to X$ is orbitally continuous at $x_0 \in X$).*

Example 1.3. In Example 1.2, for each $x \in X$, there exists a Tx.

The traditional Banach contraction principle appeared in thousands of publications should be corrected as Theorem Q.

From now on, the numbers attached to theorems, corollaries, remarks, etc. in this paper are possibly the same one in the original sources.

1.6 Early examples of the RHR maps

In this section, we introduce theorems of Rus and Hicks–Rhoades and their early extensions or relatives not mentioned in Section 1.3. From now on, the RHR maps can be applied Theorems P and H(γ1) on quasi-metric spaces.

We note that many already known results were given under some redundant assumptions. In this paper, the present author's opinions are expressed in *Comments*.

Rus [60] in 1973

The origin of our story begins from Rus:

Theorem. *Let f be a continuous selfmap of a complete metric space (X, d) satisfying*

$$d(fx, f^2x) \leq \alpha\, d(x, fx) \quad \textit{for every } x \in X,$$

where $0 < \alpha < 1$. Then f has a fixed point.

Comments

This is the origin of our Theorems P and Q.

Ćirić [12] in 1974

Theorem 1.1. *Let (M, d) be a metric space, and let $T : M \to M$ be a given mapping. Suppose that the following conditions are satisfied:*

(i) *T is orbitally continuous on M;*

(ii) *(M, d) is T-orbitally complete;*

(iii) *There exists a constant $a \in (0, 1)$ such that*

$$\min\{d(Tx, Ty), d(x, Tx), d(y, Ty)\} - \min\{d(x, Ty), d(y, Tx)\} \leq a\, d(x, y),$$

for all $x, y \in M$.

Then, for every $x \in M$, the Picard sequence $\{T^n x\}$ converges to a fixed point of T.

Comments

Among many of the works of Ćirić, we choose only two of them.

For $y = Tx$, we can consider the following:

$$\min\{d(Tx, T^2x), d(x, Tx), d(Tx, T^2x)\} - \min\{d(x, T^2x), d(Tx, Tx)\} \le \alpha\, d(x, Tx),$$

for all $x, y = Tx \in M$.
Case 1: If $d(Tx, T^2x) \le d(x, Tx)$, then $d(Tx, T^2x) \le \alpha\, d(x, Tx)$ and T is an RHR map.
Case 2: If $d(Tx, T^2x) \ge d(x, Tx)$, then $d(x, Tx) \le \alpha\, d(x, Tx)$ implies $x = Tx$.

In any case, Theorem H(γ1) works.

Up to this point, the contractive definitions slowly become more general, most of the corresponding fixed point theorems are RHR type and the proofs of them are basically the same except only a few of them.

Ćirić [13] in 1974

Ćirić proved a celebrated fixed-point theorem as follows:

Theorem 1. *Let T be a self map of a complete metric space (X, d). If there is a constant $\alpha \in (0, 1)$ such that*

$$d(Tx, Ty) \le \alpha \max\{d(x,y), d(x, Tx), d(y, Ty), d(x, Ty), d(y, Tx)\},$$

for all $x, y \in X$, then T has a unique point $z \in X$ and $d(z, T^n x_0) \to 0$ as $n \to \infty$, for all $x_0 \in X$.

Comment

This is an example of non RHR map. However, T is an RHR map whenever $0 < \alpha < 1/2$.

Dass and Gupta [16] in 1975

Theorem 1. *Let (X, d) be a complete metric space and $T : X \to X$ be a mapping. If there exist $k_1, k_2 \in [0, 1)$, with $k_1 + k_2 < 1$ such that*

$$d(Tx, Ty) \le k_1 \cdot d(y, Ty) \frac{1 + d(x, Tx)}{1 + d(x,y)} + k_2 \cdot d(x,y),$$

for all $x, y \in X$, then T has a unique fixed point $u \in X$ and the sequence $\{T^n x\}$ converges to the fixed point u for all $x \in X$.

Comments

Note that T is an RHR map and hence Theorem P can be applied. The uniqueness follows from the contractive condition.

Jaggi [25] in 1977

Theorem 2. *Let (X,d) be a complete metric space and $T : X \to X$ be a continuous mapping. If there exist $k_1, k_2 \in [0,1)$, with $k_1 + k_2 < 1$ such that*

$$d(Tx,Ty) \le k_1 \cdot \frac{d(x,Tx)}{d(y,Ty)} d(x,y) + k_2 \cdot d(x,y)$$

for all distinct $x, y \in X$, then T possesses a unique fixed point in X.

Comments

For $y = Tx$, we have the following

$$d(Tx,T^2x) \le k_1 \cdot \frac{d(x,Tx)}{d(Tx,T^2x)} d(x,Tx) + k_2 \cdot d(x,Tx),$$

which implies $d(Tx,T^2x) \le (k_1 + k_2)d(x,Tx)$. Hence T is an RHR map and Theorem P works.

Aubin [5] in 1979

Theorem 5. *Let V be a complete metric space and $f : V \to V$ be a map such that there exists an $L \in [0,1)$ satisfying*

$$d(fx,f^2x) \le L\, d(x,fx) \quad \forall x \in V.$$

If $F(x) = d(x,fx)$ on V is l. s. c., then

(1) *$\lim f^n x = p$ exists for all $x \in V$,*

$$d(f^n x,p) \le \frac{L^n}{1-L} d(x,fx),$$

and p is a fixed point of f, and

(2) *for any $u \in V$ and $\varepsilon > 0$ satisfying*

$$F(u) \le (1-L)\varepsilon,$$

f has a fixed point in $\overline{B}(u,\varepsilon)$. Further, if f is a quasi-Lipshitzian with constant k, then either u is a fixed point of f or f has a fixed point in $\overline{B}(u,\varepsilon)\backslash B(u,s)$, where $s = F(u)(1+k)^{-1}$.

Comment
This could be our first encounter to RHR maps.

Hicks and Rhoades [22] in 1979

Theorem (Hicks–Rhoades). *Let (X,d) be a complete metric space, $g : X \to X$ and $0 \leq h < 1$. Suppose there exists an x such that*

$$d(gy, g^2y) \leq h\, d(y, gy) \quad \text{for every } y \in \{x, gx, g^2x, \dots\}.$$

Then,
(i) $\lim_n g^n x = q$ *exists;*
(ii) $d(g^n x, q) \leq \frac{h^n}{1-h} d(x, gx)$;
(iii) *q is a fixed point of g if and only if $G(x) = d(x, gx)$ is g-orbitally lower semi-continuous at q.*

Comment
This is the origin of our Theorem P.

Park and Rhoades [53] in 1980

Let f be a selfmap of a topological space X. The orbit $O(x)$ of $x \in X$ under f is defined by $O(x) = \{x, f(x), f^2(x), \dots\}$. A function $G : X \to [0, \infty)$ is said to be f-orbitally lower semicontinuous at a point $p \in X$ if, for every $x_0 \in X$, $x_{n_k} \to p$ implies $G(p) \leq \liminf_k G(x_{n_k})$, where $\{x_{n_k}\}_{k=1}^{\infty}$ is a subsequence of $\{x_n\}_{n=1}^{\infty}$, which is defined by $x_{n+1} = f(x_n)$, i. e., $\{x_n\}_{n=1}^{\infty} = O(x_0)$.

Theorem 1. *Let d be a nonnegative real-valued function defined on $X \times X$ such that $d(x,y) = d(y,x)$ and $d(x,y) = 0$ iff $x = y$. If there exists a point $u \in X$ such that $\lim_n d(f^{n+1}(u), f^n(u)) = 0$, and if $\{f^n(u)\}$ has a convergent subsequence with limit $p \in X$, then p is a fixed point of f iff $G(x) = d(x, f(x))$ is f-orbitally lower semicontinuous at p.*

Theorem 2. *Let f be a selfmap of a metric space (X,d) satisfying:*
(1) *$\delta(O(x)) < \infty$ for each $x \in X$, where δ denotes the diameter.*
(2) *There exists a $u \in X$ such that $O(u)$ has a cluster point $p \in X$.*
(3) *There exists a function $\varphi : [0, \infty) \to [0, \infty)$ which is nondecreasing, continuous from the right and satisfies $\phi(t) < t$ for each $t > 0$ and the inequality*

$$d(f(x), f^2(x)) \leq \varphi(\delta(O(x) \cup O(f(y)))) \quad \text{for each } x, y \in X.$$

Then p is the unique fixed point of f and $\lim_n f^n(u) = p$.

These results extend works of Pal-Maiti, Park, Hegedüs and Daneš. A 2-metric space version of Theorem 2 is added.

Comment
In the present paper, we add new information related to the above theorems.

Park [44] in 1980

We gave a Banach-type fixed-point theorem with respect to contraction pairs of selfmaps on complete metric spaces:

Theorem 3.1. *Let g and h be selfmaps of a metric space X. If there exists a sequence $\{u_i : i \in \omega\}$ in X such that*

$$u_{2i+1} = gu_{2i}, \quad u_{2i+2} = hu_{2i+1} \quad \text{for } i \in \omega$$

and $\overline{\{u_i\}}$ is complete, and if there exists a $\lambda \in [0,1)$ such that

$$(*) \quad d(gx, hy) \leq \lambda\, d(x, y)$$

holds for any distinct $x, y \in \overline{\{u_i\}}$ satisfying either $x = hy$ or $y = gx$, then either
(1) *g or h has a fixed point in $\{u_i\}$, or*
(2) *$\{u_i\}$ converges to some $\xi \in X$, and*

$$d(u_i, \xi) \leq \frac{\lambda^i}{1-\lambda} d(u_0, u_i) \quad \text{for } i > 0.$$

Further, if one of g or h is continuous at ξ and $(*)$ holds for any distinct $x, y \in \overline{\{u_i\}}$, then ξ is a common fixed point of g and h.

Comments
This seems to be artificial, but this was made to include as many as existing results of similar nature. In fact, by putting $h = g$ and $y = gx$, we have $d(gx, g^2x) \leq \lambda\, d(x, gx)$. Hence Theorem 3.1 extends the RHR type theorem.

Berinde [6] in 2003

The class of weak contractive type mappings, introduced and studied in Berinde [7] in 2004 is compared to some other well-known contractive-type mappings in Rhoades' classification [58].

Theorem B is a version of Theorem Q of our present paper for complete metric space. Despite these important features, Theorem B suffers from one drawback—the Banach contraction forces T be continuous on X.

Berinde noted that his weak contractions "implies the so called Banach orbital condition $d(Tx, T^2x) \leq a\, d(x, Tx)$, for all $x \in X$, studied by various authors in the context of fixed point theorems, see for example Kasahara, Hicks–Rhoades, Ivanov, Rus, and Taskovic."

Comments

Until 2022, the present author was not aware of such works.

Berinde [7] in 2004

Definition 1. Let (X, d) be a metric space. A map $T : X \to X$ is called a *weak contraction* or (δ, L)-*contraction* if there exist a constant $\delta \in (0,1)$ and some $L \leq 0$ such that

$$d(Tx, Ty) \leq \delta \cdot d(x,y) + L\, d(y, Tx), \quad \text{for all } x,y \in X.$$

Remark 1. Due to the symmetry of the distance, the weak contraction condition implicitly includes the following dual one

$$d(Tx, Ty) \leq \delta \cdot d(x,y) + L \cdot d(x, Ty), \quad \text{for all } x,y \in X.$$

Comments

For $y = Tx$, a weak contraction is an RHR map.

Ran and Reurings [58] in 2004

In this paper, the following fixed-point theorem in a partially ordered metric space is proved:

Theorem 1.2 (Ran–Reurings). *Let $(M, \leq)$ be an ordered set and d be a metric on M such that (M, d) is a complete metric space. Let $U : M \to M$ be a nondecreasing mapping, i. e., $Ux \leq Uy$, for every $x,y \in M$ with $x \leq y$. Suppose that there exists $x_0 \in M$ with $x_0 \leq Ux_0$ and $L \in [0,1)$ such that*

$$d(Ux, Uy) \leq L\, d(x,y), \quad \textit{for every } x,y \in M \textit{ with } x \leq y.$$

If U is continuous, then it has a fixed point in M.

Comments

For $y = Tx$, U is an RHR map on (M, d) and Theorem 1.2 follows from Theorem H(γ1). The order is not necessary.

Nieto and Rodriguez-López [42] in 2005

Theorem. *Let $(X, d, \preccurlyeq)$ be a partially ordered complete metric spaces. If $f : X \to X$ is a continuous and monotone function such that there exist $k \in [0,1[$ and $x_0 \in X$ with $x_0 \preccurlyeq f(x_0)$ and*

$$d(f(x), f(y)) \leq k\, d(x, y)$$

for all $x, y \in X$ such that $y \preccurlyeq x$, then f has a fixed point.

Comments

Note that f is an RHR map and hence, theorem is a consequence of Theorem H(γ1). The partial order is not necessary.

Chen and Li [11] in 2007

Definition 2.1. The ordered pair (f, g) of two self-maps f and g of a metric space X is called a *Banach operator pair*, if the set $F(g)$ of fixed points of g is f-invariant, namely $f(F(g)) \subset F(g)$.

If the self-maps f and g of X satisfy

$$d(gf(x), f(x)) \leq k\, d(g(x), x), \quad \text{for all } x \in X$$

for some $k \geq 0$, $gf(x) = f(x)$ whenever $x \in F(g)$, i. e., $f(x) \in F(g)$, and hence (f, g) is a Banach operator pair. In particular, when $f = g$ and X is a normed space, it can be rewritten as

$$\|f^2(x) - f(x)\| \leq k\|f(x) - x\|, \quad \text{for all } x \in X;$$

such f is called a Banach operator of type k in the literature. This is the origin of the name “Banach operator pair.”

Comments

Here, we have an example of RHR maps.

Pathak and Shahzad [54] in 2008

Definition. Let (X, d) be a metric space. A mapping $T : Y \subset X \to X$ is called a *metric p-contraction* (or simply *p-contraction*) mapping if Y is T-invariant and it satisfies the following inequality:

$$d(Tx, T^2x) \le p(x)d(x, Tx) \text{ for all } x \in Y,$$

where $p : Y \to [0,1]$ is a function such that $p(x) < 1$ for all $x \in Y$ and $\sup_{x\in Y} p(Tx) = \alpha < 1$.

Further, if $\bigcap_{n=0}^{\infty} T^n(Y)$ is a singleton set, where $T^n(Y) = T(T^{n-1}(Y))$ for each $n \in \mathbb{N}$ and $T^0(Y) = Y$, then T is said to be a *strong p-contraction*.

Remark. 1) If $p(x) \le 1$ for all $x \in Y$, then the p-contraction map is called a *fundamental contraction*, which is also known as a Banach operator.
2) If $Y = X$ and $y = Tx$, then a Banach contraction map is a fundamental contraction.
3) If $p(x) \le 1$ for all $x \in Y$ and $\sup_{x\in Y} p(Tx) = 1$, then the p-contraction map is said to be fundamentally p-non-expansive. In particular, when $p(x) = 1$ for all $x \in Y$, then the fundamentally p-non-expansive mapping is said to be *fundamentally non-expansive*.
4) If $\sup_{x\in Y} p(x) < 1$ (or ≤ 1), then $\sup_{x\in Y} p(Tx) < 1$ (or ≤ 1) since $T(Y) \in Y$.

Comments

Here, we can find certain RHR maps. In fact, the authors gave various examples of RHR-type maps in spaces more general than metric spaces.

Kamran [26] in 2009

From Abstract: Mizoguchi–Takahashi's fixed-point theorem is a real generalization of Nadler's fixed point theorem. We refine/generalize Mizoguchi–Takahashi's fixed point theorem. Our result improves a recent result by Klim and Wadowski [2007] and extends Hicks and Rhoades [1979] fixed-point theorem to multivalued maps.

Comments

Our Theorem H has similar things and gives unified proofs of generalizations of theorems of Banach, Rus, Berinde, Hicks–Rhoades, Nadler, Covitz–Nadler, and others.

1.7 Suzuki types of the RHR maps

From 2008, Suzuki found fixed point theorems for certain RHR type maps with very sophisticated proofs. He and his colleagues continued to publish more papers within a few years. Their papers are very popular and many followers published scores of papers of the same nature. However, we can give very short proofs of some of their works; see [50].

Suzuki [65] in 2001

For metric spaces, there have appeared thousands of generalizations, modifications, imitations, extensions, and their applications. One of them is the following:

Theorem 1. *Let X be a complete metric space and let T be a mapping from X into itself. Suppose that there exist a τ-distance p on X and $r \in [0,1)$ such that $p(Tx, T^2x) \leq r \cdot p(x, Tx)$ for all $x \in X$. Assume that either of the following holds:*

(i) *If $\lim_n \sup\{p(x_n, x_m) : m > n\} = 0$, and $\lim_n p(x_n, Tx_n) = 0$, and $\lim_n p(x_n, y) = 0$, then $Ty = y$;*
(ii) *If $\{x_n\}$ and $\{Tx_n\}$ converge to y, then $Ty = y$;*
(iii) *T is continuous.*

Then there exists $x_0 \in X$ such that $Tx_0 = x_0$ and $p(x_0, x_0) = x_0$.

Comments

For $y = Tx$, a weak contraction is an RHR map. Theorems of Rus and Hicks–Rhoades are particular case of (iii).

There may be similar theorems for thousands of artificial metric type spaces. For quasi-metric spaces, either of (i) or (ii) means $\{x_n\}$ is Cauchy and T is orbitally continuous. Theorem due to Rus [1973] is a particular case of (iii). Moreover, if p is a quasi-metric, Theorem 1 follows from our Theorem H(γ1).

Suzuki [66] in 2005

In this paper, Suzuki extends the generalized Caristi's fixed point theorems due to Bae in 2003 and Bae et al. in 1994.

From the original Caristi's theorem, Suzuki deduced the following in [66]:

Theorem 2 (Suzuki). *Let X be a complete metric space with metric d. Let T be a mapping from X into itself and let f be a lower semicontinuous function from X into $[0, \infty)$. Let φ be a function from X into $[0, \infty)$ satisfying*

$$\sup\Big\{\varphi(x) : x \in X,\ f(x) \leq \inf_{w \in X} f(w) + \eta\Big\} < \infty$$

for some $\eta > 0$. Assume that

$$d(x, Tx) \leq \varphi(x)\big(f(x) - f(Tx)\big)$$

for all $x \in X$. Then there exists a fixed point $x_0 \in X$ of T.

Comments

Since the lower semicontinuity in the Caristi theorem was extended recently to the one *from above*, Suzuki's theorem and all its consequences due to Bae et al. can be also extended. Moreover, they may be extended to quasi-metric spaces. Consequently, all eleven theorems in Suzuki [66] can be improved.

Suzuki [67] in 2008

Suzuki generalized the Banach contraction principle as follows:

Theorem 2 (Suzuki). *Let (X, d) be a complete metric space and T be a mapping on X. Define a nonincreasing function θ from $[0, 1)$ onto $(1/2, 1]$ by*

$$\theta(r) = \begin{cases} 1 & \text{if } 0 \le r \le (\sqrt{5}-1)/2, \\ (1-r)r^{-2} & \text{if } (\sqrt{5}-1)/2 \le r \le 2^{-1/2}, \\ (1+r)^{-1} & \text{if } 2^{-1/2} \le r < 1. \end{cases} \tag{1.1}$$

Assume there exists $r \in [0, 1)$ such that

$$\theta(r)d(x, Tx) \le d(x, y) \quad \text{implies } d(Tx, Ty) \le rd(x, y)$$

for all $x, y \in X$. Then there exists a unique fixed point z of T. Moreover, $\lim_n T^n x = z$ for all $x \in X$.

Suzuki noted that the following theorem says that $\theta(r)$ is the best constant for every $r \in [0, 1)$.

Theorem 3. *Define a function θ as in Theorem 2. Then for each $r \in [0, 1)$, there exist a complete metric space (X, d) and a mapping T on X such that T does not have a fixed point and*

$$\theta(r)d(x, Tx) < d(x, y) \quad \text{implies } d(Tx, Ty) \le rd(x, y)$$

for all $x, y \in X$.

Comments

In Theorem 2, T is an RHR map and a Picard operator, hence Theorem 2 follows from Theorem H(γ1). From now on T is called the Suzuki type, since its many followers used it. Note that, in Theorem 3, $\theta(r)d(x, Tx) < d(x, y)$ means T can not have a fixed point $x = y = Tx$.

Recall that there are hundreds of equivalent conditions for metric completeness, e. g., Kirk, Park, Cobzaş as typical examples. Comments for them are given by Park and Rhoades in 1986. For details, see the recent article [51].

Kikkawa and Suzuki [34] in 2008

The authors state that the following is in Suzuki [67], where $\theta(r)$ is given:

Theorem 1 (Suzuki). *For a metric space (X,d), the following are equivalent:*
(i) *X is complete.*
(ii) *Every mapping T on X satisfying the following has a fixed point:*
  – *There exists $r \in [0,1)$ such that $\theta(r)d(x,Tx) \le d(x,y)$ implies $d(Tx,Ty) \le rd(x,y)$ for all $x,y \in X$.*
(iii) *There exists $r \in (0,1)$ such that every mapping T on X satisfying the following has a fixed point:*
  – *$\frac{1}{10000}d(x,Tx) \le d(x,y)$ implies $d(Tx,Ty) \le rd(x,y)$ for all $x,y \in X$.*

Two generalizations of Theorem 1 to multimaps are obtained:

Theorem 2. *Define a strictly decreasing function η from $[0,1)$ onto $(1/2,1]$ by*

$$\eta(r) = \frac{1}{1+r}.$$

Let (X,d) be a complete metric space and let T be a mapping from X into $CB(X)$. Assume that there exists $r \in [0,1)$ such that

$$\eta(r)d(x,Tx) \le d(x,y) \quad \textit{implies}\ H(Tx,Ty) \le r\, d(x,y)$$

for all $x,y \in X$. Then there exists $z \in X$ such that $z \in Tz$.

New Proof. This follows from our Theorem H(δ1). In fact, if $\{x\} = Tx$ for some $x \in X$, then let $z = x$. For any $x \in X\backslash T(x)$ and any $y \in X\backslash\{x\}$, we have

$$\eta(r)d(x,Tx) = \eta(r)\inf\{d(x,z) : z \in Tx\} \le d(x,y) \quad \text{implies } H(Tx,Ty) \le r\, d(x,y).$$

Therefore, by Theorem H(δ1), T has a fixed point $z \in X$. □

Comments

Theorem 1 can be extended as follows:

$$\text{(i)} \Longleftrightarrow \text{Theorem H}(\gamma 1)\text{ for metric spaces} \Longleftrightarrow \text{(ii)} \Longleftrightarrow \text{(iii)}.$$

This is the basis of (0) $\Longleftrightarrow$ (γ1) in the proof of Theorem H.

In Theorem 2, several things can be improved, that is, metric space, CB, and the Suzuki condition can be replaced by more general ones.

Kikkawa and Suzuki [35] in 2008

The following is a Kannan version of the Suzuki theorem [67]:

Theorem 2.2 (Kikkawa–Suzuki). *Let T be a mapping on complete metric space (X,d) and let φ be a non-increasing function from $[0,1)$ into $(1/2,1]$ defined by*

$$\varphi(r) = \begin{cases} 1, & \text{if } 0 \le r \le \frac{1}{\sqrt{2}}, \\ \frac{1}{1+r}, & \text{if } \frac{1}{\sqrt{2}} \le r < 1. \end{cases} \tag{1.2}$$

Let $\alpha \in [0,1/2)$ and $r = \alpha/(1-\alpha) \in [0,1)$. Suppose that

$$\varphi(r)d(x,Tx) \le d(x,y) \quad \text{implies } d(Tx,Ty) \le \alpha d(x,Tx) + \alpha d(y,Ty)$$

for all $x,y \in X$. Then, T has a unique fixed point z, and $\lim_n T^n x = z$ holds for every $x \in X$.

Comments

This gives an example of RHR maps, which can be applied Theorem P.

They said the following theorem shows that $\varphi(r)$ is the best constant for every r:

Theorem 2.4. *Define a function φ as in Theorem 2.2. For every $\alpha \in [0,1/2)$, putting $r = \alpha/(1-\alpha)$, there exist a complete metric space (X,d) and a mapping T on X such that T has no fixed points and*

$$\varphi(r)d(x,Tx) < d(x,y) \quad \text{implies } d(Tx,Ty) \le \alpha\, d(x,Tx) + \alpha\, d(y,Ty)$$

for all $x,y \in X$.

New Proof. Note that $\varphi(r)d(x,Tx) < d(x,y)$ means T can not have a fixed point $x = y = Tx$. □

Comments

In this paper, it is noted that $d(Tx,Ty) \le r\, d(x,y)$ in the original Suzuki type map in [67] can be replaced by one of the following in Theorem 3.1:

$$d(Tx,Ty) \le r \max\{d(x,Tx), d(y,Ty)\},$$
$$d(Tx,Ty) \le \max\{\alpha d(x,Tx), \beta d(y,Ty)\}, \quad \alpha,\beta \in [0,1).$$

Similarly, Theorem 3.3 is modified by applying such conditions.

In fact, they gave unnecessarily lengthy proofs for their theorems. Moreover, it can be replaced by the Hardy–Rogers condition.

Kikkawa and Suzuki [36] in 2008

Suzuki introduced the notion of τ-distance in order to generalize the results of Kada et al., Tataru, Zhong and others. Let (X, d) be a metric space and let p be a τ-distance on X.

The authors defined several mapping classes on X as in the following:

Theorem 8. *Let (X, d) be a metric space. Then*

$$TC_0(X) = TC(X) = TM_0(X) = TM(X)$$

holds.

In the proof, the selfmap T on (X, p) satisfies the following:

$$p(Tx, T^2x) \le r\, p(Tx, x) \quad \text{and} \quad p(T^2x, Tx) \le r\, p(x, Tx) \quad \text{for } x \in X,$$

where $r \in [0, 1)$.

Comment

Note that T is of the RHR type. Therefore, our arguments on metric spaces can be applied to several theorems in [36].

Kikkawa and Suzuki [37] in 2009

The authors give some notes on recent fixed-point theorems with constants, which are generalizations of the Banach contraction principle.

Let (X, d) be a metric space. Let $\mathrm{CB}(X)$ be the family of all nonempty closed bounded subsets of X. Let $H(\cdot, \cdot)$ be the Hausdorff metric, i. e.,

$$H(A, B) = \max\{\delta(A, B), \delta(B, A)\} \quad \text{for } A,\ B \in \mathrm{CB}(X),$$

where $\delta(A, B) = \sup_{x \in A} \inf_{y \in B} d(x, y)$.

Comment

In this paper, $H(A, B)$ in their nonlinear analysis paper [34] is replaced by $\delta(A, B)$ and similar results are obtained.

Enjouji, Nakanishi, and Suzuki [19] in 2009

Let $\Delta = \{(\alpha, \beta) : \alpha \ge 0, \beta \ge 0, \alpha + \beta < 1\}$. Define a nonincreasing function $\psi : \Delta \to (1/2, 1]$. Let T be a map on a complete metric space such that there exists $\alpha, \beta \in \Delta$ satisfying

$$\psi(\alpha,\beta)d(x,Tx) \le d(x,y) \quad \text{implies } d(Tx,Ty) \le \alpha d(x,Tx) + \beta d(y,Ty)$$

for all $x, y \in X$.

Comment
Note that T is an RHR map and that the condition $\psi(\alpha,\beta)d(x,Tx) < d(x,y)$ implies nonexistence of a fixed point of T.

Nakanishi and Suzuki [41] in 2010

In this paper, in order to observe the condition of Kannan maps more deeply, they prove a generalization of Kannan's fixed-point theorem.

Moreover, they assumed something like $\theta(\alpha)d(x,Tx) < d(x,y)$, which is false for $x = y = Tx$. Hence, from the beginning, such T can not have a fixed point.

Further, for $(\alpha,\beta) \in \Delta = [0,1)^2$ and a function $\varphi : \Delta \to (1/2,1]$, let T be a selfmap of a complete metric space (X,d) satisfying

$$\varphi(\alpha,\beta)d(x,Tx) \le d(x,y) \quad \text{implies } d(Tx,Ty) \le \max\{\alpha d(x,Tx), \beta d(y,Ty)\},$$

for all $x, y \in X$.

Comment
Note that T is an RHR map and that the condition $\varphi(\alpha,\beta)d(x,Tx) < d(x,y)$ does not hold for $x = y = Tx$.

1.8 Recent Rus–Hicks–Rhoades-type maps

After appearances of the works of Suzuki and his colleagues, their influence became serious problems. In the present section, we list related articles of other authors appeared after Suzuki's 2008 paper.

We tried to improve or correct them and to give them short proofs.

Altun and Erduran [2] in 2011

The aim of this paper is to generalize several previous results including that of Suzuki [67] in 2008 using the implicit relation technique in such a way that

$$F(d(Tx,Ty), d(x,y), d(x,Tx), d(y,Ty), d(x,Ty), d(y,Tx)) \le 0,$$

for $x, y \in X$, where $F : [0,\infty)^6 \to \mathbb{R}$ is a function as given as follows:

Let Ψ be the set of all continuous functions $F : [0,\infty)^6 \to \mathbb{R}$ satisfying the following conditions:

F1: $F(t_1,\dots,t_6)$ is nonincreasing in variables $t_2,\dots,t_6$,

F2: there exists $r \in [0,1)$, such that $F(u,v,v,u,u+v,0) \le 0$ or $F(u,v,0,u+v,u,v) \le 0$ or $F(u,v,v,v,v,v) \le 0$ implies $u \le rv$,

F3: $F(u,0,0,u,u,0) > 0$, for all $u > 0$.

Comment

From this, the authors showed $d(Tx,T^2x) \le d(x,Tx)$, that is, T is a variant of the RHR maps.

Damjanović and Dorić [15] in 2011

The authors obtain multi-valed mapping generalizations of two recent theorems of Kikkawa and Suzuki [34] in 2008 and the main theorem of Enjouji–Nakanishi–Suzuki [19] in 2009,

Let (X,d) be a metric space. We denote by $CB(X)$ the family of all non-empty closed bounded subsets of X. Let $H(\cdot,\cdot)$ be the Hausdorff metric. The following is the main result.

Theorem 2.1. *Define a non-increasing function $\varphi : [0,1) \to (0,1]$ by*

$$\varphi(r) = \begin{cases} 1 & \text{if } 0 \le r < \frac{\sqrt{5}-1}{2}, \\ 1-r & \text{if } \frac{\sqrt{5}-1}{2} \le r < 1. \end{cases} \tag{1.3}$$

Let (X,d) be a complete metric space and let $T : X \to CB(X)$. Assume that

$$\varphi(r)d(x,Tx) \le d(x,y) \text{ implies } H(Tx,Ty) \le r\max\{d(x,Tx),d(y,Ty)\}$$

for all $x,y \in X$. Then there exists $z \in X$ such that $z \in Tz$.

Comments

This has two and half page proof.

For $y \in Tx$, since

$$d(y,Ty) \le H(Tx,Ty) \le r\ \max\{d(x,Tx),d(y,Ty)\},$$

$d(y,Ty)$ can not be maximal. Hence, we have to assume $H(Tx,Ty) \le r\,d(x,Tx)$ from the beginning. Consequently, the conclusion follows from our Theorem H(ϵ1), the singleton case of H(ϵ).

Dorić and Lazović [17] in 2011

The authors obtain a Suzuki-type generalization of a fixed point theorem for generalized multivalued mappings of Ćirić in 1972. The obtained results extend furthermore the recently developed Kikkawa–Suzuki-type contractions. Applications to certain functional equations arising in dynamic programming are also considered.

The main theorem is the following:

Theorem 2.1. *In Theorem 2.1 of the previous section, the contractive condition is replaced by more general*

$$H(Tx, Ty) \leq r \cdot \max\{d(x,y), d(x, Tx), d(y, Ty), [d(x, Ty) + d(y, Tx)]/2\}.$$

Comments

This is shown by three page proof. From this, three routine corollaries and an application are given.

Karapinar and Taş [31] in 2011

In this manuscript, the notion of C-condition due to Suzuki [68] in 2008 is generalized. Some new fixed-point theorems are obtained.

Many types of selfmaps T on a subset K of a Banach space X satisfying the following condition are studied:

$$\|Tx - T^2x\| \leq \|x - Tx\| \quad \text{for } x \in K.$$

Comment

This seems to be a relative of RHR maps.

Singh and Mishra [64] in 2011

The following result is a generalization of Kikkawa and Suzuki [34, Theorem 2].

Theorem 2.1. *Let (X, d) be a complete metric space and $P : X \to CL(X)$. Assume there exists $r \in [0, 1)$ such that*

$$d(x, Px) \leq (1 + r)d(x, y) \quad \textit{implies } H(Px, Py) \leq r\, d(x, y)$$

for all $x, y \in X$. Then P has a fixed point, i. e., there exists a $z \in X$ such that $z \in Pz$.

Comment

We add our short proof:

Proof. This follows from our Theorem H(δ1). In fact, if $\{x\} = Px$ for some $x \in X$, then let $z = x$. For any $x \in X \backslash P(x)$ and any $y \in X \backslash \{x\}$, we have

$$d(x, Px) = \inf\{d(x, z) : z \in Px\} \le (1 + r)d(x, y) \quad \text{implies } H(Px, Py) \le r\, d(x, y).$$

Therefore, by Theorem H(δ1), P has a fixed point $z \in X$. □

Asgari and Abbasbigi [4] in 2012]

Abstract: In this paper, we develop the Banach contraction principle and Kannan fixed-point theorem on generalized cone metric spaces. We prove a version of Suzuki and Kannan-type generalizations of fixed point theorems in generalized cone metric spaces.

Comment

There are thousands of generalizations of metric spaces and any work on metric spaces might be extended to such ones. We show only some of them in the sequel.

Ćirić, Abbas, Rajović, and Ali [14] in 2012

It was obtained Suzuki-type fixed-point results for a generalized multi-valued mapping on a set equipped with two b-metrics. As a consequence, existence and uniqueness of solution of functional equation arising in dynamical programming is also derived.

Ghosh and Ray [21] in 2012

Theorem 2.1. *Let (X, d) be a complete generalized metric space (GMS), and T a mapping on X. Define a non-increasing function θ from $[0, 1)$ onto $(1/2, 1]$ by*

$$\theta(r) = \begin{cases} 1, & \text{if } 0 \le r \le 1/3 \\ \frac{2(1-r)}{3r(r+1)}, & \text{if } 1/3 \le r < 1. \end{cases} \tag{1.4}$$

Assume that, there exists $r \in [0, 1)$ such that

$$\theta(r)d(x, Tx) \le d(x, y) \Longrightarrow d(Tx, Ty) \le r\, d(x, y)$$

for all $x, y \in X$. Then there exists a unique fixed point z of T, and $\lim_{n\to\infty} T^n x = z$ for all $x \in X$.

Comments
By putting $y = Tx$, T becomes an RHR map. Hence, we can apply Theorems P. Other theorems in this paper can do the same to T.

Popescu [56] in 2013

The aim of this paper is to introduce a new type of multivalued operators similar to those of Kikkawa–Suzuki-type and to present some basic problems of the fixed point and strict fixed point for them.

Theorem 2.5. *Let (X, d) be a complete metric space and $T : X \to X$ be an (s, r)-contractive single-valued operator:*

$$x, y \in X \quad \text{with } d(y, Tx) \le s\, d(y, x) \quad \text{implies } d(Tx, Ty) \le r\, M_T(x, y)$$

where

$$M_T(x, y) = \max\left\{d(x, y), d(x, Tx), d(y, Ty), \frac{1}{2}[d(x, Ty) + d(y, Tx)]\right\}.$$

Then T has a fixed point. Moreover, if $s \ge 1$ then T has a unique fixed point.

Comment
Note that T is an RHR map and hence our Theorems P can be applicable. For the uniqueness, put $x = Tx, y = Ty$ in

$$d(x, y) = d(Tx, Ty) \le r\, M_T(x, y) = r \max\{d(x, y)\} = r\, d(x, y),$$

which implies $x = y$.

Salimi and Karapinar [61] in 2013

We prove the existence and uniqueness of a fixed point of certain type mapping, extension of Suzuki–Edelstein mapping, in a partially ordered complete metric space. Our results extend, improve, and generalize the existence results on the topic in the literature. We state some examples to illustrate our results.

Shobkolaei, Sedghi, Roshan, and Hussain [63] in 2013

We demonstrate a fundamental lemma for the convergence of sequences in metric-like spaces, and by using it we prove some Suzuki-type fixed point results in the setup of metric-like spaces. As an immediate consequence of our results, we obtain certain recent results in partial metric spaces as corollaries. Finally, three examples are presented to verify the effectiveness and applicability of our main results.

Comments
Metric like spaces are almost same to metric spaces. Suzuki's lengthy proof is imitated for authors' metric like spaces. If possible, its existing short proof can be reformulated for the metric like spaces.

Hussain, Kutbi, Khaleghizadeh, and Salimi [23] in 2014

As an application of the exciting results, we obtain the following Suzuki-type fixed-point theorem in 2008.

Theorem 30. *Let (X, d) be a complete metric space and let T be a self-mapping on X. Assume that there exists $r \in [0, 1)$ such that*

$$\frac{1}{1+r} d(x, Tx) \leq d(x, y) \quad \textit{implies } d(Tx, Ty) \leq rM(x, y) + Ld(y, Ty)$$

for all $x, y \in X$, where

$$M(x, y) = \max\left\{d(x, y), d(x, Tx), d(y, Ty), \frac{1}{2}[d(x, Ty) + d(y, Tx)]\right\}.$$

Then T has a fixed point.

Comment
For $y = Tx$, we have

$$\begin{aligned} d(Tx, T^2x) &\leq r \max\left\{d(x, Tx), d(Tx, T^2x), \frac{1}{2}[d(x, Tx) + d(Tx, T^2x)]\right\} \\ &= r\frac{1}{2}[d(x, Tx) + d(Tx, T^2x)] = \frac{r}{2-r}(x, Tx). \end{aligned}$$

Hence T is an RHR map and Theorem H(γ1) can be applicable. Most results in this paper can be improved following such method.

Khojasteh, Abbas, and Costache [33] in 2014

Theorem 1. *Let (X, d) be a complete metric space and let T be a mapping from X into itself. Suppose that T satisfies the following condition:*

$$d(Tx, Ty) \leq \frac{d(x, Ty) + d(y, Tx)}{d(x, Tx) + d(y, Ty) + 1} d(x, y)$$

for all $x, y \in X$. Then

(a) *T has at least one fixed point $x_0 \in X$;*

(b) *$\{T^n x\}$ converges to a fixed point, for all $x \in X$;*

(c) *If x_0 and y_0 are distinct fixed points of T, then $d(x_0, y_0) \leq 1/2$.*

Comment

Note that T is an RHR map and can be applicable Theorems P and H(γ1). The (c) follows from the contractive condition.

Kim, Sedghi and Shobkolaei [38] in 2015

Kim et al. obtained a common fixed-point theorem for two maps T and S on a complete metric spaces X satisfying the following generalized contractive condition:

$$\theta(r) \min\{d(x, Tx), d(x, Sx)\} \leq d(x, y) \implies$$
$$\max d(Sx, Sy), d(Tx, Ty), d(Sx, Ty), d(Sy, Tx) \leq r\, d(x, y).$$

for each $x, y \in X$, where θ is defined as in Suzuki in 2008.

Comments

By putting $y = Sx$ or $y = Tx$, S or T is an RHR map. Hence, by our Theorem H(δ)1, S or T has a fixed point. Let $z = Sz$ and put $x = y = z = Sz$. Then, we have

$$\max d(Sz, Sz), d(Tz, Tz), d(Sz, Tz), d(Sz, Tz) \leq r\, d(z, z) = 0,$$

and hence we have $d(Sz, Tz) = d(z, Tz) = 0$. Therefore, z is a common fixed point.

Kumam, Dung, and Sitytithakerngkiet [39] in 2015

Definition 2.4. Let $T : X \to X$ be a mapping on metric space X. The mapping T is said to be a *generalized quasi-contraction* iff there exists $q \in [0, 1)$ such that for all $x, y \in X$,

$$d(Tx, Ty) \leq q \max\{d(x, y), d(x, Tx), d(y, Ty), d(x, Ty), d(y, Tx),$$
$$d(x, T^2x), d(Tx, T^2x), d(x, T^2y), d(T^2x, Ty)\}$$

Some corollaries and multi-valued versions of them are added.

Comment

The contractive condition implies the Hegedüs condition $d(Tx, Ty) \leq q\,\mathrm{diam}\{O_T(x) \cup O_T(y)\}$. Hence, the main theorem was already known.

Karapinar [27] in 2018

Definition 2.1. Let (X, d) be a metric space. We say that a selfmap $T : X \to X$ is an *interpolative Kannan-type contraction*, if there exist a constant $\lambda \in [0, 1)$ and $\alpha \in (0, 1)$ such that

$$d(Tx, Ty) \leq \lambda[d(x, Tx)]^{\alpha}[d(y, Ty)]^{1-\alpha}$$

for all $x, y \in X$ with $x \neq Tx$.

Theorem 2.2. *Let (X, d) be a complete metric space and T be an interpolative Kannan-type contraction. Then T has a unique fixed point in X.*

Comments

By putting $y = Tx$, $[d(Tx, T^2x)]^{\alpha} \leq k[d(x, Tx)]^{\alpha}$, that is,

$$d'(Tx, T^2x) \leq k\, d'(x, Tx) \quad \text{with } d^{\alpha} = d'.$$

Therefore, T is an RHR map. The uniqueness in Theorem 2.2 is removed in the next article.

Karapinar, Agarwal, and Aydi [29] in 2018

As a correction of Theorem 1 [Theorem 2.2 in Karapinar [27] in 2018), the authors should state

Theorem 2. *Let (X, ρ) be a complete metric space. A self-map $T : X \to X$ possesses a fixed point in X, if there exist constants $\lambda \in [0, 1)$ and $\alpha \in (0, 1)$ such that*

$$\rho(T\zeta, T\eta) \leq \lambda[\rho(\zeta, T\zeta)]^{\alpha} \cdot [\rho(\eta, T\eta)]^{1-\alpha}$$

for all $\zeta, \eta \in X \backslash \mathrm{Fix}(T)$.

The following theorem was proved by Reich, Rus and Ćirić (1971–1979) independently to combine and improve both Banach and Kannan fixed-point theorems.

Theorem 3. *In the framework of a complete metric space (X, ρ), if $T : X \to X$ forms a Reich–Rus–Ćirić contraction map, i. e.,*

$$\rho(T\zeta, T\eta) \leq \lambda[\rho(\zeta,\eta) + \rho(\zeta, T\zeta) + \rho(\eta, T\eta)]$$

for all $\zeta, \eta \in X$, where $\lambda \in [0, 1/3)$, then T possesses a unique fixed point.

Notice that several variations of Reich contractions can be stated. We may state the following:

$$\rho(T\zeta, T\eta) \leq a\rho(\zeta,\eta) + b\rho(\zeta, T\zeta) + c\rho(\eta, T\eta),$$

where $a, b, c \in (0, \infty)$ such that $0 \leq a + b + c < 1$.

In this paper, we shall investigate the validity of the interpolation approach for Reich contractions in the context of partial metric spaces that was introduced by Matthews (1994).

The main contribution of the paper to ensure the existence of fixed points for interpolative Reich–Rus–Ćirić-type contraction mappings on partial metric spaces.

Comment
In this paper, we can find some RHR maps.

Karapinar, Alqahtani, and Aydi [30] in 2018

Definition 2 ([30]). Let (X, d) be a complete metric space. The mapping $T : X \to X$ is called an interpolative Hardy–Rogers-type contraction if there exist $\lambda \in [0, 1)$ and positive reals $\beta, \alpha, \gamma > 0$, with $\beta + \alpha + \gamma < 1$, such that

$$d(Tx, Ty) \leq \lambda\Big([d(x,y)]^{\beta} \cdot [d(x, Tx)]^{\alpha} \cdot [d(y, Ty)]^{\gamma} \cdot [1/2(d(x, Ty) + d(y, Tx))]^{1-\alpha-\beta-\gamma}\Big)$$

for each $x, y \in X \setminus \mathrm{Fix}(T)$.

Theorem 4 ([30]). *Let (X, d) be a complete metric space and T be an interpolative Hardy–Rogers-type contraction. Then, T has a fixed point in X.*

Comment
Note that
Case 1: $d(x, Tx) \leq d(Tx, T^2x)$ *implies* $[d(Tx, T^2x)]^{\alpha+\beta} \leq \lambda[d(x, Tx)]^{\alpha+\beta}$.
Case 2: $d(Tx, T^2x) \leq d(x, Tx)$ *implies* $d(Tx, T^2x) \leq \lambda d(x, Tx)$.

Hence, T is an RHR map.

Karapinar [28] in 2019

Given nonunique fixed-point theorems are due to Ćirić (1974), Achari (1976), Pachpatte (1979), Ćirić–Jotić (1998), and Karapinar (2019). For example,

Theorem 5. *Let $T : X \to X$ be an orbitally continuous self-map on the T-orbitally complete metric space (X, d). Suppose there exist real numbers a_1, a_2, a_3, a_4, a_5 and a self mapping $T : X \to X$ which satisfies $E(x,y) \leq a_4 d(x,y) + a_5 d(x, T^2 x)$, where*

$$E(x,y) := a_1 d(Tx, Ty) + a_2[d(x, Tx) + d(y, Ty)] + a_3[d(y, Tx) + d(x, Ty)],$$

for all $x, y \in X$. Then, T has at least one fixed point.

Comment

Here $\{a_i\}_{i=1}^5$ are chosen to hold the RHR condition $d(Tx, T^2x) \leq r\, d(x, Tx)$ with $r \in [0,1)$ for all $x \in X$. Note that Theorem 5 holds for quasi-metric spaces.

Aouine and Aliouche [3] in 2021

The main result of Khojasteh et al. [33] is extended to the following example and others:

Theorem 4. *Let (X, d) be a complete metric space and T a mapping from X into itself satisfying the following condition*

$$d(Tx, Ty) \leq \frac{d(x, Ty) + d(y, Tx)}{d(x, Tx) + d(y, Ty) + 1} \max\{d(x, Tx), d(y, Ty)\}.$$

for all $x, y \in X$. Then,

a) *T has a unique fixed point $z \in X$;*
b) *For any sequence $\{y_n\}$ in X such that $\lim_{n\to\infty} d(Ty_n, y_n) = 0$, we have $\lim_{n\to\infty} d(y_n, z) = 0$; and*
c) *T is continuous at z.*

Comment

Some applications are added. Note that T is an RHR map and that c) seems to hold for any $x \in X$.

Popescu [56, 57] in 2021

Theorem. *Suppose that (X, d) is a complete metric space and $T : X \to X$ is a mapping satisfying*

$$d(Tx, Ty) \leq M[d(x,y) + d(x, Tx) + d(y, Ty)],$$

where $M \in [0,1)$ and the following condition:

(C) *Assume that there exist real constants a, b with $a \in [0,1)$ and $b > 0$ such that for arbitrary $x \in X$ there exists $u \in X$ satisfying*
   (i) $d(u, Tu) \leq a\, d(x, Tx)$;
   (ii) $d(u, x) \leq b\, d(x, Tx)$.

Then T has a fixed point.

Comment
Note that T is not an RHR map. However, if $b \geq 1$ and $u = Tx$, then (C) implies that T is an RHR map.

Yeşilkaya [69] in 2021

Describe using Ψ the set of all nondecreasing self-mappings ψ on $[0, \infty)$ such that $\sum_{n=1}^{\infty} \psi^n(t) < \infty$ for each $t > 0$. Regard that for $\psi \in \Psi$, we have $\psi(0) = 0$ and $\psi(t) < t$ for each $t > 0$.

Corollary 2.5. *Let (X, d) be a complete metric space, $\lambda \in [0,1)$, and T be a self mapping on X such that*

$$\frac{1}{2}d(x, Tx) \leq d(x, y) \quad \textit{implies}$$

$$d(Tx, Ty) \leq \lambda\left([d(x,y)]^{\beta} \cdot [d(x, Tx)]^{\alpha} \cdot [d(y, Ty)]^{\gamma} \cdot \left[\frac{1}{2}(d(x, Ty) + d(y, Tx))\right]^{1-\alpha-\beta-\gamma}\right)$$

for each $x.y \in X \backslash \mathrm{Fix}(T)$, where $\psi \in \Psi$ and positive real $\beta, \alpha, \gamma > 0$, with $\beta + \alpha + \gamma < 1$. Then T posses a fixed point.

Comment
Note that T is an RHR map as in Theorem 1.4 of Karapinar et al. [27] in 2018.

Aouine [3] in 2022

Definition 4 (Pathak–Shahzad [54]). Let (X, d) be a metric space. A mapping $T : Y \subset X \to X$ is said to be a *metric p-contraction* (or simply *p-contraction* mapping if Y is T-invariant and it satisfies the following inequality:

$$d(T(x), T^2(x)) \leq p(x) d(x, T(x)) \quad \text{for all } x \in Y,$$

where $p : Y \to [0,1]$ is a function such that $p(x) < 1$ for all $x \in Y$ and $\sup_{x \in Y} p(Tx) = \alpha < 1$.

Further, if $\bigcap_{n=0}^{\infty} T^n(Y)$ is a singleton set, where $T^n(Y) = T(T^{n-1}(Y))$ for each $n \in \mathbb{N}$ and $T^0(Y) = Y$, then T is said to be a *strong p-contraction.*

The aim of this paper is to prove a fixed-point result for self-mapping, which satisfies p-contraction condition in partially ordered metric spaces.

Comment

Note that T is an RHR map. Our theory can be applicable in this situation.

Berinde and Pacurar [8] in 2022

In this paper, an RHR map is called a graphic contraction (orbital contraction).

Lemma 2.12. *Let (X, d) be a metric space. Any graphic contraction $T : X \to X$ admits an approximate fixed point sequence, that is, a sequence $\{x_n\} \subset X$ satisfying $d(x_n, Tx_n) \to 0$ as $n \to \infty$.*

Note that this holds for a quasi-metric spaces and that the Picard iteration associated with a graphic contraction T, i. e., the sequence $\{x_n\}$ defined by $x_{n+1} = Tx_n$, $n \geq 0$, for some $x_0 \in X$, is an approximate fixed point sequence with respect to T.

The map T is called a *weakly Picard operator* if it has a fixed point and any Picard iteration converges to a fixed point p for any $x_0 \in X$. If T is a weakly Picard operator and p is the unique fixed point, then T is called a *Picard operator.*

Then the authors showed that the examples of graphic contractions from Banach to Zamfirescu are Picard operator when X is complete.

Finally, another Picard operators were introduced from the Maia type map and its general version.

Comment

Early in 2004, Berinde [7] defined the weak contractions containing the above examples and proved that they are Picard operators. It holds for quasi-metric spaces (without symmetry) instead of metric spaces, see Park [48].

Chandra, Joshi, and Joshi [10] in 2022

Let (M, d) be a metric space, and $T : M \to M$. Then, for all $x, y \in M$, we denote

$$m(Tx, Ty) = ad(x, y) + b\max\{d(x, Tx), d(y, Ty)\} + c[d(x, Ty) + d(y, Tx)],$$

where a, b and c are non-negative reals such that $a + b + 2c = r$ with $r \in [0, 1)$. Now, we consider the following generalized contractive condition

$$\theta(r)\min\{d(x,Tx),d(x,Ty)\}\le d(x,y)\quad \text{implies } d(Tx,Ty)\le m(Tx,Ty).$$

(Here, θ is as in Suzuki [67].) It is remarkable that this condition is a generalization of the condition (22) and several other conditions mentioned in Transaction Paper of Rhoades [58] in 1977.

Comment
Note that T is an RHR map and applicable Theorem P and H(γ1).

Khantwal, Aneja, and Cairola [32] in 2022

It is presented a generalization of Suzuki's fixed-point theorem and the Matkowski contraction principle for a system of transformations on the finite product of metric spaces.

In fact, the authors use the following general Suzuki-type condition:

$$\theta(h)d_i(x_i,T_ix)\le d_i(x_i,y_i)\Longrightarrow d_i(T_ix,T_iy)\le a_{ik}d_k(x_k,y_k).$$

For $y_i = T_ix$, this reduces to $d_i(T_ix,T_i^2x)\le a_{ik}d_k(x_k,T_i^2x_k)$.

Comment
This is a generalization of the RHR map. This paper can be rewritten for the generalization.

Romaguerra [59] in 2022

For a selfmap F of a quasi-metric space (X,ρ), $c\in(0,1)$ and $x,y\in X$, one of the following conditions hold:

$$\begin{aligned}\rho(x,Fx)\le 2\rho(x,y)&\Longrightarrow \rho(Fx,Fy)\le c\rho(x,y),\\ \min\{\rho(x,Fx),\rho(y,Fy)\}\le 2\rho(x,y)&\Longrightarrow \rho(Fx,Fy)\le c\rho(x,y),\\ \min\{\rho(x,Fx),\rho(y,Fy),\rho(Fy,y)\}\le 2\rho(x,y)&\Longrightarrow \rho(Fx,Fy)\le c\rho(x,y).\end{aligned}$$

Comment
Note that such F's are RHR maps. Since all of our basic tools, Theorems H, P, and H(γ1) are applicable to quasi-metric spaces. Most of articles mentioned in this paper ignore this fact except Romaguerra [59] in 2022.

Ali, Ali, Nazir, and Ali [1] in 2023

Theorem 1. *Let $(\mathcal{T}, d)$ be a complete metric space and $\Psi : \mathcal{T} \to \mathcal{T}$. Assume that there exists $\alpha \in [0,1)$ such that*

$$d(\ell, \Psi\ell) \leq 2d(\ell, \hbar) \Longrightarrow d(\Psi\ell, \Psi\hbar) \leq \alpha\, d(\ell, \hbar)$$

for all $\ell, \hbar \in \mathcal{T}$. Then there exists a unique fixed point z of Ψ. Moreover,

$$\lim_{k\to\infty} \Psi^k \ell = z \quad \text{for all } \ell \in \mathcal{T}.$$

Comment
For $\hbar = \Psi\ell$, Ψ becomes an RHR map. For other theorems, the same remark works.

Ekayanti, Muslikh, Fitri, and Marjono [18] in 2023

We begin with the following theorem that gives the existence of a fixed point in partially ordered partial metric space.

Theorem 3.1. *Let $(X, \preccurlyeq)$ be a partially ordered set and suppose that there exists a metric d in X such that (X, p) is a complete partial metric space. Let $f : X \to X$ be a continuous and non-decreasing mapping such that there exists $k \in [0,1)$ with $p(f(x), f(y)) \leq k\, p(x,y)$ for all $x \succcurlyeq y$. If there exists $x_0 \in X$ with $x_0 \preccurlyeq f(x_0)$, then f has a fixed point, that is x^*. Furthermore, $p(x^*, x^*) = 0$.*

Comment
For quasi-metric spaces, this simply follows from Theorem H(γ1).

Fierro and Pizarro [20] in 2023

Let (X, d) be a complete metric space. The authors state that the following corollary is an equivalent version of the main result of Hicks and Rhoades [22] in 1979:

Corollary 3.3. *Let $\xi : X \to X$ be a function and $k \in [0,1)$. Suppose there exists $x_0 \in X$ such that, for all $x \in O(x_0, \xi)$, $d(\xi(x), \xi^2(x)) \leq k\, d(x, \xi(x))$. Then, there exists $x^* \in X$ such that the following two conditions hold:*
(i) $\lim_{n\to\infty} d(x^*, \xi^n(x_0)) = 0$ *and*
(ii) $d(x^*, \xi^n(x_0)) \leq \frac{k^n}{1-k} d(x_0, \xi(x_0))$, *for all $n \in \mathbb{N}$.*

Moreover, $x^* = \xi(x^*)$, if and only if, the function $x \in X \mapsto d(x, \xi(x)) \in \mathbb{R}$ is (x_0, ξ)-orbitally lower semicontinuous at x^*.

Comment
This is almost same to our Theorem P in Section 1.5. It is possible to unify this with Theorem P.

Kumar, Garakoti, Chandra, and Joshi [40] in 2023

Theorem 1. *Let (X, p) be a complete partial metric space and let $T : X \to X$ be a mapping satisfying*

$$p(Tx, Ty) \leq \alpha p(x,y) + \beta[p(x, Tx) + p(y, Ty)] + \gamma[p(x, Ty) + p(y, Tx)] + \delta[M_p(x,y) + h\, m_p(x,y)],$$

for all $x, y \in X$, where $0 < h < 1$ and $\alpha \geq 0$, $\beta, \gamma, \delta > 0$ with $\alpha + 2\beta + 2\gamma + 2\delta < 1$, and

$$M_p(x,y) = \max\{p(x, Ty), p(y, Tx)\}, \quad m_p(x,y) = \min\{p(x, Ty), p(y, Tx)\}.$$

Then T has a unique fixed point.

Comment
We show that T is an RHR map for a quasi-metric space (X, p): For $y = Tx$, we have $M_p(x,y) = p(x, Ty)$ and $m_p(x,y) = p(y, Tx) = 0$. Hence

$$p(Tx, T^2x) \leq \alpha p(x, Tx) + \beta[p(x, Tx) + p(Tx, T^2x)] + \gamma[p(x, T^2x) + p(Tx, Tx)] + \delta[p(x, Tx)],$$

for all $x, y = Tx \in X$. Then, we have

$$(1 - \beta - \gamma)p(Tx, T^2x) \leq (\alpha + \beta + \gamma + \delta)p(x, Tx).$$

Since

$$\alpha + \beta + \gamma + \delta < 1 - \beta - \gamma - \delta < 1 - \beta - \gamma,$$

T is an RHR map.

Pant and Khantwal [43] in 2023

Let $\psi : [0, \infty) \to [0, \infty)$ be upper semi-continuous from the right such that $\psi(t) < t$ for all $t > 0$.

Theorem 2.2. *Suppose (E, ρ) is a metric space. Let $f : E \to E$ a mapping such that for some $v_0 \in E$,*

$$\frac{1}{2}\rho(x, f(x)) \leq \rho(x,y) \Longrightarrow \rho(f(x), f(y)) \leq \psi(N(x,y)) \quad \text{for all } x, y \in \overline{O(v_0, f)} \text{ with } x \neq y,$$

where $N(x,y) = \max\{\rho(x,y), \rho(x, f(x)), \rho(y, f(y)), \frac{1}{2}[\rho(y, f(x)) + \rho(x, f(y))]\}$.

If E is f-orbitally complete then the sequence of iterations $(f^n(v_0))$ is Cauchy in E and converges to the unique fixed point of f in $\overline{O(v_0,f)}$.

Comment

Easily, we can show $\rho(f(x),f^2(x)) \le \psi(\rho(x,f(x)))$ and f is a relative RHR map.

Petrov [55] in 2023

Petrov defined generalized Kannan-type maps and obtained:

Proposition 1.3. *Let (X,d) be a metric space and let $T : X \to X$ be a generalized Kannan-type metric with some $\lambda \in [0,2/3)$. If x is an accumulation point of X and T is continuous at x, then the inequality*

$$d(Tx,Ty) \le \lambda[d(x,Tx) + d(y,Ty)/2]$$

holds for all points $y \in X$.

Theorem 2.1. *Let (X,d), $|X| > 3$, be a complete metric space and let the mapping $T : X \to X$ satisfy the following two conditions:*
(i) *$T(T(x)) \ne x$ for all $x \in X$ such that $Tx \ne x$.*
(ii) *T is a generalized Kannan-type mapping on X.*

Then T has a fixed point. The number of fixed points is at most two.

Comments

For any point $x \in X$, the inequality in Proposition 1.3 for $y = Tx$ implies

$$\left(1-\frac{\lambda}{2}\right)d(Tx,T^2x) \le \lambda\, d(x,Tx) \quad \text{or} \quad d(Tx,T^2) \le \alpha\, d(x,Tx) \quad \text{where } \alpha = \frac{\lambda}{2-\lambda} \in [0,1).$$

Hence, T is an RHR map and has a fixed point by Theorem P.

Suppose that there exists at least three pairwise distinct fixed points x, y and z. Then $Tx = x$, $Ty = y$ and $Tz = z$, which contradicts to the definition of generalized Kannan-type map.

The concept of generalized Kannan-type maps can be replaced by RHR maps.

Saravanan, Piramanantham and Theivaraman [62] in 2023

The authors' typical rational contraction mapping is the following:

Let (K, d) be a metric space and $T : K \to K$. Then there exists $\nu, \mu \in [0, 1)$ with $\nu + \mu < 1$ such that $d(t, Tt) + d(r, Tr) \neq 0$ and for all $t, r \in K$ such that

$$d(Tt, Tr) \leq \frac{\nu[d(t, Tt)d(t, Tr) + d(r, Tr)d(r, Tt)]}{d(t, Tr) + d(r, Tt)} + \mu d(t, r).$$

Comment

We show T is an RHR map. In fact, by putting $r = Tt$, we have

$$d(Tt, T^2t) \leq \frac{\nu[d(t, Tt)d(t, T^2t) + d(Tt, T^2t)d(Tt, Tt)]}{d(t, T^2t) + d(Tt, Tt)} + \mu d(t, Tt),$$

or

$$d(Tt, T^2t) \leq \frac{\nu[d(t, Tt)d(t, T^2t)]}{d(t, T^2t)} + \mu d(t, Tt) = (\nu + \mu)d(t, Tt).$$

1.9 Epilogue

Recall that Banach's original fixed-point theorem in 1922 was stated for normed vector spaces. Later, several researchers formulated it to the form of the Banach Contraction Principle for complete metric spaces. In the last one hundred years, there have been appeared hundreds of contraction type conditions and almost one thousand spaces which generalize, extend, or modify the complete metric spaces. Recall that the Banach contraction does not characterize the metric completeness.

Since then, a few researchers studied the Rus–Hicks–Rhoades (RHR) maps by using several different names. From 2023, one hundred years later to the Banach contraction, the present author began to study on RHR maps. We found a large number of examples of RHR maps and the so-called RHR Contraction Principle extending the Banach one. Moreover, we found that the RHR theorem is equivalent to variants of the Nadler or Covitz–Nadler theorem for multivalued contractions. Furthermore, we found that the RHR theorem characterizes the metric completeness.

Such studies were done in 2022–24. One of the significance of our recent works on metric fixed-point theory is to clarify some incorrect results with unnecessarily long proofs given by several authors. In fact, our aim of study in metric fixed-point theory since 2022 is to improve every thing there without making new spaces or new contractive condition.

Bibliography

[1] Ali, B., Ali, H., Nazir, T. and Ali, Z., *Existence of fixed points of Suzuki-type contractions of quasi-metric spaces*, Mathematics 11 (2023), 4445, https://doi.org/10.3390/math11214445.

[2] Altun, I. and Erduran, A., *A Suzuki type fixed-point theorem*, Int. J. Math. Math. Sci. 2011 (2011) 736063, 9 pages. https://doi.org/10.1155/2011/736063.

[3] Aouine, A. C., *A fixed point theorem for p-contraction mappings in partially ordered metric spaces and application to ordinary differential equations*, Bul. Acad. Ştiinţe A Repub. Mold. Mat. 3(100) (2022), 15–21. E-ISSN 2587-4322. ISSN 1024-7696.

[4] Asgari, M. S. and Abbasbigi, Z., *The generalizations of the Suzuki and Kannan fixed point theorems in G-cone metric spaces*, Anal. Theory Appl. 28(3) (2012), 1–14, https://doi.org/10.3969/j.issn.1672-4070.2012.03.009.

[5] Aubin, J. P., *Applied Functional Analysis*, John Wiley & Sons, New York, 1979.

[6] Berinde, V., *On the approximation of fixed points of weak contractive mappings*, Carpath. J. Math. 19(1) (2003), 7–22.

[7] Berinde, V., *Approximating fixed points of weak contractions using the Picard iteration*, Nonlinear Anal. Forum 9(1) (2004), 43–53.

[8] Berinde, V. and Pacurar, M., *Alternative proofs of some classical metric fixed point theorems by using approximate fixed point sequences*, Arab. J. Math. (2022). https://doi.org/10.1007/s40065-022-00398-6.

[9] Berinde, V., Petruşel, A. and Rus, I. A., *Remarks of the mappings in fixed point iterative methods in metric spaces*, Fixed Point Theory 24(2) (2023), 525–540, https://doi.org/10.24193/fpt-ro.2023.2.05.

[10] Chandra, N., Joshi, B. and Joshi, M. C., *Generalized fixed point theorems on metric spaces*, Math. Morav. 26(2) (2022), 85–101.

[11] Chen, J. and Li, Z., *Common fixed-points for Banach operator pairs in best approximation*, J. Math. Anal. Appl. 336 (2007), 1466–1475.

[12] Ćirić, Lj., *On some maps with a non-unique fixed point*, Publ. Inst. Math. 17 (1974), 52–58.

[13] Ćirić, Lj., *A generalization of Banach's contraction principle*, Proc. Am. Math. Soc. 45 (1974), 267–273.

[14] Ćirić, Lj. B., Abbas, M., Rajović, M. and Ali, B., *Suzuki type fixed point theorems for generalized multi-valued mappings on a set endowed with two b-metrics*, Appl. Math. Comput. 219 (2012), 1712–1723.

[15] Damjanović, B. and Dorić, D., *Multivalued generalizations of the Kannan fixed point theorem*, Filomat 25(1) (2011), 125–131, https://doi.org/10.2298/FIL1101125D.

[16] Dass, B. K. and Gupta, S., *An extension of Banach contraction principle through rational expressions*, Indian J. Pure Appl. Math. 6 (1975), 1455–1458.

[17] Dorić, D. and Lazović, R., *Some Suzuki-type fixed point theorems for generalized multivalued mappings and applications*, Fixed Point Theory Appl. 2011 (2011), 40.

[18] Ekayanti, A., Muslikh, M., Fitri, S. and Marjono, *Fixed point theorems in partially ordered metric space*, Adv. Fixed Point Theory 2023(13) (2023), 19, https://doi.org/10.28919/afpt/8188. ISSN 1927-6303.

[19] Enjouji, Y., Nakanishi, M. and Suzuki, T., *A generalization of Kannan's fixed point theorem*, Fixed Point Theory Appl. 1 (2009), 192872, 10 pages.

[20] Fierro, R. and Pizarro, S., *Fixed points of set-valued mappings satisfying a Banach orbital condition*, CUBO 25(1) (2023), 151–159.

[21] Ghosh, P. and Ray, A. D., *Characterization of completeness of generalized metric spaces using generalized Banach contraction principle*, Demonstr. Math. XLV(3) (2012), 717–724.

[22] Hicks, T. L. and Rhoades, B. E., *A Banach type fixed point theorem*, Math. Jpn. 24 (1979), 327–330.

[23] Hussain, N., Kutbi, M. A., Khaleghizadeh, S. and Salimi, P., *Discussions on recent results for α-ψ-contractive mappings*, Abstr. Appl. Anal. 2014 (2014), 456482, 13 pages. https://doi.org/10.1155/2014/456482.

[24] Jachymski, J., *Converses to fixed point theorems of Zermelo and Caristi*, Nonlinear Anal. 52 (2003), 1455–1463.

[25] Jaggi, D. S., *Some unique fixed point theorems*, Indian J. Pure Appl. Math. 8 (1977), 223–230.

[26] Kamran, T., *Mizoguchi–Takahashi's type fixed point theorem*, Comput. Math. Appl. 57 (2009), 507–511.

[27] Karapinar, E., *Revisiting the Kannan type contractions via interpolation*, Adv. Theory Nonlinear Anal. Appl. 2(2) (2018), 85–87, https://doi.org/10.31197/atnaa.431135.

[28] Karapinar, E., *Ćirić type nonunique fixed points results: A review*, Appl. Comput. Math. 18(1) (2019), 3–21.

[29] Karapinar, E., Agarwal, R. P. and Aydi, H., *Interpolative Reich–Rus–Ćirić type contractions on partial metric spaces*, Mathematics 6(11) (2018), 256.

[30] Karapinar, E., Alqahtani, O. and Aydi, H., *On interpolative Hardy–Rogers type contractions*, Symmetry 11 (2018), 8. https://doi.org/10.3390/sym11010008.

[31] Karapinar, E. and Taş, E., *Generalized (C)-conditions and related fixed point theorems*, Comput. Math. Appl. 61 (2011), 3370–3380.

[32] Khantwal, D., Aneja, S. and Cairola, U. C., *A generalization of Matkowski's and Suzuki's fixed point theorems*, Asian-Eur. J. Math. 15(9) (2022), 2250169, 12 pages.

[33] Khojasteh, F., Abbas, M. and Costache, S., *Two new types of fixed point theorems in complete metric spaces*, Abstr. Appl. Anal. (2014), 325840, 5 pages.

[34] Kikkawa, M. and Suzuki, T., *Three fixed point theorems for generalized contractions with constants in complete metric spaces*, Nonlinear Anal. 69 (2008), 2942–2949.

[35] Kikkawa, M. and Suzuki, T., *Some similarity between contractions and Kannan mappings*, Fixed Point Theory Appl. 649749 (2008), 8.

[36] Kikkawa, M. and Suzuki, T., *Some similarity between contractions and Kannan mappings, II*, Bull. Kyushu Inst. Technol., Pure Appl. Math. 55 (2008), 1–13.

[37] Kikkawa, M. and Suzuki, T., *Some notes on fixed point theorems with constants*, Bull. Kyushu Inst. Technol., Pure Appl. Math. 56 (2009), 11–18.

[38] Kim, J. K., Sedghi, S. and Shobkolaei, N., *Suzuki-type of common fixed point theorems in metric spaces*, J. Nonlinear Convex Anal. 16 (2015), 1779–1786.

[39] Kumam, P., Dung, N. V. and Sitytithakerngkiet, K., *A generalization of Ćirić fixed point theorems*, Filomat 29 (2015), 1549–1556.

[40] Kumar, R., Garakoti, N., Naveen Chandra, N. and Joshi, M. C., *Fixed and coincidence point theorems on partial metric spaces with an application*, Math. Morav. 27(2) (2023), 33–53.

[41] Nakanishi, M. and Suzuki, T., *An observation on Kannan mappings*, Cent. Eur. J. Math. 8(1) (2010), 170–178. https://doi.org/10.2478/s11533-009-0065-9.

[42] Nieto, J. J. and Rodriguez-López, R., *Contractive mapping theorems in partially ordered sets and applications to ordinary differential equations*, Order 22 (2005), 223–239.

[43] Pant, R. and Khantwal, D., *Fixed points of single and multivalued mappings in metric spaces with applications*, 2023. Preprint.

[44] Park, S., *Fixed points and periodic points of contractive pairs of maps*, Proc. Coll. Nat. Sci. SNU 5(1) (1980), 9–22.

[45] Park, S., *Foundations of ordered fixed point theory*, J. Nat. Acad. Sci. Repub. Korea, Nat. Sci. Ser. 61(2) (2022), 1–51.

[46] Park, S., *Equivalents of various theorems of Zermelo, Zorn, Ekeland, Caristi and others*, J. Adv. Math. Comput. Sci. 38(5) (2023), 60–73. https://doi.org/10.9734/jamcs/2023/v38i5176117.

[47] Park, S., *Relatives of a theorem of Rus–Hicks–Rhoades*, Lett. Nonlinear Anal. Appl. 1(2) (2023), 57–63.

[48] Park, S., *Almost all about Rus–Hicks–Rhoades maps in quasi-metric spaces*, Adv. Theory Nonlinear Anal. Appl. 7(2) (2023), 455–471. https://doi.org/0.31197/atnaa.1185449.

[49] Park, S., *The use of quasi-metric in the metric fixed-point theory*, J. Nonlinear Convex Anal. 25(7) (2024), 1553–1564.

[50] Park, S., *The realm of the Rus–Hicks–Rhoades maps in the metric fixed-point theory*, J. Nat. Acad. Sci. Repub. Korea, Nat. Sci. Ser. 63(1) (2024), 1–45.

[51] Park, S., *Several recent episodes on the metric completeness*, Edited by Debnath et al., to appear. RG on Jan. 11, 2024.

[52] Park, S., *Improving many metric fixed point theorems*, Lett. Nonlinear Anal. Appl. 2(2) (2024), 35–61.

[53] Park, S. and Rhoades, B. E., *Some general fixed point theorems*, Acta Sci. Math. 42 (1980), 299–304. Notices Amer. Math. Soc. Abstract 79T-B95.
[54] Pathak, H. K. and Shahzad, N., *Fixed points for generalized contractions and applications to control theory*, Nonlinear Anal. 68 (2008), 2181–2193.
[55] Petrov, E., *Fixed point theorem for generalized Kannan type mappings*, arXiv:2308.05419v1 [math.GN], 10 Aug 2023.
[56] Popescu, O., *A new type of contractive multivalued operators*, Bull. Sci. Math. 137 (2013), 30–44.
[57] Popescu, O., *A new class of contractive mappings*, Acta Math. Hung. 164 (2021), 570–579.
[58] Rhoades, B. E., *A comparison of various definitions of contractive definitions*, Trans. Am. Math. Soc. 226 (1977), 257–290.
[59] Romaguera, S., *Contractions of Kannan-type and Chatterjea-type on fuzzy quasi-metric spaces*, Results Nonlinear Anal. 5 (2022), 347–359.
[60] Rus, I. A., *Teoria Punctului fix, II*, Univ. Babes-Bolyai, Cluj, 1973.
[61] Salimi, P. and Karapinar, E., *Suzuki–Edelstein type contractions via auxiliary functions*, Math. Probl. Eng. 2013 (2013), 648528, 8 pages. https://doi.org/10.1155/2013/648528.
[62] Saravanan, K., Piramanantham, V. and Theivaraman, V. R., *New approximate fixed point results for rational contraction mappings*, Ratio Math. 492023 (2023), 69–87.
[63] Shobkolaei, N., Sedghi, S., Roshan, J. R. and Hussain, N., *Suzuki-type fixed point results in metric-like spaces*, J. Funct. Spaces Appl. 2013 (2013), 143686, 9 pages. https://doi.org/10.1155/2013/143686.
[64] Singh, S. L. and Mishra, S. N., *Fixed point theorems for single-valued and multi-valued maps*, Nonlinear Anal. 74 (2011), 2243–2248.
[65] Suzuki, T., *Generalized distance and existence theorems in complete metric spaces*, J. Math. Anal. Appl. 253 (2001), 440–458.
[66] Suzuki, T., *Generalized Caristi's fixed point theorems by Bae and others*, J. Math. Anal. Appl. 302 (2005), 502–508.
[67] Suzuki, T., *A generalized Banach contraction principle that characterizes metric completeness*, Proc. Am. Math. Soc. 136 (2008), 1861–1869.
[68] Suzuki, T., *Fixed point theorems and convergence theorems for some generalized nonexpansive mappings*, J. Math. Anal. Appl. 340 (2008), 1088–1095.
[69] Yeşilkaya, S. S., *On interpolative Hardy-Rogers contractive of Suzuki type mappings*, Topol. Algebra Appl. 9 (2021), 13–19.

Jyoti Saikia and Pradip Debnath

2 On the fixed points of Hardy–Rogers and Ćirić–Reich–Rus-type interpolative cyclic contraction in *b*-metric spaces and rectangular *b*-metric spaces

Abstract: Fixed-point theory in b-metric spaces has been a subject of increasing interest due to its scope of non-trivial generalization of classical metric space results. However, existing research has largely focused on standard contraction mappings, leaving a gap in the study of interpolative cyclic contractions in these generalized spaces. To address this gap, we establish two significant fixed-point theorems for Hardy–Rogers-type and Cirić–Reich–Rus-type cyclic contractions within both b-metric spaces and rectangular b-metric spaces. By employing an interpolative approach, our results provide a broader and more flexible framework for studying fixed points, extending beyond the constraints of traditional metric spaces. Furthermore, we validate our theoretical findings through concrete examples, demonstrating their applicability. Since b-metric spaces represent a non-trivial extension of classical metric spaces, this study significantly expands the scope of fixed-point theory by incorporating a wider class of contraction mappings. Our results not only strengthen the mathematical foundation of cyclic contractions in generalized spaces, but also pave the way for future research in nonlinear analysis and applied mathematics.

Keywords: Fixed point, contraction map, b-metric space, complete metric space, Hardy–Rogers-type cyclic contraction, Ćirić–Reich–Rus-type cyclic contraction

MSC 2020: 47H10, 54H25, 54E50

Conflict of interests: The authors declare that the work presented in this study was not impacted by any of their known conflicting financial interests or personal relationships.

Data availability: No new information was generated or used in this study.

Authors' contributions: Author J. S. contributed in Conceptualization, Investigation, Validation, Methodology, Writing the original draft; Author P. D. contributed in Conceptualization, Methodology, Investigation, Supervision and Editing.

Funding: This study was not supported by outside funds.

Jyoti Saikia, Department of Mathematics, Tingkhong College, Dibrugarh, Assam - 786612, India, e-mail: jyotisaikia885@gmail.com
Pradip Debnath, Department of Mathematical Sciences, Tezpur University, Napaam, Assam - 784028, India, e-mail: pradip@tezu.ernet.in

https://doi.org/10.1515/9783112215760-002

2.1 Introduction

The study of cyclic contractions plays a fundamental role in modern mathematics, with applications spanning fixed-point algorithms, functional analysis, and nonlinear analysis. While classical contraction mappings have been extensively explored, the investigation of interpolative cyclic contractions remains relatively underdeveloped, particularly within the framework of b-metric spaces and rectangular b-metric spaces. Addressing this gap, our study establishes new fixed-point theorems for two significant classes of interpolative cyclic contractions: Hardy–Rogers-type and Cirić–Reich–Rus-type contractions.

Fixed-point theory has been a pivotal mathematical tool since its early development by Joseph Liouville (1837) and Charles Émile Picard (1890), with applications in differential equations, optimization, and computational mathematics. Rooted in the method of successive approximations, this theory provides a foundation for determining the existence of solutions in various mathematical and engineering problems. Since Bakhtin (1989) introduced b-metric spaces, later expanded by Czerwik [6], researchers have extensively studied fixed-point results in this generalized setting. Several important contributions in b-metric spaces can be found in the literature [9, 1, 2, 11, 12, 15–18], highlighting the increasing interest in this field.

However, existing research has primarily focused on standard contraction mappings, leaving interpolative cyclic contractions largely unexplored in b-metric spaces. Motivated by prior studies on Hardy–Rogers and Cirić–Reich–Rus-type contractions [8, 13, 7, 3, 4], we extend these results by introducing and analyzing their interpolative counterparts within b-metric and rectangular b-metric spaces. These generalizations allow for a broader class of mappings, bridging the gap between classical and cyclic contractions, while enhancing the theoretical framework of fixed-point theory in generalized metric spaces.

By establishing the existence and uniqueness of fixed points for these newly defined contractions and supporting our findings with concrete examples, this study provides a significant advancement in the field. Our results not only extend the applicability of fixed-point theory but also open new avenues for future research in nonlinear analysis, optimization, and computational mathematics.

2.2 Preliminaries

We begin by listing certain key definitions that are helpful in our findings. Throughout the work, the symbols $\mathbb{N}$, $\mathbb{R}$, and $\mathbb{R}+$ represent the sets of natural numbers, real numbers, and non-negative real numbers, respectively.

Definition 2.1 ([5, 6]). Let $\mathfrak{B} \neq \phi$ be a set and the mapping $\mathfrak{b} : \mathfrak{B} \times \mathfrak{B} \to [0, \infty)$ with:

(i) $\mathfrak{b}(x_1, x_2) = 0 \iff x_1 = x_2$;

(ii) $\mathfrak{b}(x_1, x_2) = \mathfrak{b}(x_2, x_1)\ \forall x_1, x_2 \in \mathfrak{B}$;
(iii) $\exists s \in \mathbb{R}$ with $s \geq 1$ such that $\mathfrak{b}(x_1, x_3) \leq s[\mathfrak{b}(x_1, x_2) + \mathfrak{b}(x_1, x_3)]\ \forall x_1, x_2, x_3 \in \mathfrak{B}$.

Then $\mathfrak{b}$ is referred to as a $\mathfrak{b}$-metric on $\mathfrak{B}$ and $(\mathfrak{B}, \mathfrak{b})$. is referred to as a $\mathfrak{b}$-metric space with coefficient s.

Definition 2.2 ([10]). Let $\mathfrak{B} \neq \phi$ be a set and the mapping $\mathfrak{b} : \mathfrak{B} \times \mathfrak{B} \to [0, \infty)$ with:
(i) $\mathfrak{b}(x_1, x_2) = 0 \iff x_1 = x_2$;
(ii) $\mathfrak{b}(x_1, x_2) = \mathfrak{b}(x_2, x_1)\ \forall x_1, x_2 \in \mathfrak{B}$;
(iii) $\exists$ a real number $s \geq 1$ such that $\mathfrak{b}(x_1, x_2) \geq s[\mathfrak{b}(x_1, x_3) + \mathfrak{b}(x_3, x_4) + \mathfrak{b}(x_4, x_2)]\ \forall x_1, x_2 \in \mathfrak{B}$ and all distinct points $x_3, x_4 \in \mathfrak{B} \backslash \{x_1, x_2\}$.

Then $(\mathfrak{B}, \mathfrak{b})$ is a rectangular $\mathfrak{b}$-metric space with coefficient s, and $\mathfrak{b}$ is a rectangular $\mathfrak{b}$-metric on $\mathfrak{B}$.

Definition 2.3 ([10]). Consider a rectangular $\mathfrak{b}$-metric space $(\mathfrak{B}, \mathfrak{b})$, a sequence $\{x_n\}$ in $\mathfrak{B}$, and $x_1 \in \mathfrak{B}$. Then,
(i) The sequence $\{x_n\}$ is converges to x_1 in $(\mathfrak{B}, \mathfrak{b})$, if each $\epsilon > 0\ \exists\ n_0 \in \mathbb{N}$ such that $\mathfrak{b}(x_n, x_1) < \epsilon\ \ \forall\ n > n_0$ and $\lim_{x \longrightarrow \infty} x_n = x_1$;
(ii) The sequence $\{x_n\}$ in $(\mathfrak{B}, \mathfrak{b})$ is Cauchy sequence if for any $\epsilon > 0\ \exists\ n_0 \in \mathbb{N}$ such that $\mathfrak{b}(x_n, x_{n+p}) < \epsilon\ \ \forall\ n > n_0,\ p > 0$;
(iii) If every Cauchy sequence in $\mathfrak{B}$ converges to some $x_1 \in \mathfrak{B}$. Then $(\mathfrak{B}, \mathfrak{b})$ is said to be a complete(rectangular) $\mathfrak{b}$-metric space.

2.3 Results in $\mathfrak{b}$-metric spaces

In this section, we establish interpolative Hardy–Rogers-type cyclic contractions in $\mathfrak{b}$-metric spaces.

Definition 2.4. Let X_1 and X_2 be non-empty subsets of $\mathfrak{B}$ and let $(\mathfrak{B}, \mathfrak{b})$ be a b-metric space. A cyclic map $I : X_1 \cup X_2 \to X_1 \cup X_2$ is an interpolative Hardy–Rogers-type cyclic contraction if $\exists k \in [0, 1)$ and $\alpha_1, \alpha_2, \alpha_3 \in [0, 1)$, with $\alpha_1 + \alpha_2 + \alpha_3 < 1$ such that

$$\mathfrak{b}(Ix_1, Ix_2) \leq k[\mathfrak{b}(x_1, x_2)]^{\alpha_2}[\mathfrak{b}(Ix_1, x_1)]^{\alpha_1}[\mathfrak{b}(Ix_2, x_2)]^{\alpha_3}\left[\frac{1}{2s}(\mathfrak{b}(Ix_1, x_2) + \mathfrak{b}(Ix_2.x_1))\right]^{1-(\alpha_1+\alpha_2+\alpha_3)}. \tag{2.1}$$

$\forall (x_1, x_2) \epsilon X_1 \times X_2$ and $x_1, x_2 \notin Fix(I)$.

Theorem 2.5. *Let $I : X_1 \cup X_2 \to X_1 \cup X_2$ be an interpolative Hardy–Rogers-type cyclic contraction, let $(\mathfrak{B}, \mathfrak{b})$ be a complete b-metric space, and let X_1 and X_2 be non-empty subsets of $\mathfrak{B}$. Then, in $X_1 \cap X_2$, I has a unique fixed point.*

Proof. Fix $x_1 \in X_1$, by definition (2.1), we have

$$\mathfrak{b}(I^2x_1, Ix_1) \le k[\mathfrak{b}(Ix_1, x_1)]^{\alpha_2}[\mathfrak{b}(I^2x_1, Ix_1)]^{\alpha_1}[\mathfrak{b}(Ix_1, x_1)]^{\alpha_3}\left[\frac{1}{2s}\mathfrak{b}(I^2x_1, x_1) + \mathfrak{b}(Ix_1, Ix_1)\right]^{1-(\alpha_1+\alpha_2+\alpha_3)} \tag{2.2}$$

$$\implies [\mathfrak{b}(I^2x_1, Ix_1)]^{1-\alpha_1} \le k[\mathfrak{b}(Ix_1, x_1)]^{\alpha_2+\alpha_1}\left[\frac{1}{2s}(\mathfrak{b}(I^2x_1, x_1))\right]^{1-\alpha_1-\alpha_2-\alpha_3} \tag{2.3}$$

(by the third property of b-metric)

$$[\mathfrak{b}(I^2x_1, Ix_1)]^{1-\alpha_1} \le k[\mathfrak{b}(Ix_1, x_1)]^{\alpha_2+\alpha_1}\left[\frac{1}{2s}.s(\mathfrak{b}(I^2x_1, Ix_1) + \mathfrak{b}(Ix_1, x_1))\right]^{1-\alpha_1-\alpha_2-\alpha_3}. \tag{2.4}$$

Let us suppose $\mathfrak{b}(Ix_1, x_1) < \mathfrak{b}(I^2x_1, Ix_1)$.

Thus

$$\frac{1}{2}(\mathfrak{b}(I^2x_1, Ix_1) + \mathfrak{b}(Ix_1, x_1)) \le \mathfrak{b}(I^2x_1, Ix_1)$$

From (2.4)

$$\begin{aligned} [\mathfrak{b}(I^2x_1, Ix_1)]^{1-\alpha_1} &\le k[\mathfrak{b}(Ix_1, x_1)]^{\alpha_2+\alpha_1}[\mathfrak{b}(I^2x_1, Ix_1)]^{1-\alpha_1-\alpha_2-\alpha_3} \\ \implies [\mathfrak{b}(I^2x_1, Ix_1)]^{\alpha_2+\alpha_3} &\le k[\mathfrak{b}(Ix_1, x_1)]^{\alpha_2+\alpha_3}. \end{aligned}$$

Thus, $\mathfrak{b}(Ix_1, x_1) \ge \mathfrak{b}(I^2x_1, Ix_1)$, which is a contradiction.

Therefore,

$$\frac{1}{2}(\mathfrak{b}(I^2x_1, Ix_1) + \mathfrak{b}(Ix_1, x_1)) \le \frac{1}{2}(\mathfrak{b}(Ix_1, x_1) + \mathfrak{b}(Ix_1, x_1)) = \mathfrak{b}(Ix_1, x_1).$$

Therefore, (2.4) yields

$$\begin{aligned} [\mathfrak{b}(I^2x_1, Ix_1)]^{1-\alpha_1} &\le k[\mathfrak{b}(Ix_1, x_1)]^{\alpha_2+\alpha_3}[\mathfrak{b}(Ix_1, x_1)]^{1-\alpha_1-\alpha_2-\alpha_3} \\ \implies [\mathfrak{b}(I^2x_1, Ix_1)]^{1-\alpha_1} &\le k[\mathfrak{b}(Ix_1, x_1)]^{1-\alpha_1}. \end{aligned}$$

Proceeding in this way $\forall n \in \mathbb{N}$, we have

$$\mathfrak{b}(I^{n+1}x_1, I^nx_1) \le t^n\mathfrak{b}(Ix_1, x_1),$$

where $t = k^{\frac{1}{1-\alpha_1}}$ and $t \in (0, 1]$ so $t^n \in [0, 1)$.

From [14] we know that a sequence in a b-metric space $(\mathfrak{B}, \mathfrak{b})$, which satisfies the above inequality is a Cauchy sequence.

Therefore, I^nx_1 is a Cauchy sequence $\implies$ I^nx_1 is a convergent sequence, converges to $z \in X_1 \cup X_2$ such that $T^nx \to z$ as $n \to \infty$.

Again, the sequences $\{T^{2n}x_1\}$ and $\{T^{2n+1}x_1\}$ in X_1 and X_2 respectively have the same limit z. We conclude that $z \in X_1 \cap X_2$ is non-empty since X_1 and X_2 are closed.

We claim that $Iz = z$

$$\begin{aligned}
\mathfrak{b}(Iz, z) &= \lim_{x_1 \longrightarrow \infty} \mathfrak{b}(Iz, I^{2n}x_1) \\
&= \lim_{x_1 \longrightarrow \infty} \mathfrak{b}(Iz, I.I^{2n-1}x_1) \\
&\leq \lim_{x_1 \longrightarrow \infty} k[\mathfrak{b}(z, I^{2n-1}x_1)]^{\alpha_2}[\mathfrak{b}(I^{2n}x_1, I^{2n-1}x_1)]^{\alpha_1} \\
&\quad \times [\mathfrak{b}(Iz, z)]^{\alpha_3}\left[\frac{1}{2s}(\mathfrak{b}(Iz, x_1) + b(I^{2n}x_1, z))\right]^{1-\alpha_1-\alpha_2-\alpha_3}.
\end{aligned}$$

Taking $n \to \infty$ in the above inequality, we got $b(Iz, z) = 0 \implies Iz = z$.

To show that the fixed point z is unique, consider that there is $l \in X_1 \cup X_2$ such that $z \neq l$ and $Il = l$.

Considering that I is acyclic, we obtain $l \in X_1 \cap X_2$.

Thus,

$$\begin{aligned}
\mathfrak{b}(z, l) &= \mathfrak{b}(Iz.Il) \\
&\leq k[\mathfrak{b}(z, l)]^{\alpha_2}[\mathfrak{b}(Iz, z)]^{\alpha_1}[\mathfrak{b}(Il, l)]^{\alpha_3}\left[\frac{1}{2s}(\mathfrak{b}(Iz, l) + \mathfrak{b}(Il, z))\right]^{1-\alpha_1-\alpha_2-\alpha_3} \\
&= k[\mathfrak{b}(z, l)]^{\alpha_2}[\mathfrak{b}(z, z)]^{\alpha_1}[\mathfrak{b}(l, l)]^{\alpha_3}\left[\frac{1}{2s}(\mathfrak{b}(Iz, l) + \mathfrak{b}(Il, z))\right]^{1-\alpha_1-\alpha_2-\alpha_3} \\
&= 0 \\
\implies & \mathfrak{b}(z, l) \leq 0 \implies \mathfrak{b}(z, l) = 0.
\end{aligned}$$

Since $z = l$, we obtained that z is a unique fixed point of I. □

Corollary 2.6. *Consider two non-empty closed subsets of b-metric space $(\mathfrak{B}, \mathfrak{b})$, X_1 and X_2. Suppose that $I_1 : X_1 \to X_2$ and $I_2 : X_2 \to X_1$ are two functions.*

Assume that $\exists k \in [0, 1)$ and $\alpha_1, \alpha_2, \alpha_3 \in (0, 1)$ with sum of α_1, α_2 and α_3 is less than 1 such that

$$\mathfrak{b}(I_1x_1, I_2x_2) \leq [\mathfrak{b}(x_1, x_2)]^{\alpha_2}[\mathfrak{b}(I_1x_1, x_1)]^{\alpha_1}[\mathfrak{b}(I_2x_2, x_2)]^{\alpha_3}\left[\frac{1}{2s}(\mathfrak{b}(I_1x_1, x_2) + \mathfrak{b}(I_2x_2, x_1))\right]^{1-\alpha_1-\alpha_2-\alpha_3}$$

for all $(x_1, x_2) \in X_1 \times X_2$ with $x_1, x_2 \notin Fix(I_i)$, $i = 1, 2$.

Then, there exists a unique $z \in X_1 \cap X_2$ such that $I_1(z) = I_2(z) = z$

Proof. Let $I : X_1 \cup X_2 \to X_1 \cup X_2$ given by

$$I(x) = \begin{cases} I_1(x_1) & \text{if } x_1 \in X_1, \\ I_2(x_1) & \text{if } x_1 \in X_2. \end{cases}$$

Then, in b-metric space $(\mathfrak{B}, b)$, I is an interpolative Hardy–Rogers-type cyclic contraction. Using Theorem 2.5, we can now conclude that I has a fixed point $l \in X_1 \cap X_2$ such that $T(l) = l$.

Again,

$$\begin{aligned} & l \in X_1 \cap X_2 \\ \implies & l \in X_1 \text{ and } l \in X_2 \\ \implies & I_1(l) = l \text{ and } I_2(l) = l. \end{aligned}$$

□

Example 2.7. Let $E = [0, 2]$ and $I : X_1 \cup X_2 \to X_1 \cup X_2$ defined by

$$\mathfrak{b}(x_1, x_2) = \begin{cases} 0, & \text{if } x_1 = x_2, \\ 1, & \text{if } x_1 \neq x_2, \end{cases}$$

and

$$Ix = \begin{cases} \frac{7}{8}, & \text{if } x_1 \in [0, 1], \\ \frac{x}{3}, & \text{if } x_1 \in (1, 2]. \end{cases}$$

Suppose that $X_1 = X_2 = [0, 2]$ and $\mathfrak{b} : E \times E \to [0, +\infty)$ is the $\mathfrak{b}$-metric.

Case 1: If $x_1, x_2 \in [0.1]$, we get $\mathfrak{b}(Ix_1, Ix_2) = 0$, so Definition 2.4 holds.

Case 2: If $x_1 \in [0, 1]$ and $x_2 \in (1, 2]$ (vice-versa), then

$$\mathfrak{b}(Ix_1, Ix_2) = \mathfrak{b}\left(\frac{7}{8}, \frac{x_2}{3}\right) = 1 \quad \left(1 < x_2 \leq 2 \implies \frac{7}{8} \neq \frac{x_2}{3}\right)$$

and let us choose $k = \frac{1}{2}$, $\alpha_1 = \frac{1}{2}$, $\alpha_2 = \frac{1}{2}$, $\alpha_3 = \alpha_2 = \frac{1}{2}$ and $s = 2^2$. Then, we have

$$\begin{aligned} & \frac{1}{2}[\mathfrak{b}(x_1, x_2)]^{\frac{1}{2}}[\mathfrak{b}(Ix_1, x_1)]^{\frac{1}{2}}[\mathfrak{b}(Ix_2, x_2)]^{\frac{1}{2}}\left[\frac{1}{2 \cdot 2^2}(\mathfrak{b}(Ix_1, x_2) + \mathfrak{b}(Ix_2, x_1))\right]^{\frac{-1}{2}} \\ & = \frac{1}{2}(1)^{\frac{1}{2}}\left[\mathfrak{b}\left(\frac{7}{8}, x_1\right)\right]^{\frac{1}{2}}\left[\mathfrak{b}\left(\frac{y}{3}, x_2\right)\right]^{\frac{1}{2}}\left[\frac{1}{2 \cdot 2^2}\left(\mathfrak{b}\left(\frac{7}{8}, x_2\right) + \mathfrak{b}\left(\frac{x_2}{3}, x_1\right)\right)\right]^{\frac{-1}{2}} \\ & = \frac{1}{2} \cdot 1 \cdot 1 \cdot 1 \cdot \left(\frac{1}{2^2}\right)^{\frac{-1}{2}} \\ & = 1. \end{aligned}$$

Since $(0 \leq x_1 \leq 1; 1 < x_2 \leq 2, \implies x_2 \neq \frac{x_2}{3}, x_1 \neq \frac{7}{8}, x_1 \neq \frac{x_2}{3})$, therefore, the conditions of Definition 2.4 hold.

Case 3: Let $x_1, x_2 \in (1, 2]$.
Here $x_1 = x_2$, so we obtain $\mathfrak{b}(Ix_1, Ix_2) = 0$. Thus, the Definition 2.4 is satisfied.

Again, if $x_1 \neq x_2$, by Case 2, we will have the result.

Thus, we get that I is an interpolative Hardy–Rogers-type cyclic contraction with unique fixed point $x_1 = \frac{7}{8}$.

2.4 Results in rectangular $\mathfrak{b}$-metric spaces

We define and study interpolative Ćirić–Reich–Rus-type cyclic contraction in the framework of rectangular $\mathfrak{b}$-metric spaces in this section.

Definition 2.8. Let $(\mathfrak{B}, \mathfrak{b}, s)$ be a rectangular $\mathfrak{b}$-metric space. If there is a constant $k \in [0,1)$ and $\alpha_1, \alpha_2 \in (0,1)$ with $\alpha_1 + \alpha_2 < 1$, then a cyclic map $I : X_1 \cup X_2 \to X_1 \cup X_2$ is called an interpolative Ćirić–Reich–Rus-type cyclic contraction such that

$$\mathfrak{b}(Ix_1, Ix_2) \le k[\mathfrak{b}(x_1, x_2)]^{\alpha_2}[\mathfrak{b}(x_1, Ix_1)]^{\alpha_1}[\mathfrak{b}(x_2, Ix_2)]^{1-\alpha_1-\alpha_2}, \quad \forall x_1, x_2 \in \mathfrak{B}\backslash Fix(I). \tag{2.5}$$

Theorem 2.9. *Let $(\mathfrak{B}, \mathfrak{b}, s)$ be a complete rectangular $\mathfrak{b}$-metric space. If $I : X_1 \cup X_2 \to X_1 \cup X_2$ is an interpolative Ćirić–Reich–Rus-type cyclic contraction, then I has a fixed point in $X_1 \cap X_2$.*

Proof. Let $x_0 \in X_1$ and $\{\sigma_n\}$ be given by $\sigma_n = I^n x_0$ $\forall n \in \mathbb{N}$. If $\exists n_0 \in \mathbb{N}$ such that $\sigma_{n_0} = \sigma_{n_0+1}$, then σ_{n_0} is a fixed point of I and the proof is complete as $(\sigma_{n_0+1} = I\sigma n_0 = I\sigma_{n_0+1}) \implies \sigma_{n_0+1} = I\sigma_{n_0+1}$.

So let $\sigma_n \ne \sigma_{n+1}$ $\forall n \ge 0$.

Substituting x_1 by σ_n and x_2 by σ_{n-1}, and by Definition 2.8, we got

$$\begin{aligned}
\mathfrak{b}(I\sigma_n, I\sigma_{n-1}) &= \mathfrak{b}(\sigma_{n+1}, \sigma_n)\\
&\le k[\mathfrak{b}(\sigma_n, \sigma_{n-1})]^{\alpha_2}[\mathfrak{b}(\sigma_n, I\sigma_n)]^{\alpha_1}[\mathfrak{b}(\sigma_{n-1}, I\sigma_{n-1})]^{1-\alpha_1-\alpha_2}\\
&= k[\mathfrak{b}(\sigma_n, \sigma_{n-1})]^{\alpha_2}[\mathfrak{b}(\sigma_n, \sigma_{n+1})]^{\alpha_1}[\mathfrak{b}(\sigma_{n-1}, \sigma_n)]^{1-\alpha_1-\alpha_2}\\
&\le k[\mathfrak{b}(\sigma_n, \sigma_{n-1})]^{1-\alpha_1}[\mathfrak{b}(\sigma_n, \sigma_{n+1})]^{\alpha_1}\\
\implies & [\mathfrak{b}(\sigma_n, \sigma_{n+1})]^{1-\alpha_1} \le k[\mathfrak{b}(\sigma_n, \sigma_{n-1})]^{1-\alpha_1}\\
\implies & \mathfrak{b}(\sigma_n, \sigma_{n+1}) \le k^{\frac{1}{1-\alpha_1}}\mathfrak{b}(\sigma_n, \sigma_{n-1})\\
\implies & \mathfrak{b}(\sigma_n, \sigma_{n+1}) \le \mu\mathfrak{b}(\sigma_n, \sigma_{n-1}) = \mu\mathfrak{b}(\sigma_{n-1}, \sigma_n),
\end{aligned} \tag{2.6}$$

where $\mu = k^{\frac{1}{1-\alpha_1}} \in [0,1)$.

So, $\sigma_n \ne \sigma_{n+p}$ for $n \ge 0, p \ge 1$.

If $\sigma_n = \sigma_{n+p}$ for some $n \ge 0$ and $p \ge 1$, we get $\sigma_{n+1} = \sigma_{n+p+1}$. And inequality (2.6) implies

$$\begin{aligned}
\mathfrak{b}(\sigma_{n+1}, \sigma_n) &= \mathfrak{b}(\sigma_{n+p+1}, \sigma_{n+p})\\
&\le \mu^p\mathfrak{b}(\sigma_{n+1}, \sigma_n)\\
&< \mathfrak{b}(\sigma_{n+1}, \sigma_n)\\
\implies \mathfrak{b}(\sigma_{n+1}, \sigma_n) &< \mathfrak{b}(\sigma_{n+1}, \sigma_n),
\end{aligned}$$

which is a contraction.

From Definition 2.8, we obtain

$$\mathfrak{b}(\sigma_m, \sigma_n) \le k[\mathfrak{b}(\sigma_{m-1}, \sigma_{n-1})]^{\alpha_2}[\mathfrak{b}(\sigma_{m-1}, \sigma_m)]^{\alpha_1}[\mathfrak{b}(\sigma_{n-1}, \sigma_n)]^{1-\alpha_1-\alpha_2}, \quad \forall m, n \in \mathbb{N}. \tag{2.7}$$

Using classical inequality

$$u_1^{k_1} u_2^{k_2} u_3^{1-k_1-k_2} \le k_1 u_1 + k_2 u_2 + (1 - k_1 - k_2)u_3,$$

$\forall u_1, u_2, u_3 \in [0, \infty)$ and $k_1, k_2 \in [0, \infty)$ and $k_1 + k_2 \le 1$ from inequality (2.7) we have

$$\mathfrak{b}(\sigma_m, \sigma_n) \le k[\alpha_2 \mathfrak{b}(\sigma_{m-1}, \sigma_{n-1}) + \alpha_1 \mathfrak{b}(\sigma_{m-1}, \sigma_m) + (1 - \alpha_1 - \alpha_2)\mathfrak{b}(\sigma_{n-1}, \sigma_n)],$$

$\forall m, n \in \mathbb{N}$.

Therefore, from inequalities (2.6) and (2.7), we obtain

$$\mathfrak{b}(\sigma_m, \sigma_n) \le k\alpha_2 \mathfrak{b}(\sigma_{m-1}, \sigma_{n-1}) + \alpha_1 k^m \mathfrak{b}(\sigma_0, \sigma 1) + (1 - \alpha_1 - \alpha_2)k^n \mathfrak{b}(\sigma_0, \sigma_1), \quad \forall m, n \in \mathbb{N}.$$

So, from [14] we obtain that every sequences in a $\mathfrak{b}$-metric space $(\mathfrak{B}, \mathfrak{b}, s)$ is a Cauchy sequence if its satisfies the above property. Therefore, $\{\sigma_n\}$ is a Cauchy sequence. By completeness of $(\mathfrak{B}, \mathfrak{b}, s)$ $\exists x^* \in X$ such that $\lim_{x \longrightarrow \infty} \sigma_n = x^*$.

And as $\{\sigma_n\}$ is a sequence in X_1 and $\{\sigma_{n+1}\}$ is a sequence in X_2 with the same limit, we have $x^* \in X_1 \cap X_2$.

Also, x^* is now determined to be the fixed point of I. For any $n \in \mathbb{N}$.

$$\mathfrak{b}(x^*, Ix^*) \le s[\mathfrak{b}(x^*, \sigma_{n+1}) + \mathfrak{b}(\sigma_{n+1}, \sigma_{n+2}) + \mathfrak{b}(\sigma_{n+2}, Ix^*)].$$

Now, using Definition 2.8 and inequality (2.7), we have

$$\begin{aligned} \mathfrak{b}(x^*, Ix^*) &\le s[\mathfrak{b}(x^*, \sigma_{n+1}) + \mathfrak{b}(\sigma_{n+1}, \sigma_{n+2})] \\ &\quad + s[\mathfrak{b}(\sigma_{n+1}, x^*)]^{\alpha_2}[\mathfrak{b}(\sigma_{n+1}, \sigma_{n+2})]^{\alpha_1}[\mathfrak{b}(x^*, Ix^*)]^{1-\alpha_1-\alpha_2}. \end{aligned} \tag{2.8}$$

Taking limit as $n \to \infty$ in inequality (2.8). We have

$$\mathfrak{b}(x^*, Ix^*) = 0.$$

Hence, $Ix^* = x^*$.

Now, let $z \in X_1 \cap X_2$ and $I(z) = z$ such that $z \ne x^*$. Then from definition (2.5),

$$\begin{aligned} \mathfrak{b}(z, x^*) &= \mathfrak{b}(Iz, Ix^*) \\ &\le [\mathfrak{b}(z, x^*)]^{\alpha_2}[\mathfrak{b}(z, Iz)]^{\alpha_1}[\mathfrak{b}(x^*, Ix^*)]^{1-\alpha_1-\alpha_2}. \end{aligned}$$

Hence, $\mathfrak{b}(z, x^*) = 0$, i. e. $z = x^*$.

Therefore, x^* is the unique fixed point of $I : X_1 \cup X_2 \to X_1 \cup X_2$ in the complete rectangular $\mathfrak{b}$-metric space $(\mathfrak{B}, \mathfrak{b}, s)$. □

Example 2.10. Let $E = \mathbb{R}$ and let $\mathfrak{b}(x_1, x_2) = (x_1 - x_2)^4$ for each $x_1, x_2 \in E$. Then, $(\mathfrak{B}, \mathfrak{b})$ is a complete rectangular $\mathfrak{b}$-metric space with $s = 2$. Set

$$X_1 = \{0, 2\}, \quad X_2 = \{0, 3\}.$$

Define $I : X_1 \cup X_2 \to X_1 \cup X_2$ by $I0 = 0, I2 = 0, I3 = 1$ and let us choose $k = \frac{1}{2}, \alpha_1 = \frac{1}{2}$ $\alpha_2 = \frac{1}{2}$.

We have $b(x_1, x_1) = 0\ \forall x_1 \in X_1 \cup X_2$. So, definition (2.5) holds.

$$\mathfrak{b}(I2, I3) = \mathfrak{b}(0, 1) = 1 < \frac{1}{2}[\mathfrak{b}(2, 3)^4]^{\frac{1}{2}}[\mathfrak{b}(2, I2)^4]^{\frac{1}{2}}[\mathfrak{b}(3, I3)^4]^{1-\frac{1}{2}-\frac{1}{2}},$$
$$\mathfrak{b}(I3, I2) = \mathfrak{b}(1, 0) = 1 < \frac{1}{2}[\mathfrak{b}(3, 2)^4]^{\frac{1}{2}}[\mathfrak{b}(3, I3)^4]^{\frac{1}{2}}[\mathfrak{b}(2, I2)^4]^{1-\frac{1}{2}-\frac{1}{2}}.$$

Consequently, I is an interpolative Ćirić–Reich–Rus-type cyclic contraction with a unique fixed point $x = 0$, since it satisfies every criterion of Theorem 2.9.

2.5 Conclusion and future work

(i) This study introduces and explores Hardy–Rogers-type and Cirić–Reich–Rus-type cyclic contractions within the framework of b-metric spaces, expanding the existing literature of contraction mappings.

(ii) The study bridges existing contraction principles with cyclic mappings in b-metric spaces, contributing new theoretical insights that have not been explored in previous works.

(iii) We support our theoretical contributions with illustrative examples, ensuring that the established results are not merely theoretical constructs, but hold practical mathematical significance.

(iv) By considering interpolative contractions, our approach provides a generalization of classical fixed-point theorems, thus expanding their applicability to a wider range of mathematical and applied problems.

Future research directions

(i) Extension to partial metric and G-metric spaces—investigate whether similar fixed-point results hold in partial metric spaces or G-metric spaces, potentially leading to a broader theoretical framework.

(ii) Application in computational and applied mathematics—explore how these cyclic contractions can be applied in numerical methods, optimization problems, or differential equations, bridging theory and practical applications.

(iii) Fixed-point stability and iterative approximations—analyze the stability of fixed points under perturbations and develop iterative approximation techniques for computational implementations.
(iv) Multi-valued and stochastic extensions—extend the study to multi-valued mappings or stochastic contraction conditions, which could be relevant for probabilistic and fuzzy fixed-point theory.

Bibliography

[1] Amini-Harandi, A., *Fixed point theory for quasi-contraction maps in b-metric spaces*, Fixed Point Theory 15 (2014), 351–358.

[2] Aydi, H., Bota, M. F., Karapınar, E. and Mitrović, S., *A fixed point theorem for set-valued quasi-contractions in b-metric spaces*, Fixed Point Theory Appl. 2012, 88 (2012).

[3] Aydi, H., Chen, E. and Karapinar, E., *Interpolative Ćirić–Reich–Rus type contractions via the Branciri distance*, Mathematics 7(1) (2019), 84.

[4] Aydi, H., Karapinar, E. and Hierro, A. F. R., *ω-interpolative Ćirić–Reich–Rus type contractions*, Mathematics 7(57) (2019), 2–8.

[5] Bakhtin, I. A., *The contraction mapping in almost metric spaces*, Funct. Ana. Gos. Pedagog. Inst. Unianowsk. 30 (1989), 26–37.

[6] Czerwik, S., *Contraction mappings in b-metric spaces*, Acta Math. Univ. Ostrav. 1(1) (1993), 5–11.

[7] Debnath, P., Mitrović, Z. and Radenović, S., *Interpolative Hardy–Rogers and Reich–Rus–Ćirić type contractions in b-metric spaces and rectangular b-metric spaces*, Mat. Vesn. 72(4) (2020), 368–374.

[8] Edraoui, M., El Koufi, A. and Aamiri, M., *On interpolative Hardy–Rogers type cyclic contractions*, Appl. Gen. Topol. 25(1) (2024), 117–124.

[9] Ege, O., *Complex valued rectangular b-metric spaces and an application to linear equations*, J. Nonlinear Sci. Appl. 8(6) (2015), 1014–1021.

[10] George, R., Radenović, S., Reshma, K. P. and Shukla, S., *Rectangular b-metric spaces and contraction principle*, J. Nonlinear Sci. Appl. 8(6) (2015), 1005–1013.

[11] Hussain, N., Parvaneh, V., Roshan, J. R. and Kadelburg, Z., *Fixed points of cyclic weakly (ψ, ϕ, L, A, B)-contractive mappings in ordered b-metric spaces with applications*, Fixed Point Theory Appl. 256(2013) (2013), 1–18.

[12] Hussain, N., Saadati, R. and Agrawal, R. P., *On the topology and wt-distance on metric type spaces*, Fixed Point Theory Appl. 88(2014) (2014), 1–14.

[13] Karapinar, E., Alqahtani, O. and Aydi, H., *On interpolative Hardy–Rogers type contractions*, Symmetry 11(1) (2018), 8.

[14] Miculescu, R. and Mihail, A., *New fixed point theorems for set-valued contractions in b-metric spaces*, J. Fixed Point Theory Appl. 19 (2017), 2153–2163.

[15] Mukheimer, A. A., *α-ϕ-contractive mappings on b-metric space*, J. Comput. Anal. Appl. 18(4) (2015), 636–644.

[16] Nashine, H. K. and Kadelburg, Z., *Cyclic generalized φ-contractions in b-metric spaces and an application to integral equations*, Filomat 28(10) (2014), 2047–2057.

[17] Radenović, S., Došenović, T., Lampert, T. A. and Golubović, Z., *A note on some recent fixed point results for cyclic contractions in b-metric spaces and an application to integral equations*, Appl. Math. Comput. 273 (2016), 155–164.

[18] Roshan, J. R., Shobkolaei, N., Sedghi, S. and Abbas, M., *Common fixed point of four maps in b-metric spaces*, Hacet. J. Math. Stat. 43(4) (2014), 613–624.

Nihal Taş and Elif Kaplan

3 Hybrid-type fixed-point results on S-metric spaces

Abstract: Fixed-point theory is one of the most studied areas in recent times. This theory continues to be studied from different perspectives. One of these methods is hybrid contractive conditions. In this section, we obtain new fixed-point results by defining hybrid contractive condition on S-metric spaces. We also give an application to the fixed-circle problem. Finally, by leaving an open problem for G-metric spaces, new study topics is shed light.

Keywords: Hybrid contractive condition, S-metric space, fixed-circle problem

MSC 2020: Primary 54H25, Secondary 47H09, 47H10

3.1 Introduction

Fixed-point (FP) theory is one of the most studied areas of mathematics. It is an important subject to study, both theoretically and in terms of researching its applications. Metric FP theory basically provides methods to guarantee the existence and uniqueness of the FP of a given function. We research and study various FP theorems for this purpose. Different methods generalize the obtained FP results in this context, and the literature continues to introduce new results.

Using different techniques to generalize the contractive condition is the most common method. FP theory has introduced numerous contractive conditions for this purpose. For example, one of the generalized contractive conditions used is hybrid type contractive conditions (for example, see [1, 5, 6, 9, 19]).

In cases where generalization of the contractive condition is not sufficient, generalization of the metric space (MS) structure is often used. For example, the concept of an S-metric space (S-MS) has been defined as a generalization of MS [17], and various FP results have been studied on this space (see [2–4, 7, 10, 11, 18] and the references therein).

One of the most current generalization approaches is the fixed-circle problem, which is considered a geometric approach. This problem becomes important when the number of FPs of a given function is more than one. Both theoretical studies and research in various fields should take this importance into consideration. In order to obtain solutions to this problem, various theorems and results continue to be studied

Nihal Taş, Department of Mathematics, Balıkesir University, 10145 Balıkesir, Turkey, e-mail: nihaltas@balikesir.edu.tr
Elif Kaplan, Department of Mathematics, Ondokuz Mayıs University, Samsun, Turkey, e-mail: elifaydin@omu.edu.tr

https://doi.org/10.1515/9783112215760-003

both on MSs and generalized MSs, and the theory obtained is supported by the examples and applications given (see [12–15, 20]).

In the light of the studies mentioned above, in this section we obtain new FP results by using hybrid-type contractive conditions on S-MSs. First, we define the notion of Jaggi-type hybrid F_S-contraction (Cn). We then provide an example to show the consistency of the obtained FP theorem. Basic results regarding the proven main theorem are given, and hybrid-type FP results are obtained for b-MSs by using the relationship between S-metric and b-metric. Additionally, new results are provided to the fixed-circle problem as an application of the obtained results in a geometric sense.

3.2 Preliminaries

In this section, we recall the basic notions and results related to S-MSs.

Definition 3.1 ([17]). Let $\chi \neq \varnothing$ and $S : \chi \times \chi \times \chi \to [0, \infty)$ be a function. If S satisfies the following conditions for all $\alpha, \beta, \gamma, \eta \in \chi$:
(S1) $S(\alpha, \beta, \gamma) = 0 \Longleftrightarrow \alpha = \beta = \gamma$;
(S2) $S(\alpha, \beta, \gamma) \leq S(\alpha, \alpha, \eta) + S(\beta, \beta, \eta) + S(\gamma, \gamma, \eta)$;

then S is called an S-metric on χ and the pair (χ, S) is called an S-MS.

Lemma 3.1 ([17]). *Let (χ, S) be an S-MS. Then,*

$$S(\alpha, \alpha, \beta) = S(\beta, \beta, \alpha).$$

The above Lemma 3.1 can be considered as a symmetry condition on an S-MSs.

In [11] was investigated the relationships between a d metric and an S-metric. The formula for an S-MS that is generated by the metric d is as follows:

Supposing that (χ, d) is a MS. Therefore, the S-metric on χ is the function $S_d : \chi \times \chi \times \chi \to [0, \infty)$ defined by

$$S_d(\alpha, \beta, \gamma) = d(\alpha, \gamma) + d(\beta, \gamma),$$

for all $\alpha, \beta, \gamma \in \chi$. The S-metric S_d is termed the S-metric generated by d. Furthermore, there exists an instance of an S-metric that cannot be obtained from any metric d. (for more details, see [11]).

Definition 3.2 ([17]). Let (χ, S) be an S-MS and $A \subset \chi$.
(1) A sequence $\{\alpha_n\}$ in χ converges to α if and only if $S(\alpha_n, \alpha_n, \alpha) \to 0$ as $n \to \infty$. That is, there exists $n_0 \in \mathbb{N}$ such that for all $n \geq n_0$,

$$S(\alpha_n, \alpha_n, \alpha) < \varepsilon,$$

for each $\varepsilon > 0$. We denote this by

$$\lim_{n\to\infty} \alpha_n = \alpha$$

or

$$\lim_{n\to\infty} S(\alpha_n, \alpha_n, \alpha) = 0.$$

(2) A sequence $\{\alpha_n\}$ in χ is called a Cauchy sequence if

$$S(\alpha_n, \alpha_n, \alpha_m) \to 0$$

as $n, m \to \infty$. That is, there exists $n_0 \in \mathbb{N}$ such that for all $n, m \geq n_0$,

$$S(\alpha_n, \alpha_n, \alpha_m) < \varepsilon$$

for each $\varepsilon > 0$.

(3) The S-MS (χ, S) is called complete if every Cauchy sequence is convergent.

Lemma 3.2 ([17]). *Let (χ, S) be an S-MS. If $\{\alpha_n\}$ and $\{\beta_n\}$ are sequences in χ such that $\alpha_n \to \alpha$ and $\beta_n \to \beta$, then*

$$S(\alpha_n, \alpha_n, \beta_n) \to S(\alpha, \alpha, \beta).$$

Remark 3.1 ([18]). Every S-MS is topologically equivalent to a b-MS.

Corollary 3.1 ([18]). *Let $T : \chi \to \beta$ be a map from an S-MS χ to an S-MS β. Then T is continuous at $\alpha \in \chi$ if and only if $T\alpha_n \to T\alpha$, whenever $\alpha_n \to \alpha$.*

Now, we recall some necessary sets of mappings given in [16, 19, 21] and the references therein.

Let Ω_F be the set of all mappings $F : (0, \infty) \to \mathbb{R}$ satisfying the following conditions:

($F1$) F is strictly increasing, that is, for $\eta, \mu \in (0, \infty)$ such that

$$\eta < \mu \Longrightarrow F(\eta) < F(\mu).$$

($F2$) For each sequence $\{\alpha_n\}$ of positive numbers

$$\lim_{n\to\infty} \alpha_n = 0 \Longleftrightarrow \lim_{n\to\infty} F(\alpha_n) = -\infty.$$

($F3$) There exists $k \in (0, 1)$ such that

$$\lim_{\eta\to 0^+} \eta^k F(\eta) = 0.$$

$\psi : [0, \infty) \to [0, \infty)$ is called (c)-comparison function if its nondecreasing and there exist $n_0 \in \mathbb{N}$, $k \in (0, 1)$ and a convergent series

$$\sum_{n=1}^{\infty} \gamma_n$$

such that

$$\gamma_n \geq 0$$

and

$$\psi^{n+1}(t) \leq k\psi^n(t) + \gamma_n,$$

for $n \geq n_0$ and $t \geq 0$. Ψ represents the set of all functions that (c)-comparison functions.

Lemma 3.3 ([16]). *Let $\psi \in \Psi$. Then ψ satisfies the followings:*
(i) $\psi(t) < t$ *for any* $t \in \mathbb{R}^+$.
(ii) ψ *is continuous at* 0.
(iii) *The series* $\sum_{n=1}^{\infty} \psi^n(t)$ *is convergent for* $t \geq 0$.

3.3 Main results

In this section, we define the notion of a Jaggi-type hybrid F_S-Cn and prove some FP results on S-MSs.

Definition 3.3. Let (χ, S) be an S-MS and $T : \chi \to \chi$ be a self-mapping. T is called a Jaggi-type hybrid F_S-Cn if there exist $\tau > 0$, $\psi \in \Psi$ and $F \in \Omega_F$ such that

$$S(T\alpha, T\alpha, T\beta) > 0 \Longrightarrow \tau + F(S(T\alpha, T\alpha, T\beta)) \leq F(\psi(J^n_{T_S}(\alpha, \beta))),$$

for all $\alpha, \beta \in \chi$, $\alpha \neq \beta$, where $k \geq 0$,

$$J^k_{T_S}(\alpha, \beta) = \begin{cases} [\eta(\frac{S(\alpha,\alpha,T\alpha)S(\beta,\beta,T\beta)}{S(\alpha,\alpha,\beta)})^k + \mu(S(\alpha,\alpha,\beta))^k]^{\frac{1}{k}} & \text{if } k > 0,\ \alpha, \beta \in \chi,\ \alpha \neq \beta, \\ [S(\alpha,\alpha,T\alpha)]^\eta [S(\beta,\beta,T\beta)]^\mu & \text{if } k = 0, \alpha, \beta \in \chi - F_\chi(T), \end{cases}$$

$\eta, \mu \geq 0$ such that $\eta + \mu = 1$ and

$$F_\chi(T) = \{\alpha \in \chi : T\alpha = \alpha\}.$$

Theorem 3.1. *Let (χ, S) be a complete S-MS and $T : \chi \to \chi$ be a continuous Jaggi-type hybrid F_S-Cn. Then T has a unique FP χ. Also, the sequence $\{T^n\alpha_0\}$ converges to α for any $\alpha_0 \in \chi$.*

Proof. Let $\alpha_0 \in \chi$ such that $\alpha_0 \notin F_\chi(T)$ and let us define a sequence $\{\alpha_n\}$ by

$$T^n\alpha_0 = T\alpha_{n-1} = \alpha_n.$$

Assume that $S(a_n, a_n, a_{n+1}) > 0$ for all $n \in \mathbb{N} \cup \{0\}$. On the contrary, if there is $n_0 \in \mathbb{N} \cup \{0\}$ such that

$$S(a_{n_0}, a_{n_0}, a_{n_0+1}) = 0,$$

then we get

$$Ta_{n_0} = a_{n_0+1} = a_{n_0},$$

that is, a_{n_0} is a FP of T.

Let us consider the following two cases:

Case 1: Assume $k > 0$

Let us take

$$a = a_{n-1} \quad \text{and} \quad \beta = Ta_{n-1} = a_n.$$

Using the hypothesis, we get

$$\tau + F(S(Ta_{n-1}, Ta_{n-1}, T^2 a_{n-1})) \le F(\psi(J^k_{T_S}(a_{n-1}, Ta_{n-1}))),$$

where

$$\begin{aligned} J^k_{T_S}&(a_{n-1}, Ta_{n-1}) \\ &= \left[\eta\left(\frac{S(a_{n-1}, a_{n-1}, Ta_{n-1})S(Ta_{n-1}, Ta_{n-1}, T^2 a_{n-1})}{S(a_{n-1}, a_{n-1}, Ta_{n-1})}\right)^k + \mu(S(a_{n-1}, a_{n-1}, Ta_{n-1}))^k\right]^{\frac{1}{k}} \\ &= [\eta(S(a_n, a_n, a_{n+1}))^k + \mu(S(a_{n-1}, a_{n-1}, Ta_{n-1}))^k]^{\frac{1}{k}}. \end{aligned}$$

Hence, by the above two inequality, we obtain

$$\tau + F(S(a_n, a_n, a_{n+1})) \le F(\psi[\eta(S(a_n, a_n, a_{n+1}))^k + \mu(S(a_{n-1}, a_{n-1}, a_n))^k]^{\frac{1}{k}}).$$

Let $S(a_n, a_n, a_{n+1}) \ge S(a_{n-1}, a_{n-1}, a_n)$. By the non-decreasing property of ψ, we get

$$\begin{aligned} \tau + F(S(a_n, a_n, a_{n+1})) &\le F(\psi[\eta(S(a_n, a_n, a_{n+1}))^k + \mu(S(a_{n-1}, a_{n-1}, a_n))^k]^{\frac{1}{k}}) \\ &\le F(\psi[\eta(S(a_n, a_n, a_{n+1}))^k + \mu(S(a_n, a_n, a_{n+1}))^k]^{\frac{1}{k}}) \\ &= F(\psi[(\eta + \mu)(S(a_n, a_n, a_{n+1}))^k]^{\frac{1}{k}}) \\ &= F(\psi[S(a_n, a_n, a_{n+1})]) \\ &< F(S(a_n, a_n, a_{n+1})) \end{aligned}$$

which is a contradiction since $\tau > 0$. Therefore, we say that

$$S(\alpha_n, \alpha_n, \alpha_{n+1}) \le S(\alpha_{n-1}, \alpha_{n-1}, \alpha_n).$$

Hence, we obtain

$$\tau + F(S(\alpha_n, \alpha_n, \alpha_{n+1})) \le F(\psi[S(\alpha_{n-1}, \alpha_{n-1}, \alpha_n)])$$

and

$$F(S(\alpha_n, \alpha_n, \alpha_{n+1})) \le F(\psi[S(\alpha_{n-1}, \alpha_{n-1}, \alpha_n)]) - \tau.$$

Using the mathematical induction, we have

$$\begin{aligned} F(S(\alpha_n, \alpha_n, \alpha_{n+1})) &\le F(\psi^n(S(\alpha_0, \alpha_0, \alpha_1))) - n\tau \\ &\le F(\psi^n(S(\alpha_0, \alpha_0, \alpha_1))). \end{aligned}$$

Let

$$S(\alpha_n, \alpha_n, \alpha_{n+1}) = \alpha_n.$$

Then, we write

$$F(\alpha_n) \le F(\psi^n(\alpha_0)) \tag{3.1}$$

Let us take $n \to \infty$, we obtain

$$\lim_{n \to \infty} F(\alpha_n) = -\infty$$

and by (F_2), we have

$$\lim_{n \to \infty} \alpha_n = 0.$$

By (F_3), there exist $k \in (0, 1)$ such that

$$\lim_{n \to \infty} \alpha_n^k F(\alpha_n) = 0.$$

Using the inequality (3.1), we get

$$\alpha_n^k F(\alpha_n) - \alpha_n^k F(\psi^n(\alpha_0)) \le \alpha_n^k (F(\psi^n(\alpha_0)) - n\tau) - \alpha_n^k F(\psi^n(\alpha_0)),$$

for all $n \in \mathbb{N}$. Then, we have

$$\alpha_n^k F(\alpha_n) - \alpha_n^k F(\psi^n(\alpha_0)) \le -\alpha_n^k n\tau$$

and

$$\frac{1}{\tau}[\alpha_n^k F(\psi^n(\alpha_0)) - \alpha_n^k F(\alpha_n)] \geq n\alpha_n^k,$$

that is,

$$n\alpha_n^k \leq \frac{1}{\tau}[\alpha_n^k F(\psi^n(\alpha_0)) - \alpha_n^k F(\alpha_n)].$$

Let us take $n \to \infty$, we have

$$\lim_{n\to\infty} n\alpha_n^k = 0.$$

There is $n_1 \in \mathbb{N}$ such that

$$n\alpha_n^k \leq 1,$$

for all $n \geq n_1$. Then, we get

$$\alpha_n \leq \frac{1}{n^{\frac{1}{k}}},$$

for all $n \geq n_1$. Our goal now is to demonstrate that $\{\alpha_n\}$ is a Cauchy sequence. To achieve this, let us take $m, n \in \mathbb{N}$ such that

$$m > n \geq n_1.$$

Hence, using the triangle inequality and Lemma 3.1, we obtain

$$\begin{aligned}
S(\alpha_n, \alpha_n, \alpha_m) &\leq S(\alpha_n, \alpha_n, \alpha_{n+1}) + S(\alpha_n, \alpha_n, \alpha_{n+1}) + S(\alpha_m, \alpha_m, \alpha_{n+1}) \\
&= 2S(\alpha_n, \alpha_n, \alpha_{n+1}) + S(\alpha_{n+1}, \alpha_{n+1}, \alpha_m) \\
&\leq 2(\alpha_n + \alpha_{n+1} + \cdots + \alpha_{m-1}) \\
&= 2\sum_{i=n}^{m-1} \alpha_i \\
&< 2\sum_{i=n}^{\infty} \alpha_i \\
&\leq \sum_{i=n}^{\infty} \frac{2}{i^{\frac{1}{k}}}
\end{aligned}$$

and so $\{\alpha_n\}$ is Cauchy.

From the completeness hypothesis, there exists $\gamma \in \chi$ such that

$$\lim_{n\to\infty} \alpha_n = \gamma.$$

By the continuity hypothesis, we get

$$\begin{aligned} S(T\gamma, T\gamma, \gamma) &= \lim_{n\to\infty} S(T\alpha_n, T\alpha_n, \alpha_n) \\ &= \lim_{n\to\infty} S(\alpha_{n+1}, \alpha_{n+1}, \alpha_n) \\ &= 0, \end{aligned}$$

that is,

$$T\gamma = \gamma.$$

Case 2: Assume $k = 0$
Let us take

$$\alpha = \alpha_{n-1} \quad \text{and} \quad \beta = T\alpha_{n-1} = \alpha_n.$$

Using the contractive property hypothesis, we get

$$\tau + F(S(T\alpha_{n-1}, T\alpha_{n-1}, T^2\alpha_{n-1})) \le F(\psi(J^k_{T_S}(\alpha_{n-1}, T\alpha_{n-1}))),$$

where

$$\begin{aligned} J^k_{T_S}(\alpha_{n-1}, T\alpha_{n-1}) &= (S(\alpha_{n-1}, \alpha_{n-1}, T\alpha_{n-1}))^\eta (S(T\alpha_{n-1}, T\alpha_{n-1}, T^2\alpha_{n-1}))^\mu \\ &= (S(\alpha_{n-1}, \alpha_{n-1}, \alpha_n))^\eta (S(\alpha_n, \alpha_n, \alpha_{n+1}))^\mu. \end{aligned}$$

Hence, we have

$$\tau + F(S(\alpha_n, \alpha_n, \alpha_{n+1})) \le F(\psi[(S(\alpha_{n-1}, \alpha_{n-1}, \alpha_n))^\eta (S(\alpha_n, \alpha_n, \alpha_{n+1}))^\mu]).$$

Suppose that

$$S(\alpha_n, \alpha_n, \alpha_{n+1}) \ge S(\alpha_{n-1}, \alpha_{n-1}, \alpha_n).$$

By the above inequality, we get

$$\begin{aligned} \tau + F(S(\alpha_n, \alpha_n, \alpha_{n+1})) &\le F(\psi[(S(\alpha_{n-1}, \alpha_{n-1}, \alpha_n))^\eta (S(\alpha_n, \alpha_n, \alpha_{n+1}))^\mu]) \\ &\le F(\psi[(S(\alpha_n, \alpha_n, \alpha_{n+1}))^{\eta+\mu}]) \\ &= F(\psi(S(\alpha_n, \alpha_n, \alpha_{n+1}))) \\ &< F(S(\alpha_n, \alpha_n, \alpha_{n+1})), \end{aligned}$$

which is a contradiction since $\tau > 0$. Therefore, we have

$$S(\alpha_n, \alpha_n, \alpha_{n+1}) \le S(\alpha_{n-1}, \alpha_{n-1}, \alpha_n)$$

and

$$\tau + F(S(\alpha_n, \alpha_n, \alpha_{n+1})) \le F(\psi(S(\alpha_{n-1}, \alpha_{n-1}, \alpha_n))).$$

Using the mathematical induction, we have

$$F(S(\alpha_n, \alpha_n, \alpha_{n+1})) \le F(\psi^n(S(\alpha_0, \alpha_0, \alpha_1))) - n\tau.$$

We demonstrate that $\{\alpha_n\}$ is Cauchy using the similar reasoning as in Case 1. From the completeness hypothesis, there exists $\gamma \in \chi$ such that

$$\lim_{n\to\infty} \alpha_n = \gamma.$$

Hence, γ is a FP of T.

Now, we prove that T has a unique FP. On the contrary, let w be an another FP of T such that $w \neq \gamma$. Using the hypothesis, we obtain

$$\tau + F(S(T\gamma, T\gamma, Tw)) \le F(\psi(J^k_{T_S}(\gamma, w))). \tag{3.2}$$

Case 1: Let $k > 0$

Then we have

$$\begin{aligned} J^k_{T_S}(\gamma, w) &= \left[\eta\left(\frac{S(\gamma,\gamma,T\gamma)S(w,w,Tw)}{S(\gamma,\gamma,w)}\right)^k + \mu(S(\gamma,\gamma,w))^k\right]^{\frac{1}{k}} \\ &= \mu^{\frac{1}{k}} S(\gamma,\gamma,w) \\ &\le S(\gamma,\gamma,w). \end{aligned}$$

From the inequality (3.2), we obtain

$$\begin{aligned} \tau + F(S(\gamma,\gamma,w)) &\le F(\psi(S(\gamma,\gamma,w))) \\ &< F(S(\gamma,\gamma,w)), \end{aligned}$$

a contradiction since $\tau > 0$.

Case 2: Let $k = 0$

Then we have

$$\begin{aligned} J^k_{T_S}(\gamma, w) &= (S(\gamma,\gamma,T\gamma))^\eta (S(w,w,Tw))^\mu \\ &= (S(\gamma,\gamma,\gamma))^\eta (S(w,w,w))^\mu \\ &= 0. \end{aligned}$$

Using the hypothesis, we get

$$\tau + F(S(y, y, w)) \leq F(\psi(0)) < F(0),$$

a contradiction with the definition of F.

Consequently, y is a unique FP of T. □

Remark 3.2. (1) If the S-metric is generated by any metric d, then the notion of a Jaggi-type hybrid F_S-Cn and the notion of a Jaggi-type hybrid F-Cn are coincidence.
(2) Our main theorem generalize Theorem 3.2, proved in [19]. Also, if the used S-metric is generated by any metric d, the these two theorems are coincidence.
(3) Our main theorem gives us the existence and uniqueness theorem for a FP of a self-mapping T. In this context, our theorem also becomes important in items of generalizing the Banach FP theorem in S-MSs.

We inspire Example 3.3 given in [19] and give a new example for the S-metric, which is not generated by any metric d.

Example 3.1. Let $\chi = [0, \infty)$ and the function $S : \chi \times \chi \times \chi \to [0, \infty)$ be defined as

$$S(\alpha, \beta, \gamma) = |\alpha - \gamma| + |\alpha + \gamma - 2\beta|,$$

for all $\alpha, \beta, \gamma \in \chi$ [11]. This S-metric is not generated by any metric d (for more details, see [11]). Then (χ, S) is a complete S-MS. Let us consider the self-mapping $T : \chi \to \chi$ be defined by

$$T\alpha = \begin{cases} \frac{1}{8}\alpha e^{-\tau} & \text{if } \alpha \in [0, 1], \\ \frac{1}{8}e^{-\tau} & \text{if } \alpha > 1, \end{cases}$$

for all $\alpha \in \chi$ where $\tau > 0$. The self-mapping T is a continuous function. Let us define the function $\psi : \mathbb{R}^+ \to \mathbb{R}^+$ by

$$\psi(t) = \frac{t}{8},$$

for all $t > 0$ and the function $F : (0, \infty) \to \mathbb{R}$ by

$$F(\eta) = \ln(\eta^2 + \eta),$$

where $\eta > 0$. If $\alpha, \beta \in (1, \infty)$, then the proof is clear. Hence, we suppose that $\alpha, \beta \in [0, 1)$ such that $T\alpha \neq T\beta$ and $\eta = 0$, $\mu = 1$. Under the following cases. We show that T is a Jaggi-type hybrid F_S-Cn.

Case 1:
For $k > 0$, let us take $k = 1$. Then we get

$$\begin{aligned}S(T\alpha, T\alpha, T\beta) &= 2|T\alpha - T\beta| = 2\frac{1}{8}e^{-\tau}|\alpha - \beta| = e^{-\tau}\frac{S(\alpha,\alpha,\beta)}{8}\\ &= e^{-\tau}\psi(S(\alpha,\alpha,\beta)) \le e^{-\tau}\psi(J^k_{T_S}(\alpha,\beta))\end{aligned}$$

and

$$\begin{aligned}S(T\alpha, T\alpha, T\beta)[S(T\alpha, T\alpha, T\beta) + 1] &\le e^{-\tau}\psi(S(\alpha,\alpha,\beta))[e^{-\tau}\psi(S(\alpha,\alpha,\beta) + 1)]\\ &\le e^{-\tau}[\psi(J^k_{T_S}(\alpha,\beta))(\psi(J^k_{T_S}(\alpha,\beta)) + 1)].\end{aligned}$$

Hence, we obtain

$$\begin{aligned}\tau + F(S(T\alpha, T\alpha, T\beta)) &= \tau + \ln[(S(T\alpha, T\alpha, T\beta))^2 + S(T\alpha, T\alpha, T\beta)]\\ &\le \tau + \ln[e^{-\tau}(\psi(J^k_{T_S}(\alpha,\beta))^2 + \psi(J^k_{T_S}(\alpha,\beta)))]\\ &= \ln[\psi(J^k_{T_S}(\alpha,\beta))^2 + \psi(J^k_{T_S}(\alpha,\beta))]\\ &= F(\psi(J^k_{T_S}(\alpha,\beta))).\end{aligned}$$

Case 2:
For $s = 0$, let us take $\eta = 0$, $\mu = 1$ and $\alpha = T\beta$ for all $\alpha, \beta \in \chi$. Then we get

$$S(T\alpha, T\alpha, T\beta) \le e^{-\tau}\psi(J^k_{T_S}(\alpha,\beta)).$$

By the similar arguments used in Case 1, we have

$$\tau + F(S(T\alpha, T\alpha, T\beta)) \le F(\psi(J^k_{T_S}(\alpha,\beta))).$$

Consequently, T satisfies all conditions of our main theorem. Then, $\alpha = 0$ is the unique FP of T, that is,

$$F_\chi(T) = \{0\}.$$

Now, we give the following corollaries:

Corollary 3.2. *Let (χ, S) be a complete S-MS and $T : \chi \to \chi$ be a continuous self-mapping. If there exist $\tau > 0$ and $F \in \Omega_F$ such that*

$$\tau + F(S(T\alpha, T\alpha, T\beta)) \le F(\eta(A^k_{T_S}(\alpha,\beta)))$$

for all $\alpha, \beta \in \chi$, where

$$A^k_{T_S}(\alpha,\beta) = \left[\eta\left(\frac{S(\alpha,\alpha,T\alpha)S(\beta,\beta,T\beta)}{S(\alpha,\alpha,\beta)}\right)^k + \mu(S(\alpha,\alpha,\beta))^k\right]^{\frac{1}{k}},$$

$\eta,\mu \geq 0$ with $\eta+\mu=1$, $k>0$ and $\eta \in (0,1)$, then T has a unique FP in χ.

Proof. Let us define the mapping $\psi : \mathbb{R}^+ \to \mathbb{R}^+$ as

$$\psi(t) = \eta^t,$$

for all $t > 0$ and $\eta \in (0,1)$. Then, from our main theorem, T has a unique FP γ in χ. □

Corollary 3.3. *Let (χ, S) be a complete S-MS and $T : \chi \to \chi$ be a continuous self-mapping. If there exist $\tau > 0$ and $F \in \Omega_F$ such that*

$$\tau + F(S(T\alpha, T\alpha, T\beta)) \leq F(\psi(B^k_{T_S}(\alpha,\beta)))$$

for all $\alpha, \beta \in \chi$, where

$$B^k_{T_S}(\alpha,\beta) = [S(\alpha,\alpha,T\alpha)]^\eta [S(\beta,\beta,T\beta)]^\mu,$$

$\eta,\mu \geq 0$ with $\eta+\mu=1$, then T has a unique FP in χ.

Proof. Let us consider $k=0$ in our main theorem. Then the proof can be easily seen. □

Corollary 3.4. *Let (χ, S) be a complete S-MS and $T : \chi \to \chi$ be a continuous self-mapping. If there exist $\tau > 0$ and $F \in \Omega_F$ such that*

$$\tau + F(S(T\alpha, T\alpha, T\beta)) \leq F(\eta(C^k_{T_S}(\alpha,\beta)))$$

for all $\alpha, \beta \in \chi$, where

$$C^k_{T_S}(\alpha,\beta) = [S(\alpha,\alpha,T\alpha)S(\beta,\beta,T\beta)]^{\frac{1}{2}},$$

$\eta \in (0,1)$, then T has a unique FP in χ.

Proof. Let us take $\eta = \mu = \frac{1}{2}$ in our main theorem. Then the proof is clear. □

Remark 3.3. If the used S-metric is generated by any metric, then our corollaries coincidence with Corollaries 3.4–3.6 given in [19].

Since every S-metric generates a b-metric (see [18] for more details), the following definition, theorem and corollaries can be easily obtained.

Definition 3.4. Let (χ, d^S) be a b-MS and $T : \chi \to \chi$ be a self-mapping. T is called a Jaggi-type hybrid F_b-Cn if there are $\tau > 0$, $\psi \in \Psi$ and $F \in \Omega_F$ such that

$$d^S(T\alpha, T\beta) > 0 \Longrightarrow \tau + F(d^S(T\alpha, T\beta)) \leq F(\psi(J^k_{T_b}(\alpha,\beta))),$$

for all $\alpha, \beta \in \chi$, $\alpha \neq \beta$ where $k \geq 0$,

$$J_{T_b}^k(\alpha,\beta) = \begin{cases} [\eta(\frac{d^S(\alpha,T\alpha)d^S(\beta,T\beta)}{d^S(\alpha,\beta)}) + \mu(d^S(\alpha,\beta))^k]^{\frac{1}{k}} & \text{if } k > 0,\ \alpha,\beta \in \chi,\ \alpha \neq \beta, \\ (d^S(\alpha,T\alpha))^\eta (d^S(\beta,T\beta))^\mu & \text{if } k = 0,\ \alpha,\beta \in \chi - F_\chi(T), \end{cases}$$

$\eta, \mu \geq 0$ such that $\eta + \mu = 1$.

Theorem 3.2. *Let (χ, d^S) be a complete b-MS and $T : \chi \to \chi$ be a continuous Jaggi-type hybrid F_b-Cn. Then T has a unique FP χ.*

Corollary 3.5. *Let (χ, d^S) be a complete b-MS and $T : \chi \to \chi$ be a continuous self-mapping. If there exist $\tau > 0$ and $F \in \Omega_F$ such that*

$$\tau + F(d^S(T\alpha, T\beta)) \leq F(\eta(A_{T_b}^k(\alpha,\beta))),$$

for all $\alpha, \beta \in \chi$, where

$$A_{T_b}^k(\alpha,\beta) = \left[\eta\left(\frac{d^S(\alpha,T\alpha)d^S(\beta,T\beta)}{d^S(\alpha,\beta)}\right)^k + \mu(d^S(\alpha,\beta))^k\right]^{\frac{1}{k}},$$

$\eta, \mu \geq 0$ with $\eta + \mu = 1$, $k > 0$ and $\eta \in (0,1)$, then T has a unique FP in χ.

Corollary 3.6. *Let (χ, d^S) be a complete b-MS and $T : \chi \to \chi$ be a continuous self-mapping. If there exist $\tau > 0$ and $F \in \Omega_F$ such that*

$$\tau + F(d^S(T\alpha, T\beta)) \leq F(\psi(B_{T_b}^k(\alpha,\beta))),$$

for all $\alpha, \beta \in \chi$, where

$$B_{T_b}^k(\alpha,\beta) = [d^S(\alpha,T\alpha)]^\eta [d^S(\beta,T\beta)]^\mu,$$

$\eta, \mu \geq 0$ with $\eta + \mu = 1$, then T has a unique FP in χ.

Corollary 3.7. *Let (χ, d^S) be a complete b-MS and $T : \chi \to \chi$ be a continuous self-mapping. If there exist $\tau > 0$ and $F \in \Omega_F$ such that*

$$\tau + F(d^S(T\alpha, T\beta)) \leq F(\eta(C_{T_b}^k(\alpha,\beta)))$$

for all $\alpha, \beta \in \chi$, where

$$C_{T_b}^k(\alpha,\beta) = [d^S(\alpha,T\alpha)d^S(\beta,T\beta)]^{\frac{1}{2}},$$

$\eta \in (0,1)$, then T has a unique FP in χ.

3.4 An application to the fixed-circle problem

In this section, we give an application to the fixed-circle problem as a geometric approach. Now, we recall the notions of a circle and a disc on an S-MS given in [13, 17], respectively.

Let $(\chi, \mathcal{S})$ be an S-MS and $T : \chi \to \chi$ a self-mapping. A circle $C^S_{a_0,r}$ and a disc $D^S_{a_0,r}$ are defined as follows:

$$C^S_{a_0,r} = \{a \in \chi : \mathcal{S}(a, a, a_0) = r\}$$

and

$$D^S_{a_0,r} = \{a \in \chi : \mathcal{S}(a, a, a_0) \leq r\},$$

where $r \in [0, \infty)$.

If $Ta = a$ for all $a \in C^S_{a_0,r}$ (resp. $a \in D^S_{a_0,r}$), then the circle $C^S_{a_0,r}$ (resp. the disc $D^S_{a_0,r}$) is called as the fixed circle (resp. the fixed disc) of T.

Definition 3.5. Let (χ, S) be an S-MS and $T : \chi \to \chi$ be a self-mapping. T is called a Jaggi-type hybrid $F^S_{a_0}$-Cn if there exist $\tau > 0$, $\psi \in \Psi$, $F \in \Omega_F$ and $a_0 \in \chi$ such that

$$S(a, a, Ta) > 0 \Longrightarrow \tau + F(S(a, a, Ta)) \leq F(\psi(J^k_{T_S}(a, a_0))),$$

for all $a, \beta \in \chi$ with $a \neq \beta$.

Proposition 3.1. *Let (χ, S) be an S-MS and $T : \chi \to \chi$ be a self-mapping. T is called a Jaggi-type hybrid $F^S_{a_0}$-Cn with $a_0 \in \chi$. Then, we have $a_0 \in F_\chi(T)$, that is,*

$$S(a_0, a_0, Ta_0) > 0.$$

Proof. Using the Cn hypothesis, we consider the following two cases:

Case 1: Let $k > 0$
Then we have

$$J^k_{T_S}(a_0, a_0) = \infty,$$

a contradiction.

Case 2: Let $k = 0$
Then we have

$$J^k_{T_S}(a_0, a_0) = S(a_0, a_0, Ta_0),$$

and

$$\begin{aligned}\tau + F(S(a_0, a_0, Ta_0)) &\leq F(\psi(J^k_{T_S}(a_0, a_0)))\\ &= F(\psi(S(a_0, a_0, Ta_0)))\\ &< F(S(a_0, a_0, Ta_0))\end{aligned}$$

a contradiction with $\tau > 0$. Consequently, it should be $a_0 \in F_\chi(T)$. □

Theorem 3.3. *Let (χ, S) be an S-MS and $T : \chi \to \chi$ be a Jaggi-type hybrid $F^S_{a_0}$-Cn with $a_0 \in \chi$ and the number r be defined as*

$$r = \inf\{S(a, a, Ta) : a \notin F_\chi(T),\ a \in \chi\}.$$

Then we have $C_{a_0,r} \subseteq F_\chi(T)$.

Proof. Let $a \in C_{a_0,r}$ such that $a \notin F_\chi(T)$. Then using the hypothesis, we get

$$\tau + F(S(a, a, Ta)) \leq F(\psi(J^k_{T_S}(a, a_0))). \tag{3.3}$$

We consider the following cases:

Case 1: Let $k > 0$
From Proposition 3.1, we have

$$\begin{aligned}J^k_{T_S}(a, a_0) &= \left[\eta\left(\frac{S(a, a, Ta)S(a_0, a_0, Ta_0)}{S(a, a, a_0)}\right)^k + \mu(S(a, a, a_0))^k\right]^{\frac{1}{k}}\\ &= \mu^{\frac{1}{k}} S(a, a, a_0)\\ &= \mu^{\frac{1}{k}}.r\\ &\leq \mu^{\frac{1}{k}} S(a, a, Ta).\end{aligned}$$

From (3.3), we obtain

$$\begin{aligned}\tau + F(S(a, a, Ta)) &\leq F(\psi(\mu^{\frac{1}{k}} S(a, a, Ta)))\\ &< F(\mu^{\frac{1}{k}} S(a, a, Ta))\\ &\leq F(S(a, a, Ta)),\end{aligned}$$

a contradiction with $\tau > 0$.

Case 2: Let $k = 0$
From Proposition 3.1, we have

$$J^k_{T_S}(a, a_0) = (S(a, a, Ta))^\eta (S(a_0 a_0 Ta_0))^\mu = 0.$$

From (3.3), we have

$$\tau + F(S(\alpha, \alpha, T\alpha)) \le F(\psi(0)) = F(0),$$

a contradiction with the definition of F. Consequently, $C^S_{\alpha_0,r}$ is a fixed circle of T, that is,

$$C^S_{\alpha_0,r} \subseteq F_\alpha(T). \qquad \square$$

Example 3.2. Let $\chi = \mathbb{R}$ be the S-MS with the S-metric give in Example 3.1. Let us define the self-mapping $T : \chi \to \chi$ by

$$T\alpha = \begin{cases} 6 & \text{if } \alpha = 8, \\ \alpha & \text{if } \alpha \in \mathbb{R} - \{8\}, \end{cases}$$

for all $\alpha \in \mathbb{R}$. T is a Jaggi-type hybrid $F^S_{\alpha_0}$-Cn with $\alpha_0 = 0$, $\psi(t) = \frac{t}{2}$, $\tau = \ln 2$, $k = 1 = \mu$, $\eta > 0$ and $F = \ln \alpha$. Indeed, for $\alpha = 8$, we have

$$S(8, 8, T8) = S(8, 8, 6) = 4$$

and

$$\begin{aligned} J^k_{T_S}(8, 0) &= \left[\eta\left(\frac{S(8, 8, T8)S(0, 0, T0)}{S(8, 8, 0)}\right)^k + \mu(S(8, 8, 0)^k)\right]^{\frac{1}{k}} \\ &= \mu^{\frac{1}{k}} S(8, 8, 0) \\ &= 16. \end{aligned}$$

Hence, we get

$$\begin{aligned} \ln 8 = \ln 2 + \ln 4 = \tau + \ln(S(8, 8, 6)) &= \tau + F(S(\alpha, \alpha T\alpha)) \\ \le \ln 8 = \ln\left(\frac{16}{2}\right) &= \ln(\psi(16)) \\ = F(\psi(J^k_{T_S}(8, 0))) &= F(\psi(J^k_{T_S}(\alpha, \alpha_0))). \end{aligned}$$

Also, we have

$$r = \inf\{S(\alpha, \alpha, T\alpha) : \alpha \notin F_\chi(T)\} = 4$$

and

$$\begin{aligned} C^S_{0,4} &= \{\alpha \in \mathbb{R} : S(\alpha, \alpha, 0) = 4\} \\ &= \{\alpha \in \mathbb{R} : 2|\alpha| = 4\} \\ &= \{-2, 2\}. \end{aligned}$$

Consequently, $C^S_{0,4}$ is a fixed circle of T, that is,

$$C^S_{0,4} = \{-2, 2\} \subseteq F_\chi(T) = \mathbb{R} - \{8\}.$$

3.5 Conclusions and future works

In this chapter, we present new FP results with the fixed-circle application on S-MSs. The obtained results are supported by some examples. In these examples, we use the special S-metric, which is not generated by any metric d. For this reason, this study is important for FP theory in view of generalized MSs.

In the literature, we know that the class of S-metrics and G-metrics are distinct (for more details, see [2] and the references therein). In this context, we define the following notion on G-MSs introduced in [8]:

Definition 3.6. Let (χ, G) be a G-MS and $T : \chi \to \chi$ be a self-mapping. T is called a Jaggi-type F_G-Cn if there exist $\tau > 0$, $\psi \in \Psi$ and $F \in \Omega_F$ such that

$$G(T\alpha, T\alpha, T\beta) > 0 \Longrightarrow \tau + F(G(T\alpha, T\alpha, T\beta)) \le F(\psi(J^k_{T_G}(\alpha, \beta))),$$

for all $\alpha, \beta \in \chi$, $\alpha \neq \beta$ where $k \geq 0$,

$$J^k_{T_G}(\alpha,\beta) = \begin{cases} [\eta(\frac{G(\alpha,\alpha,T\alpha)G(\beta,\beta,T\beta)}{G(\alpha,\alpha,\beta)})^k + \mu(G(\alpha,\alpha,\beta))^k]^{\frac{1}{k}} & \text{if } k > 0,\ \alpha,\beta \in \chi,\ \alpha \neq \beta, \\ (G(\alpha,\alpha,T\alpha))^{\eta}(G(\beta,\beta,T\beta))^{\mu} & \text{if } k = 0,\ \alpha,\beta \in \chi - F_\chi(T), \end{cases}$$

$\eta, \mu \geq 0$ such that $\eta + \mu = 1$.

Using this definition, it can be investigated some answers to the following questions:
(1) Are these any existence FP theorems for the Jaggi-type hybrid F_G-Cn self-mapping?
(2) Are there any uniqueness conditions for the existence FP theorems?
(3) Are there some applications to the notion of Jaggi-type hybrid F_C-Cn?

Bibliography

[1] Alansari, M., Mohammed, S. S., Azam, A. and Hussain, N., *On multivalued hybrid contractions with applications*, J. Funct. Spaces 1 (2020), 5401403.
[2] Dung, N. V., Hieu, N. T. and Radojević, S., *Fixed point theorems for g-monotone maps on partially ordered S-metric spaces*, Filomat 28(9) (2014), 1885–1898.
[3] Gupta, A., *Cyclic contraction on S-metric space*, Int. J. Anal. Appl. 3(2) (2013), 119–130.
[4] Hieu, N. T., Ly, N. T. and Dung, N. V., *A generalization of Ćirić quasi-contractions for maps on S-metric spaces*, Thai J. Math. 13(2) (2015), 369–380.
[5] Jiddah, J. A., Alansari, M., Mohamed, O. M., Shagari, M. S. and Bakery, A. A., *Fixed point results of Jaggi-type hybrid contraction in generalized metric space*, J. Funct. Spaces 1 (2022), 2205423.
[6] Karapınar, E. and Fulga, A., *A hybrid contraction that involves Jaggi type*, Symmetry 11(5) (2019), 715.
[7] Mlaiki, N., *α-ψ-contractive mapping on S-metric space*, Math. Sci. Lett. 4(1) (2015), 9–12.
[8] Mustafa, Z. and Sims, B., *A new approach to generalized metric spaces*, J. Nonlinear Convex Anal. 7(2) (2006), 289–297.
[9] Ogbumba, R. O., Shagari, M. S., Alansari, M., Khalid, T. A., Mohamed, E. A. and Bakery, A. A., *Advancements in hybrid fixed point results and F-contractive operators*, Symmetry 15(6) (2023), 1253.

[10] Özgür, N. Y. and Taş, N., *Some fixed point theorems on S-metric spaces*, Mat. Vesn. 69(1) (2017), 39–52.

[11] Özgür, N. Y. and Taş, N., *Some new contractive mappings on S-metric spaces and their relationships with the mapping (S25)*, Math. Sci. 11(1) (2017), 7–16.

[12] Özgür, N. Y. and Taş, N., *Some fixed-circle theorems on metric spaces*, Bull. Malays. Math. Sci. Soc. 42(4) (2019), 1433–1449.

[13] Özgür, N. Y. and Taş, N., *Fixed-circle problem on S-metric spaces with a geometric viewpoint*, Facta Univ., Ser. Math. Inform. 34(3) (2019), 459–472.

[14] Özgür, N. Y., Taş, N. and Çelik, U., *New fixed-circle results on S-metric spaces*, Bull. Math. Anal. Appl. 9(2) (2017), 10–23.

[15] Özgür, N. Y., *Fixed-disc results via simulation functions*, Turk. J. Math. 43(6) (2019), 2794–2805.

[16] Rus, I. A., *Generalized Contractions and Applications*, Cluj University Press, Cluj-Napocca, Romania, 2001.

[17] Sedghi, S., Shobe, N. and Aliouche, A., *A generalization of fixed point theorems in S-metric spaces*, Mat. Vesn. 64(3) (2012), 258–266.

[18] Sedghi, S. and Dung, N. V., *Fixed point theorems on S-metric spaces*, Mat. Vesn. 66(1) (2014), 113–124.

[19] Shagari, M. S., Jehoshaphat, M., Ogbumba, R. O. and Chiroma, R., *Fixed point results of hybrid F-contractions*, J. Linear Topol. Algebra 13(2) (2024), 81–91.

[20] Taş, N., *Suzuki–Berinde type fixed-point and fixed-circle results on S-metric spaces*, J. Linear Topol. Algebra 7(3) (2018), 233–244.

[21] Wardowski, D., *Fixed points of a new type of contractive mappings in complete metric spaces*, Fixed Point Theory Appl. 2012 (2012), 94.

Silvestru Sever Dragomir

4 Refinements and reverses of Jensen tensorial inequality for twice differentiable functions of selfadjoint operators in Hilbert spaces

Abstract: Let H be a Hilbert space. In this paper, we show among others that, if $\psi : I \subseteq \mathbb{R} \to \mathbb{R}$ is a twice differentiable convex function on $\mathring{I}$, $\phi : I \to \mathbb{R}$ is twice differentiable and $m\psi''(t) \leq \phi''(t) \leq M\psi''(t)$, $t \in \mathring{I}$, for some constants $m < M$, A_i, $i \in \{1, \ldots, n\}$ are selfadjoint operators with $\operatorname{Sp}(A_i) \subset I$, $i \in \{1, \ldots, n\}$ and $w_i \geq 0$, $i \in \{1, \ldots, n\}$ with $\sum_{i=1}^{n} w_i = 1$, then

$$m\left[\sum_{i=1}^{n} w_i \widehat{\psi(\mathbf{A}_i)} - \psi\left(\sum_{i=1}^{n} w_i \widehat{\mathbf{A}_i}\right)\right] \leq \sum_{i=1}^{n} w_i \widehat{\phi(\mathbf{A}_i)} - \phi\left(\sum_{i=1}^{n} w_i \widehat{\mathbf{A}_i}\right) \leq M\left[\sum_{i=1}^{n} w_i \widehat{\psi(\mathbf{A}_i)} - \psi\left(\sum_{i=1}^{n} w_i \widehat{\mathbf{A}_i}\right)\right],$$

where $\hat{\mathbf{A}}_i$ is defined as a tensorial product of A_i in position $i = 1, \ldots, n$ and with 1 in the other positions. Let $(A_t)_{t\in\Omega}$ be a continuous field of operators in $B(H)$ with $\operatorname{Sp}(A_t) \subseteq [m, M]$ for $t \in \Omega$, then for $\nu \in [0, 1]$, we have the inequalities for the Hadamard product

$$\begin{aligned}
0 &\leq (1-\nu)\nu \exp(m)\left[\int_\Omega A_t^2 d\mu(t) \circ 1 - \int_\Omega A_t d\mu(t) \circ \int_\Omega A_t d\mu(t)\right] \\
&\leq \int_\Omega \exp(A_t) d\mu(t) \circ 1 - \int_\Omega (\exp(1-\nu)A_t) d\mu(t) \circ \int_\Omega \exp(\nu A_t) d\mu(t) \\
&\leq (1-\nu)\nu \exp(M)\left[\int_\Omega A_t^2 d\mu(t) \circ 1 - \int_\Omega A_t d\mu(t) \circ \int_\Omega A_t d\mu(t)\right].
\end{aligned}$$

Keywords: Tensorial product, Hadamard product, selfadjoint operators, convex functions

MSC 2020: 47A63, 47A99

Silvestru Sever Dragomir, Applied Mathematics Research Group, ISILC, Victoria University, PO Box 14428, Melbourne City, MC 8001, Australia; and School of Computer Science & Applied Mathematics, University of the Witwatersrand, Private Bag 3, Johannesburg 2050, South Africa, e-mail: sever.dragomir@vu.edu.au; URL: http://rgmia.org/dragomir

https://doi.org/10.1515/9783112215760-004

4.1 Introduction

Let $I_1, \ldots, I_k$ be intervals from $\mathbb{R}$ and let $f : I_1 \times \cdots \times I_k \to \mathbb{R}$ be an essentially bounded real function defined on the product of the intervals. Let $A = (A_1, \ldots, A_n)$ be a k-tuple of bounded selfadjoint operators on Hilbert spaces $H_1, \ldots, H_k$ such that the spectrum of A_i is contained in I_i for $i = 1, \ldots, k$. We say that such a k-tuple is in the domain of f. If

$$A_i = \int_{I_i} \lambda_i dE_i(\lambda_i)$$

is the spectral resolution of A_i for $i = 1, \ldots, k$; by following [2], we define

$$f(A_1, \ldots, A_k) := \int_{I_1} \cdots \int_{I_k} f(\lambda_1, \ldots, \lambda_1) dE_1(\lambda_1) \otimes \cdots \otimes dE_k(\lambda_k) \tag{4.1}$$

as a bounded selfadjoint operator on the tensorial product $H_1 \otimes \cdots \otimes H_k$.

If the Hilbert spaces are of finite dimension, then the above integrals become finite sums, and we may consider the functional calculus for arbitrary real functions. This construction [2] extends the definition of Korányi [10] for functions of two variables and has the property that

$$f(A_1, \ldots, A_k) = f_1(A_1) \otimes \cdots \otimes f_k(A_k), \tag{4.2}$$

whenever f can be separated as a product $f(t_1, \ldots, t_k) = f_1(t_1) \ldots f_k(t_k)$ of k functions each depending on only one variable.

It is know that, *if f is super-multiplicative (sub-multiplicative)* on $[0, \infty)$, namely

$$f(st) \geq (\leq) f(s)f(t) \text{ for all } s, t \in [0, \infty)$$

and if f is continuous on $[0, \infty)$, then [8, p. 173]

$$f(A \otimes B) \geq (\leq) f(A) \otimes f(B) \quad \text{for all } A, B \geq 0. \tag{4.3}$$

This follows by observing that, if

$$A = \int_{[0,\infty)} t dE(t) \quad \text{and} \quad B = \int_{[0,\infty)} s dF(s)$$

are the spectral resolutions of A and B, then

$$f(A \otimes B) = \int_{[0,\infty)} \int_{[0,\infty)} f(st) dE(t) \otimes dF(s) \tag{4.4}$$

for the continuous function f on $[0, \infty)$.

Recall the *geometric operator mean* for the positive operators $A, B > 0$

$$A\#_t B := A^{1/2}(A^{-1/2}BA^{-1/2})^t A^{1/2},$$

where $t \in [0,1]$ and

$$A\#B := A^{1/2}(A^{-1/2}BA^{-1/2})^{1/2} A^{1/2}.$$

By the definitions of # and ⊗, we have

$$A\#B = B\#A \quad \text{and} \quad (A\#B) \otimes (B\#A) = (A \otimes B)\#(B \otimes A).$$

In 2007, S. Wada [11] obtained the following *Callebaut-type inequalities* for tensorial product

$$\begin{aligned}(A\#B) \otimes (A\#B) &\leq \frac{1}{2}\left[(A\#_\alpha B) \otimes (A\#_{1-\alpha} B) + (A\#_{1-\alpha} B) \otimes (A\#_\alpha B)\right] \\ &\leq \frac{1}{2}(A \otimes B + B \otimes A) \end{aligned} \tag{4.5}$$

for $A, B > 0$ and $\alpha \in [0,1]$.

Recall that the *Hadamard product* of A and B in $B(H)$ is defined to be the operator $A \circ B \in B(H)$ satisfying

$$\langle (A \circ B)e_j, e_j \rangle = \langle Ae_j, e_j \rangle \langle Be_j, e_j \rangle$$

for all $j \in \mathbb{N}$, where $\{e_j\}_{j\in\mathbb{N}}$ is an *orthonormal basis* for the separable Hilbert space H.

It is known that, see [7], we have the representation

$$A \circ B = \mathcal{U}^*(A \otimes B)\mathcal{U}, \tag{4.6}$$

where $\mathcal{U} : H \to H \otimes H$ is the isometry defined by $\mathcal{U}e_j = e_j \otimes e_j$ for all $j \in \mathbb{N}$.

If f *is super-multiplicative operator concave (sub-multiplicative operator convex)* on $[0, \infty)$, then also [8, p. 173]

$$f(A \circ B) \geq (\leq) f(A) \circ f(B) \quad \text{for all } A, B \geq 0. \tag{4.7}$$

We recall the following elementary inequalities for the Hadamard product

$$A^{1/2} \circ B^{1/2} \leq \left(\frac{A + B}{2}\right) \circ 1 \quad \text{for } A, B \geq 0$$

and Fiedler inequality

$$A \circ A^{-1} \geq 1 \quad \text{for } A > 0.$$

As extension of Kadison's Schwarz inequality on the Hadamard product, Ando [1] showed that

$$A \circ B \leq (A^2 \circ 1)^{1/2}(B^2 \circ 1)^{1/2} \quad \text{for } A, B \geq 0$$

and Aujla and Vasudeva [3] gave an alternative upper bound

$$A \circ B \leq (A^2 \circ B^2)^{1/2} \quad \text{for } A, B \geq 0.$$

It has been shown in [9] that $(A^2 \circ 1)^{1/2}(B^2 \circ 1)^{1/2}$ and $(A^2 \circ B^2)^{1/2}$ are incomparable for 2-square positive definite matrices A and B.

Motivated by the above results, in this chapter, we show among others that, if $\psi : I \subseteq \mathbb{R} \to \mathbb{R}$ is a twice differentiable convex function on $\mathring{I}$, $\phi : I \to \mathbb{R}$ is twice differentiable and $m\psi''(t) \leq \phi''(t) \leq M\psi''(t)$, $t \in \mathring{I}$, for some constants $m < M$, A_i, $i \in \{1, \ldots, n\}$ are selfadjoint operators with $\operatorname{Sp}(A_i) \subset I$, $i \in \{1, \ldots, n\}$ and $w_i \geq 0$, $i \in \{1, \ldots, n\}$ with $\sum_{i=1}^n w_i = 1$, then

$$\begin{aligned} m\left[\sum_{i=1}^n w_i \widehat{\psi(\mathbf{A}_i)} - \psi\left(\sum_{i=1}^n w_i \widehat{\mathbf{A}}_i\right)\right] &\leq \sum_{i=1}^n w_i \widehat{\phi(\mathbf{A}_i)} - \phi\left(\sum_{i=1}^n w_i \widehat{\mathbf{A}}_i\right) \\ &\leq M\left[\sum_{i=1}^n w_i \widehat{\psi(\mathbf{A}_i)} - \psi\left(\sum_{i=1}^n w_i \widehat{\mathbf{A}}_i\right)\right], \end{aligned}$$

where $\hat{\mathbf{A}}_i$ is defined as a tensorial product of A_i in position $i = 1, \ldots, n$ and with 1 in the other positions. Let $(A_t)_{t\in\Omega}$ be a continuous field of operators in $B(H)$ with $\operatorname{Sp}(A_t) \subseteq [m, M]$ for $t \in \Omega$, then for $\nu \in [0, 1]$, we have the inequalities for the Hadamard product

$$\begin{aligned} 0 &\leq (1-\nu)\nu \exp(m)\left[\int_\Omega A_t^2 d\mu(t) \circ 1 - \int_\Omega A_t d\mu(t) \circ \int_\Omega A_t d\mu(t)\right] \\ &\leq \int_\Omega \exp(A_t) d\mu(t) \circ 1 - \int_\Omega (\exp(1-\nu)A_t) d\mu(t) \circ \int_\Omega \exp(\nu A_t) d\mu(t) \\ &\leq (1-\nu)\nu \exp(M)\left[\int_\Omega A_t^2 d\mu(t) \circ 1 - \int_\Omega A_t d\mu(t) \circ \int_\Omega A_t d\mu(t)\right]. \end{aligned}$$

4.2 Some facts related Jessen's inequality

Let L be a linear class of real-valued functions $g : E \to \mathbb{R}$ having the properties

(L1) $f, g \in L$ imply $(\alpha f + \beta g) \in L$ for all $\alpha, \beta \in \mathbb{R}$;

(L2) $\mathbf{1} \in L$, i. e., if $f_0(t) = 1$, $t \in E$ then $f_0 \in L$.

An *isotonic linear functional* $\mathcal{A} : L \to \mathbb{R}$ is a functional satisfying

(A1) $\mathcal{A}(\alpha f + \beta g) = \alpha\mathcal{A}(f) + \beta\mathcal{A}(g)$ for all $f, g \in L$ and $\alpha, \beta \in \mathbb{R}$.
(A2) If $f \in L$ and $f \geq 0$, then $\mathcal{A}(f) \geq 0$.
The mapping $\mathcal{A}$ is said to be *normalized if*
(A3) $\mathcal{A}(\mathbf{1}) = 1$.

Isotonic, that is, order-preserving, linear functionals are natural objects in analysis which enjoy a number of convenient properties. Thus, they provide, for example, Jessen's inequality, which is a functional form of Jensen's inequality.

We recall Jessen's inequality ([4], see also [6]).

Theorem 4.1. *Let $\phi : I \subseteq \mathbb{R} \to \mathbb{R}$ (I is an interval), be a convex function and $f : E \to I$ such that $\phi \circ f, f \in L$. If $\mathcal{A} : L \to \mathbb{R}$ is an isotonic linear and normalized functional, then*

$$\phi(\mathcal{A}(f)) \leq \mathcal{A}(\phi \circ f). \tag{4.8}$$

In, we obtained among others, the following refinement and reverse of Jessen's inequality:

Theorem 4.2. *Let $\psi : I \subseteq \mathbb{R} \to \mathbb{R}$ be a twice differentiable convex function on $\mathring{I}$, $f : E \to I$ such that $\psi \circ f, f \in L$ and $\mathcal{A} : L \to \mathbb{R}$ be an isotonic linear and normalized functional. If $\phi : I \to \mathbb{R}$ is twice differentiable and $m\psi''(t) \leq \phi''(t) \leq M\psi''(t)$, $t \in \mathring{I}$, then*

$$\begin{aligned} m[\mathcal{A}(\psi \circ f) - \psi(\mathcal{A}(f))] &\leq \mathcal{A}(\phi \circ f) - \phi(\mathcal{A}(f)) \\ &\leq M[\mathcal{A}(\psi \circ f) - \psi(\mathcal{A}(f))], \end{aligned} \tag{4.9}$$

provided $\phi \circ f, f \in L$.

Let $p \in (-\infty, 0) \cup (1, \infty)$ and define $g_p : I \subset (0, \infty) \to \mathbb{R}$, $g_p(t) = \phi''(t)t^{2-p}$. Assume that $-\infty < \gamma \leq g_p(t) \leq \Gamma < \infty$, $t \in I$. This implies that $\gamma t^{p-2} \leq \phi''(t) \leq \Gamma t^{p-2}$, so if we take $m = \frac{\gamma}{p(p-1)}$, $M = \frac{\Gamma}{p(p-1)}$ and $\psi(t) = \ell^p(t)$, where $\ell(t) = t$, then we have $m\psi''(t) \leq \phi''(t) \leq M\psi''(t)$, $t \in \mathring{I}$.

By Theorem 4.2. we can state the following:

Corollary 4.1. *If $\phi : I \subset (0, \infty) \to \mathbb{R}$ is twice differentiable and $\gamma t^{p-2} \leq \phi''(t) \leq \Gamma t^{p-2}$, $t \in \mathring{I}$, $p \in (-\infty, 0) \cup (1, \infty)$, then*

$$\begin{aligned} \frac{\gamma}{p(p-1)}[\mathcal{A}(f^p) - \mathcal{A}^p(f)] &\leq \mathcal{A}(\phi \circ f) - \phi(\mathcal{A}(f)) \\ &\leq \frac{\Gamma}{p(p-1)}[\mathcal{A}(f^p) - \mathcal{A}^p(f)], \end{aligned} \tag{4.10}$$

provided $\phi \circ f, f, f^p \in L$.

In particular, if $\gamma \leq \phi''(t) \leq \Gamma$, $t \in \mathring{I}$, then

$$\frac{\gamma}{2}[\mathcal{A}(f^2) - \mathcal{A}^2(f)] \leq \mathcal{A}(\phi \circ f) - \phi(\mathcal{A}(f)) \leq \frac{\Gamma}{2}[\mathcal{A}(f^2) - \mathcal{A}^2(f)], \tag{4.11}$$

provided $\phi \circ f, f, f^2 \in L$.

Also, if $\gamma t^{-3} \le \phi''(t) \le \Gamma t^{-3}$, $t \in \mathring{I}$, *then*

$$\frac{\gamma}{2}[\mathcal{A}(f^{-1}) - \mathcal{A}^{-1}(f)] \le \mathcal{A}(\phi \circ f) - \phi(\mathcal{A}(f)) \le \frac{\Gamma}{2}[\mathcal{A}(f^{-1}) - \mathcal{A}^{-1}(f)], \tag{4.12}$$

provided $\phi \circ f, f, f^{-1} \in L$.

Define $l(t) = t^2\phi''(t)$, $t \in \mathring{I} \subset (0,\infty)$ and assume that $-\infty < s \le l(t) \le S < \infty$ for $t \in \mathring{I}$. This implies that $st^{-2} \le \phi''(t) \le St^{-2}$ for $t \in \mathring{I}$, so if we take $m = s$, $M = \mathcal{A}$ and $\psi(t) = -\ln t$, then we have $m\psi''(t) \le \phi''(t) \le M\psi''(t)$, $t \in \mathring{I}$.

By Theorem 4.2, we can state the following:

Corollary 4.2. *If* $\phi : I \subset (0,\infty) \to \mathbb{R}$ *is twice differentiable and* $st^{-2} \le \phi''(t) \le St^{-2}$ *for* $t \in \mathring{I}$, *then*

$$s[\ln(\mathcal{A}(f)) - \mathcal{A}(\ln f)] \le \mathcal{A}(\phi \circ f) - \phi(\mathcal{A}(f)) \le S[\ln(\mathcal{A}(f)) - \mathcal{A}(\ln f)], \tag{4.13}$$

provided $\phi \circ f, f, \ln f \in L$.

Consider $k(t) = t\phi''(t)$, $t \in \mathring{I}$ and assume that $-\infty < \delta \le k(t) \le \Delta < \infty$ for $t \in \mathring{I}$. This implies that $\delta t^{-1} \le \phi''(t) \le t^{-1}\Delta$ for $t \in \mathring{I}$, so if we take $m = \delta$, $M = \Delta$ and $\psi(t) = t\ln t$, then we have $m\psi''(t) \le \phi''(t) \le M\psi''(t)$, $t \in \mathring{I}$.

By utilizing Theorem 4.2, we can also state the following:

Corollary 4.3. *If* $\phi : I \subset (0,\infty) \to \mathbb{R}$ *is twice differentiable and* $\delta t^{-1} \le \phi''(t) \le t^{-1}\Delta$ *for* $t \in \mathring{I}$, *then*

$$\begin{aligned}\delta[\mathcal{A}(f\ln f) - \mathcal{A}(f)\ln(\mathcal{A}(f))] &\le \mathcal{A}(\phi \circ f) - \phi(\mathcal{A}(f)) \\ &\le \Delta[\mathcal{A}(f\ln f) - \mathcal{A}(f)\ln(\mathcal{A}(f))],\end{aligned} \tag{4.14}$$

provided $\phi \circ f, f, f\ln f \in L$.

A counterpart of Jessen's result was proved by Beesack and Pečarić in [4] for compact intervals $I = [\alpha,\beta]$.

Theorem 4.3. *Let* $\phi : [\alpha,\beta] \subset \mathbb{R} \to \mathbb{R}$ *be a convex function and* $f : E \to [\alpha,\beta]$ *such that* $\phi \circ f, f \in L$. *If* $A : L \to \mathbb{R}$ *is an isotonic linear and normalized functional, then*

$$A(\phi \circ f) \le \frac{\beta - A(f)}{\beta - \alpha}\phi(\alpha) + \frac{A(f) - \alpha}{\beta - \alpha}\phi(\beta). \tag{4.15}$$

In, we obtained the following refinement and reverse of Beesack–Pečarić inequality (4.15):

Theorem 4.4. *Let* $\psi : [\alpha,\beta] \subset \mathbb{R} \to \mathbb{R}$ *be a twice differentiable convex function on* (α,β), $f : E \to [\alpha,\beta]$ *such that* $\psi \circ f, f \in L$ *and* $\mathcal{A} : L \to \mathbb{R}$ *be an isotonic linear and normalized* *functional. If $\phi : [\alpha,\beta] \to \mathbb{R}$ is twice differentiable and $m\psi''(t) \le \phi''(t) \le M\psi''(t)$, $t \in (\alpha,\beta)$, then*

$$\begin{aligned} & m\left[\frac{\beta - A(f)}{\beta-\alpha}\psi(\alpha) + \frac{A(f)-\alpha}{\beta-\alpha}\psi(\beta) - A(\psi\circ f)\right] \\ & \le \frac{\beta - A(f)}{\beta-\alpha}\phi(\alpha) + \frac{A(f)-\alpha}{\beta-\alpha}\phi(\beta) - A(\phi\circ f). \\ & \le M\left[\frac{\beta - A(f)}{\beta-\alpha}\psi(\alpha) + \frac{A(f)-\alpha}{\beta-\alpha}\psi(\beta) - A(\psi\circ f)\right]. \end{aligned} \tag{4.16}$$

We consider the p-logarithmic mean of two positive numbers given by

$$L_p(a,b) := \begin{cases} a & \text{if } b = a, \\ [\frac{b^{p+1}-a^{p+1}}{(p+1)(b-a)}]^{\frac{1}{p}} & \text{if } a \ne b \end{cases}$$

and $p \in \mathbb{R}\setminus\{-1,0\}$.

Corollary 4.4. *If $-\infty < \gamma t^{p-2} \le \phi''(t) \le t^{p-2}\Gamma < \infty$, $t \in (\alpha,\beta) \subset (0,\infty)$, then*

$$\begin{aligned} & \frac{\gamma}{p(p-1)}[pL_{p-1}^{p-1}(\alpha,\beta)A(f) - \alpha\beta(p-1)L_{p-2}^{p-2}(\alpha,\beta) - A(f^p)] \\ & \le \frac{\beta - A(f)}{\beta-\alpha}\phi(\alpha) + \frac{A(f)-\alpha}{\beta-\alpha}\phi(\beta) - A(\phi\circ f) \\ & \le \frac{\Gamma}{p(p-1)}[pL_{p-1}^{p-1}(\alpha,\beta)A(f) - \alpha\beta(p-1)L_{p-2}^{p-2}(\alpha,\beta) - A(f^p)]. \end{aligned} \tag{4.17}$$

If $-\infty < k \le \phi''(t) \le K < \infty$, $t \in (\alpha,\beta) \subset (0,\infty)$, then by taking in (4.17) $p = 2$, we get

$$\begin{aligned} \frac{k}{2}A[(f-\alpha)(\beta-f)] & = \frac{k}{2}[(\alpha+\beta)A(f) - \alpha\beta - A(f^2)] \\ & \le \frac{\beta - A(f)}{\beta-\alpha}\phi(\alpha) + \frac{A(f)-\alpha}{\beta-\alpha}\phi(\beta) - A(\phi\circ f) \\ & \le \frac{K}{2}[(\alpha+\beta)A(f) - \alpha\beta - A(f^2)] \\ & = \frac{K}{2}A[(f-\alpha)(\beta-f)]. \end{aligned} \tag{4.18}$$

Let $I(\cdot,\cdot)$ denotes the identric mean, i. e., we recall it

$$I(u,v) := \begin{cases} u & \text{if } v = u, \\ \frac{1}{e}(\frac{u^u}{v^v})^{\frac{1}{u-v}}, & v \ne u. \end{cases}$$

Corollary 4.5. *If $-\infty < st^{-2} \le \phi''(t) \le t^{-2}S < \infty$ for $t \in (\alpha,\beta) \subset (0,\infty)$, then*

$$s\left[A(\ln f) + \ln\left[I\left(\frac{1}{\alpha},\frac{1}{\beta}\right)\right] + 1 - \frac{A(f)}{L(\alpha,\beta)}\right]$$

$$\leq \frac{\beta - A(f)}{\beta - \alpha}\phi(\alpha) + \frac{A(f) - \alpha}{\beta - \alpha}\phi(\beta) - A(\phi \circ f)$$
$$\leq S\left[A(\ln f) + \ln\left[I\left(\frac{1}{\alpha}, \frac{1}{\beta}\right)\right] + 1 - \frac{A(f)}{L(\alpha,\beta)}\right]. \tag{4.19}$$

Assume that $G(\alpha,\beta) = \sqrt{ab}$ is the geometric mean and $L(\alpha,\beta)$ is the logarithmic mean, i. e., we recall it

$$L(\alpha,\beta) := \begin{cases} \alpha & \text{if } \beta = \alpha, \\ \frac{\beta-\alpha}{\ln\beta - \ln\alpha} & \text{if } \beta \neq \alpha. \end{cases}$$

Corollary 4.6. *If* $-\infty < \delta t^{-1} \leq \phi''(t) \leq t^{-1}\Delta < \infty$ *for* $t \in (\alpha,\beta)$, *then*

$$\delta\left[A(f)\ln I(\alpha,\beta) - \frac{G^2(\alpha,\beta)}{L(\alpha,\beta)} + A(f) - A(f\ln f)\right]$$
$$\leq \frac{\beta - A(f)}{\beta - \alpha}\phi(\alpha) + \frac{A(f) - \alpha}{\beta - \alpha}\phi(\beta) - A(\phi \circ f)$$
$$\leq \Delta\left[A(f)\ln I(\alpha,\beta) - \frac{G^2(\alpha,\beta)}{L(\alpha,\beta)} + A(f) - A(f\ln f)\right]. \tag{4.20}$$

4.3 Some properties for functions of tensorial products

Recall the following property of the tensorial product

$$(AC) \otimes (BD) = (A \otimes B)(C \otimes D) \tag{4.21}$$

that holds for any $A, B, C, D \in B(H)$.

If we take $C = A$ and $D = B$, then we get

$$A^2 \otimes B^2 = (A \otimes B)^2.$$

By induction and using (4.21), we derive that

$$A^n \otimes B^n = (A \otimes B)^n \quad \text{for natural } n \geq 0. \tag{4.22}$$

In particular,

$$A^n \otimes 1 = (A \otimes 1)^n \quad \text{and} \quad 1 \otimes B^n = (1 \otimes B)^n \tag{4.23}$$

for all $n \geq 0$.

We also observe that, by (4.21), the operators $A \otimes 1$ and $1 \otimes B$ are commutative and

$$(A \otimes 1)(1 \otimes B) = (1 \otimes B)(A \otimes 1) = A \otimes B. \tag{4.24}$$

Moreover, for two natural numbers m, n, we have

$$(A \otimes 1)^m (1 \otimes B)^n = (1 \otimes B)^n (A \otimes 1)^m = A^m \otimes B^n. \tag{4.25}$$

By induction over m, we derive

$$(A_1 \otimes A_2 \otimes \cdots \otimes A_m)^n = A_1^n \otimes A_2^n \otimes \cdots \otimes A_m^n \quad \text{for natural } n \geq 0 \tag{4.26}$$

and

$$A_1 \otimes A_2 \otimes \cdots \otimes A_m = (A_1 \otimes 1 \otimes \cdots \otimes 1)(1 \otimes A_2 \otimes \cdots \otimes 1) \cdots (1 \otimes 1 \otimes \cdots \otimes A_m) \tag{4.27}$$

and the m operators $(A_1 \otimes 1 \otimes \cdots \otimes 1)$, $(1 \otimes A_2 \otimes \cdots \otimes 1), \ldots$ and $(1 \otimes 1 \otimes \cdots \otimes A_m)$ are commutative between them.

We define for $A_i, B_i \in B(H)$, $i \in \{1, \ldots, n\}$, $\bigotimes_{i=1}^n B_i := B_1 \otimes \cdots \otimes B_n$,

$$\hat{\mathbf{A}}_i := 1 \otimes \cdots \otimes A_i \otimes \cdots \otimes 1, \quad i = 2, \ldots, n-1,$$

and

$$\hat{\mathbf{A}}_1 := A_1 \otimes 1 \otimes \cdots \otimes 1 \quad \text{while } \hat{\mathbf{A}}_n := 1 \otimes \cdots \otimes 1 \otimes A_n.$$

Basically, $\hat{\mathbf{A}}_i$ is defined as a tensorial product of A_i in position $i = 1, \ldots, n$ and with 1 in the other positions.

We need the following identity for the tensorial product, see also [5]:

Lemma 4.1. *Assume A_i, $i \in \{1, \ldots, m\}$ are selfadjoint operators with $\operatorname{Sp}(A_i) \subset I_i$, $i \in \{1, \ldots, m\}$ and with the spectral resolutions*

$$A_i = \int_{I_i} t_i dE(t_i), \quad i \in \{1, \ldots, m\}.$$

Let f_i, $i \in \{1, \ldots, m\}$ be continuous on I_i and φ continuous on an interval K that contains the product of the intervals $f(I_1) \ldots f(I_m)$, then

$$\varphi\left(\bigotimes_{i=1}^m f_i(A_i)\right) = \int_{I_1} \cdots \int_{I_m} \varphi\left(\prod_{i=1}^m f_i(t_i)\right) dE(t_1) \otimes \cdots \otimes dE(t_m). \tag{4.28}$$

We also have

$$\varphi(\widehat{\mathbf{f}_i(\mathbf{A}_i)}) = \widehat{\varphi(\mathbf{f}_i(\mathbf{A}_i))} \tag{4.29}$$

for all $i = 1, \ldots, m$.

Proof. By Stone–Weierstrass theorem, any continuous function can be approximated by a sequence of polynomials, therefore it suffices to prove the equality for the power function $\varphi(t) = t^n$ with n any natural number.

Then, by (4.1) and (4.26), we obtain

$$\begin{aligned}
&\int_{I_1}\cdots\int_{I_m}[f_1(t_1)\ldots f_m(t_m)]^n dE(t_1)\otimes\cdots\otimes dE(t_m)\\
&=\int_{I_1}\cdots\int_{I_m}[f_1(t_1)]^n\ldots[f_m(t_m)]^n dE(t_1)\otimes\cdots\otimes dE(t_m)\\
&=[f(A_1)]^n\otimes\cdots\otimes[f_m(A_m)]^n=[f_1(A_1)\otimes\cdots\otimes f_m(A_m)]^n,
\end{aligned}$$

which shows that the identity (4.26) is valid for the power function.

This proves the identity (4.28)

By taking $f_j \equiv 1$ for $j = 1,\ldots,m$ and $j \neq i$ in (4.28), we get

$$\begin{aligned}
\varphi(1\otimes\cdots\otimes f_i(A_i)\otimes\cdots\otimes 1)&=\int_{I_1}\cdots\int_{I_m}\varphi(f_i(t_i))dE(t_1)\otimes\cdots\otimes dE(t_m)\\
&=1\otimes\cdots\otimes\varphi(f_i(A_i))\otimes\cdots\otimes 1,
\end{aligned}$$

which proves (4.29). □

Corollary 4.7. *Assume A_i, $i \in \{1,\ldots,m\}$ are selfadjoint operators with $\operatorname{Sp}(A_i) \subset I_i$ and f_i, $i \in \{1,\ldots,m\}$ are continuous and positive on I_i, then*

$$\ln\left(\bigotimes_{i=1}^{m} f_i(A_i)\right) = \sum_{i=1}^{n} \widehat{\ln(\mathbf{f}_i(\mathbf{A}_i))}. \tag{4.30}$$

Also,

$$\ln(\widehat{\mathbf{f}_i(\mathbf{A}_i)}) = \widehat{\ln(\mathbf{f}_i(\mathbf{A}_i))} \tag{4.31}$$

for all $i = 1,\ldots,m$.

Proof. Assume that

$$A_i = \int_I t_i dE(t_i)$$

are the spectral resolutions for A_i, $i = 1,\ldots,m$.

We have for $\varphi(u) = \ln u$, $u > 0$, in (4.28) that

$$\ln(f_1(A)\otimes\cdots\otimes f_m(A_m)) = \int_{I_1}\cdots\int_{I_m}\ln(f_1(t_1)\ldots f_m(t_m))dE(t_1)\otimes\cdots\otimes dE(t_m)$$

$$
\begin{aligned}
&= \int_{I_1} \cdots \int_{I_m} [\ln f_1(t_1) + \cdots + \ln f_m(t_m)] dE(t_1) \otimes \cdots \otimes dE(t_m) \\
&= \int_{I_1} \cdots \int_{I_m} \ln f_1(t_1) dE(t_1) \otimes \cdots \otimes dE(t_m) + \cdots \\
&\quad + \int_{I_1} \cdots \int_{I_m} \ln f_m(t_m) dE(t_1) \otimes \cdots \otimes dE(t_m) \\
&= (\ln f_1(A_1)) \otimes 1 \otimes \cdots \otimes 1 + \cdots + 1 \otimes \cdots \otimes 1 \otimes (\ln f_m(A_m))
\end{aligned}
$$

and the identity (4.30) is proved. □

Corollary 4.8. *Assume A_i, $i \in \{1, \ldots, m\}$ are selfadjoint operators with $\operatorname{Sp}(A_i) \subset I_i$ and f_i, $i \in \{1, \ldots, m\}$ are continuous on I_i, then for $r > 0$*

$$
\left| \bigotimes_{i=1}^{m} f_i(A_i) \right|^r = \bigotimes_{i=1}^{m} |f_i(A_i)|^r \tag{4.32}
$$

and

$$
|\widehat{\mathbf{f}_i(\mathbf{A}_i)}|^r = \widehat{|\mathbf{f}_i(\mathbf{A}_i)|^r} \tag{4.33}
$$

for all $i = 1, \ldots, m$.

Proof. From (4.28), we have for the function $\varphi(t) = |t|^r$ that

$$
\begin{aligned}
&|f_1(A_1) \otimes \cdots \otimes f_m(A_m)|^r \\
&\quad = \int_{I_1} \cdots \int_{I_m} |f_1(t_1) \ldots f_m(t_m)|^r dE(t_1) \otimes \cdots \otimes dE(t_m) \\
&\quad = \int_{I_1} \cdots \int_{I_m} |f_1(t_1)|^r \ldots |f_m(t_m)|^r dE(t_1) \otimes \cdots \otimes dE(t_m) \\
&\quad = |f_1(A_1)|^r \otimes \cdots \otimes |f_m(A_m)|^r,
\end{aligned}
$$

which proves (4.32).

The identity (4.33) follows in a similar way. □

Corollary 4.9. *Assume A_i, $i \in \{1, \ldots, m\}$ are positive operators and $q_i \geq 0$, $i \in \{1, \ldots, m\}$, then*

$$
\varphi\left(\bigotimes_{i=1}^{m} A_i^{q_i} \right) = \int_{I_1} \cdots \int_{I_m} \varphi\left(\prod_{i=1}^{m} t_i^{q_i} \right) dE(t_1) \otimes \cdots \otimes dE(t_m). \tag{4.34}
$$

We also have the additive result [5]:

Lemma 4.2. *Assume A_i, $i \in \{1,\dots,m\}$ are selfadjoint operators with* $\mathrm{Sp}(A_i) \subset I_i$, $i \in \{1,\dots,m\}$ *and with the spectral resolutions*

$$A_i = \int_{I_i} t_i dE(t_i), \quad i \in \{1,\dots,m\}.$$

Let g_i, $i \in \{1,\dots,m\}$ be continuous on I_i and ψ continuous on an interval K that contains the sum of the intervals $g(I_1) + \cdots + g(I_m)$, then

$$\psi\left(\sum_{i=1}^{m} \widehat{\mathbf{g}_i(\mathbf{A}_i)}\right) = \int_{I_1} \cdots \int_{I_m} \psi\left(\sum_{i=1}^{m} g_i(t_i)\right) dE(t_1) \otimes \cdots \otimes dE(t_m). \tag{4.35}$$

Proof. Let f_i, continuous, positive and such that $g_i(t_i) = \ln f_i(t_i)$, $t_i \in I_i$, $i \in \{1,\dots,m\}$. Then

$$\sum_{i=1}^{m} g_i(t_i) = \sum_{i=1}^{m} \ln f_i(t_i) = \ln\left(\prod_{i=1}^{m} f_i(t_i)\right).$$

By (4.28), we get for $\varphi = \psi \circ \ln$ that

$$\begin{aligned}
&\int_{I_1} \cdots \int_{I_m} \psi\left(\sum_{i=1}^{m} g_i(t_i)\right) dE(t_1) \otimes \cdots \otimes dE(t_m) \\
&= \int_{I_1} \cdots \int_{I_m} \psi\left(\ln\left(\prod_{i=1}^{m} f_i(t_i)\right)\right) dE(t_1) \otimes \cdots \otimes dE(t_m) \\
&= \int_{I_1} \cdots \int_{I_m} (\psi \circ \ln)\left(\prod_{i=1}^{m} f_i(t_i)\right) dE(t_1) \otimes \cdots \otimes dE(t_m) \\
&= (\psi \circ \ln)\left(\bigotimes_{i=1}^{m} f_i(A_i)\right) = \psi\left(\ln\left(\bigotimes_{i=1}^{m} f_i(A_i)\right)\right).
\end{aligned}$$

By (4.30), we also have

$$\ln\left(\bigotimes_{i=1}^{m} f_i(A_i)\right) = \sum_{i=1}^{n} \widehat{\ln(\mathbf{f}_i(\mathbf{A}_i))} = \sum_{i=1}^{m} \widehat{\mathbf{g}_i(\mathbf{A}_i)}$$

and the identity (4.35) is obtained. □

Corollary 4.10. *Assume that A_i, $i \in \{1,\dots,m\}$ and g_i, $i \in \{1,\dots,m\}$ are as in Lemma 4.2 and $r > 0$, then*

$$\left|\sum_{i=1}^{m} \widehat{\mathbf{g}_i(\mathbf{A}_i)}\right|^r = \int_{I_1} \cdots \int_{I_m} \left|\sum_{i=1}^{m} g_i(t_i)\right|^r dE(t_1) \otimes \cdots \otimes dE(t_m). \tag{4.36}$$

Also, if we take $\psi = \exp$, then we get

$$\exp\left(\sum_{i=1}^{m} \widehat{\mathbf{g}_i(\mathbf{A}_i)}\right) = \int_{I_1} \cdots \int_{I_m} \exp\left(\sum_{i=1}^{m} g_i(t_i)\right) dE(t_1) \otimes \cdots \otimes dE(t_m)$$
$$= \bigotimes_{i=1}^{m} \exp[g_i(A_i)]. \tag{4.37}$$

The case of convex combination is as follows:

Corollary 4.11. *Assume A_i, $i \in \{1, \dots, m\}$ are selfadjoint operators with $\operatorname{Sp}(A_i) \subset I$, $i \in \{1, \dots, m\}$ and with the spectral resolutions*

$$A_i = \int_I t_i dE(t_i), \quad i \in \{1, \dots, m\}.$$

If $p_i \geq 0$ with $\sum_{i=1}^{n} p_i = 1$, and ψ continuous on I, then

$$\psi\left(\sum_{i=1}^{m} p_i \widehat{\mathbf{A}_i}\right) = \int_I \cdots \int_I \psi\left(\sum_{i=1}^{m} p_i t_i\right) dE(t_1) \otimes \cdots \otimes dE(t_m). \tag{4.38}$$

Follows by (4.24) for $g_i(t_i) = p_i t_i$, $i \in \{1, \dots, m\}$.

4.4 Main results

Our first main result is as follows:

Theorem 4.5. *Let $\psi : I \subseteq \mathbb{R} \to \mathbb{R}$ be a twice differentiable convex function on $\mathring{I}$. If $\phi : I \to \mathbb{R}$ is twice differentiable and $m\psi''(t) \leq \phi''(t) \leq M\psi''(t)$, $t \in \mathring{I}$, for some constants $m < M$, A_i, $i \in \{1, \dots, n\}$ are selfadjoint operators with $\operatorname{Sp}(A_i) \subset I$, $i \in \{1, \dots, n\}$ and $w_i \geq 0$, $i \in \{1, \dots, n\}$ with $\sum_{i=1}^{n} w_i = 1$, then*

$$m\left[\sum_{i=1}^{n} w_i \widehat{\psi(\mathbf{A}_i)} - \psi\left(\sum_{i=1}^{n} w_i \widehat{\mathbf{A}_i}\right)\right] \leq \sum_{i=1}^{n} w_i \widehat{\phi(\mathbf{A}_i)} - \phi\left(\sum_{i=1}^{n} w_i \widehat{\mathbf{A}_i}\right)$$
$$\leq M\left[\sum_{i=1}^{n} w_i \widehat{\psi(\mathbf{A}_i)} - \psi\left(\sum_{i=1}^{n} w_i \widehat{\mathbf{A}_i}\right)\right]. \tag{4.39}$$

Proof. If we write (4.22) for the weighted discrete sum, then we get

$$m\left[\sum_{i=1}^{n} w_i \psi(t_i) - \psi\left(\sum_{i=1}^{n} w_i t_i\right)\right] \leq \sum_{i=1}^{n} w_i \phi(t_i) - \phi\left(\sum_{i=1}^{n} w_i t_i\right)$$
$$\leq M\left[\sum_{i=1}^{n} w_i \psi(t_i) - \psi\left(\sum_{i=1}^{n} w_i t_i\right)\right], \tag{4.40}$$

for all $t_i \in I$, $i \in \{1, \dots, n\}$.

Assume that we have the spectral resolutions

$$A_i = \int_I t_i dE(t_i), \quad i \in \{1, \dots, n\}.$$

Now, if we take the integral $\int_I \cdots \int_I$ over $dE(t_1) \otimes \cdots \otimes dE(t_n)$ in (4.40), then we get

$$\begin{aligned} m&\left[\sum_{i=1}^n w_i \int_I \cdots \int_I \psi(t_i) dE(t_1) \otimes \cdots \otimes dE(t_n) - \int_I \cdots \int_I \psi\left(\sum_{i=1}^n w_i t_i\right) dE(t_1) \otimes \cdots \otimes dE(t_n)\right] \\ &\leq \sum_{i=1}^n w_i \int_I \cdots \int_I \phi(t_i) dE(t_1) \otimes \cdots \otimes dE(t_n) - \int_I \cdots \int_I \phi\left(\sum_{i=1}^n w_i t_i\right) dE(t_1) \otimes \cdots \otimes dE(t_n) \\ &\leq M\left[\sum_{i=1}^n w_i \int_I \cdots \int_I \psi(t_i) dE(t_1) \otimes \cdots \otimes dE(t_n) - \int_I \cdots \int_I \psi\left(\sum_{i=1}^n w_i t_i\right) dE(t_1) \otimes \cdots \otimes dE(t_n)\right]. \end{aligned} \tag{4.41}$$

Since, by (4.1)

$$\int_I \cdots \int_I \psi(t_i) dE(t_1) \otimes \cdots \otimes dE(t_n) = 1 \otimes \cdots \otimes \int_I \psi(t_i) dE(t_i) \otimes \cdots \otimes 1 = \widehat{\psi(\mathbf{A}_i)},$$

by (4.38)

$$\int_I \cdots \int_I \psi\left(\sum_{i=1}^n w_i t_i\right) dE(t_1) \otimes \cdots \otimes dE(t_n) = \psi\left(\sum_{i=1}^m w_i \widehat{\mathbf{A}}_i\right)$$

and the similar equalities for the function ϕ, then by (4.41), we get (4.39). □

Corollary 4.12. *If $\phi : I \subset (0, \infty) \to \mathbb{R}$ is twice differentiable and $\gamma t^{p-2} \leq \phi''(t) \leq \Gamma t^{p-2}$, $t \in \mathring{I}$, $p \in (-\infty, 0) \cup (1, \infty)$, then*

$$\begin{aligned} &\frac{\gamma}{p(p-1)}\left[\sum_{i=1}^n w_i \widehat{\mathbf{A}_i^p} - \left(\sum_{i=1}^n w_i \widehat{\mathbf{A}}_i\right)^p\right] \\ &\quad \leq \sum_{i=1}^n w_i \widehat{\phi(\mathbf{A}_i)} - \phi\left(\sum_{i=1}^n w_i \widehat{\mathbf{A}}_i\right) \\ &\quad \leq \frac{\Gamma}{p(p-1)}\left[\sum_{i=1}^n w_i \widehat{\mathbf{A}_i^p} - \left(\sum_{i=1}^n w_i \widehat{\mathbf{A}}_i\right)^p\right]. \end{aligned} \tag{4.42}$$

If $\gamma \le \phi''(t) \le \Gamma$, $t \in \mathring{I}$, then from (4.42) for $p = 2$, we get

$$\frac{\gamma}{2}\left[\sum_{i=1}^{n} w_i\widehat{\mathbf{A}_i^2} - \left(\sum_{i=1}^{n} w_i\widehat{\mathbf{A}_i}\right)^2\right] \le \sum_{i=1}^{n} w_i\widehat{\phi(\mathbf{A}_i)} - \phi\left(\sum_{i=1}^{n} w_i\widehat{\mathbf{A}_i}\right) \le \frac{\Gamma}{2}\left[\sum_{i=1}^{n} w_i\widehat{\mathbf{A}_i^2} - \left(\sum_{i=1}^{n} w_i\widehat{\mathbf{A}_i}\right)^2\right]. \tag{4.43}$$

If $\gamma t^{-3} \le \phi''(t) \le \Gamma t^{-3}$, $t \in \mathring{I}$, then for $p = -1$ in (4.42), we get

$$\frac{\gamma}{2}\left[\sum_{i=1}^{n} w_i\widehat{\mathbf{A}_i^{-1}} - \left(\sum_{i=1}^{m} w_i\widehat{\mathbf{A}_i}\right)^{-1}\right] \le \sum_{i=1}^{n} w_i\widehat{\phi(\mathbf{A}_i)} - \phi\left(\sum_{i=1}^{n} w_i\widehat{\mathbf{A}_i}\right) \le \frac{\Gamma}{2}\left[\sum_{i=1}^{n} w_i\widehat{\mathbf{A}_i^{-1}} - \left(\sum_{i=1}^{n} w_i\widehat{\mathbf{A}_i}\right)^{-1}\right]. \tag{4.44}$$

Corollary 4.13. *If $\phi : I \subset (0,\infty) \to \mathbb{R}$ is twice differentiable and $st^{-2} \le \phi''(t) \le St^{-2}$ for $t \in \mathring{I}$, then*

$$s\left[\ln\left(\sum_{i=1}^{n} w_i\widehat{\mathbf{A}_i}\right) - \ln\left(\bigotimes_{i=1}^{n} A_i^{w_i}\right)\right] \le \sum_{i=1}^{n} w_i\widehat{\phi(\mathbf{A}_i)} - \phi\left(\sum_{i=1}^{n} w_i\widehat{\mathbf{A}_i}\right) \le S\left[\ln\left(\sum_{i=1}^{n} w_i\widehat{\mathbf{A}_i}\right) - \ln\left(\bigotimes_{i=1}^{n} A_i^{w_i}\right)\right]. \tag{4.45}$$

Corollary 4.14. *If $\phi : I \subset (0,\infty) \to \mathbb{R}$ is twice differentiable and $\delta t^{-1} \le \phi''(t) \le t^{-1}\Delta$ for $t \in \mathring{I}$, then*

$$\begin{aligned} m&\left[\sum_{i=1}^{n} w_i\widehat{\mathbf{A}_i \ln(\mathbf{A}_i)} - \sum_{i=1}^{n} w_i\widehat{\mathbf{A}_i}\ln\left(\sum_{i=1}^{n} w_i\widehat{\mathbf{A}_i}\right)\right] \\ &\le \sum_{i=1}^{n} w_i\widehat{\phi(\mathbf{A}_i)} - \phi\left(\sum_{i=1}^{n} w_i\widehat{\mathbf{A}_i}\right) \\ &\le M\left[\sum_{i=1}^{n} w_i\widehat{\mathbf{A}_i \ln(\mathbf{A}_i)} - \sum_{i=1}^{n} w_i\widehat{\mathbf{A}_i}\ln\left(\sum_{i=1}^{n} w_i\widehat{\mathbf{A}_i}\right)\right]. \end{aligned} \tag{4.46}$$

Corollary 4.15. *If $\phi : I \to \mathbb{R}$ is twice differentiable and $\theta \exp t \le \phi''(t) \le \Theta \exp t$, $t \in \mathring{I}$, for some constants $\theta < \Theta$, A_i, $i \in \{1,\dots,n\}$ are selfadjoint operators with $\mathrm{Sp}(A_i) \subset I$, $i \in \{1,\dots,n\}$ and $w_i \ge 0$, $i \in \{1,\dots,n\}$ with $\sum_{i=1}^{n} w_i = 1$, then*

$$\theta\left[\sum_{i=1}^{n} w_i\,\widehat{\exp(\mathbf{A}_i)} - \bigotimes_{i=1}^{n} \exp(w_iA_i)\right]$$

$$
\begin{aligned}
&\le \sum_{i=1}^{n} w_i \widehat{\phi(\mathbf{A}_i)} - \phi\left(\sum_{i=1}^{n} w_i \widehat{\mathbf{A}_i}\right) \\
&\le \Theta\left[\sum_{i=1}^{n} w_i \widehat{\exp(\mathbf{A}_i)} - \bigotimes_{i=1}^{n} \exp(w_i A_i)\right]. \qquad (4.47)
\end{aligned}
$$

We also have:

Theorem 4.6. *Let $\psi : [\alpha,\beta] \subset \mathbb{R} \to \mathbb{R}$ be a twice differentiable convex function on (α,β). If $\phi : [\alpha,\beta] \to \mathbb{R}$ is twice differentiable and $m\psi''(t) \le \phi''(t) \le M\psi''(t)$, $t \in (\alpha,\beta)$, A_i, $i \in \{1,\dots,n\}$ are selfadjoint operators with $\operatorname{Sp}(A_i) \subset [\alpha,\beta]$, $i \in \{1,\dots,n\}$ and $w_i \ge 0$, $i \in \{1,\dots,n\}$ with $\sum_{i=1}^n w_i = 1$, then*

$$
\begin{aligned}
&m\left[\psi(\alpha)\frac{\beta - \sum_{i=1}^n w_i\widehat{\mathbf{A}_i}}{\beta-\alpha} + \psi(\beta)\frac{\sum_{i=1}^n w_i\widehat{\mathbf{A}_i} - \alpha}{\beta-\alpha} - \sum_{i=1}^n w_i\widehat{\psi(\mathbf{A}_i)}\right] \\
&\le \phi(\alpha)\frac{\beta - \sum_{i=1}^n w_i\widehat{\mathbf{A}_i}}{\beta-\alpha} + \phi(\beta)\frac{\sum_{i=1}^n w_i\widehat{\mathbf{A}_i} - \alpha}{\beta-\alpha} - \sum_{i=1}^n w_i\widehat{\phi(\mathbf{A}_i)} \\
&\le M\left[\psi(\alpha)\frac{\beta - \sum_{i=1}^n w_i\widehat{\mathbf{A}_i}}{\beta-\alpha} + \psi(\beta)\frac{\sum_{i=1}^n w_i\widehat{\mathbf{A}_i} - \alpha}{\beta-\alpha} - \sum_{i=1}^n w_i\widehat{\psi(\mathbf{A}_i)}\right]. \qquad (4.48)
\end{aligned}
$$

Proof. If we write (4.16) for the weighted discrete sum, then we get

$$
\begin{aligned}
&m\left[\frac{\beta - \sum_{i=1}^n w_i t_i}{\beta-\alpha}\psi(\alpha) + \frac{\sum_{i=1}^n w_i t_i - \alpha}{\beta-\alpha}\psi(\beta) - \sum_{i=1}^n w_i\psi(t_i)\right] \\
&\le \frac{\beta - \sum_{i=1}^n w_i t_i}{\beta-\alpha}\phi(\alpha) + \frac{\sum_{i=1}^n w_i t_i - \alpha}{\beta-\alpha}\phi(\beta) - \sum_{i=1}^n w_i\phi(t_i) \\
&\le M\left[\frac{\beta - \sum_{i=1}^n w_i t_i}{\beta-\alpha}\psi(\alpha) + \frac{\sum_{i=1}^n w_i t_i - \alpha}{\beta-\alpha}\psi(\beta) - \sum_{i=1}^n w_i\psi(t_i)\right] \qquad (4.49)
\end{aligned}
$$

for all $t_i \in I$, $i \in \{1,\dots,n\}$.

Assume that we have the spectral resolutions

$$
A_i = \int_I t_i dE(t_i), \quad i \in \{1,\dots,n\}.
$$

Now, if we take the integral $\int_I \cdots \int_I$ over $dE(t_1) \otimes \cdots \otimes dE(t_n)$ in (4.49), then we get

$$
\begin{aligned}
&m\left[\psi(\alpha)\frac{\beta - \int_I \cdots \int_I \sum_{i=1}^n w_i t_i dE(t_1) \otimes \cdots \otimes dE(t_n)}{\beta-\alpha}\right. \\
&\quad + \psi(\beta)\frac{\sum_{i=1}^n \int_I \cdots \int_I \sum_{i=1}^n w_i t_i dE(t_1) \otimes \cdots \otimes dE(t_n) - \alpha}{\beta-\alpha}
\end{aligned}
$$

$$
\begin{aligned}
&- \int_I \cdots \int_I \sum_{i=1}^{n} w_i \psi(t_i) dE(t_1) \otimes \cdots \otimes dE(t_n) \Bigg] \\
&\leq \phi(\alpha) \frac{\beta - \int_I \cdots \int_I \sum_{i=1}^{n} w_i t_i dE(t_1) \otimes \cdots \otimes dE(t_n)}{\beta - \alpha} \\
&\quad + \phi(\beta) \frac{\int_I \cdots \int_I \sum_{i=1}^{n} w_i t_i dE(t_1) \otimes \cdots \otimes dE(t_n) - \alpha}{\beta - \alpha} \\
&\quad - \int_I \cdots \int_I \sum_{i=1}^{n} w_i \phi(t_i) dE(t_1) \otimes \cdots \otimes dE(t_n) \\
&\leq M \Bigg[\frac{\beta - \int_I \cdots \int_I \sum_{i=1}^{n} w_i t_i dE(t_1) \otimes \cdots \otimes dE(t_n)}{\beta - \alpha} \psi(\alpha) \\
&\quad \times \frac{\sum_{i=1}^{n} \int_I \cdots \int_I \sum_{i=1}^{n} w_i t_i dE(t_1) \otimes \cdots \otimes dE(t_n) - \alpha}{\beta - \alpha} \psi(\beta) \\
&\quad - \int_I \cdots \int_I \sum_{i=1}^{n} w_i \psi(t_i) dE(t_1) \otimes \cdots \otimes dE(t_n) \Bigg],
\end{aligned}
$$

which, as in the proof of Theorem 4.5, gives (4.48). □

Corollary 4.16. *If $-\infty < k \leq \phi''(t) \leq K < \infty$, $t \in (\alpha, \beta) \subset (0, \infty)$, then*

$$
\begin{aligned}
&\frac{k}{2} \left[\alpha^2 \frac{\beta - \sum_{i=1}^{n} w_i \widehat{\mathbf{A}_i}}{\beta - \alpha} + \beta^2 \frac{\sum_{i=1}^{n} w_i \widehat{\mathbf{A}_i} - \alpha}{\beta - \alpha} - \sum_{i=1}^{n} w_i \widehat{\mathbf{A}_i^2} \right] \\
&\leq \phi(\alpha) \frac{\beta - \sum_{i=1}^{n} w_i \widehat{\mathbf{A}_i}}{\beta - \alpha} + \phi(\beta) \frac{\sum_{i=1}^{n} w_i \widehat{\mathbf{A}_i} - \alpha}{\beta - \alpha} - \sum_{i=1}^{n} w_i \widehat{\boldsymbol{\phi}(\mathbf{A}_i)} \\
&\leq \frac{K}{2} \left[\alpha^2 \frac{\beta - \sum_{i=1}^{n} w_i \widehat{\mathbf{A}_i}}{\beta - \alpha} + \beta^2 \frac{\sum_{i=1}^{n} w_i \widehat{\mathbf{A}_i} - \alpha}{\beta - \alpha} - \sum_{i=1}^{n} w_i \widehat{\mathbf{A}_i^2} \right].
\end{aligned}
\tag{4.50}
$$

By employing the inequalities from Corollaries 4.4, 4.5 and 4.6 for the weighted sum, one can derive other similar inequalities. The details are omitted.

We observe that the inequality (4.50) is equivalent to

$$
\begin{aligned}
&\frac{k}{2} \sum_{i=1}^{n} w_i (\widehat{\mathbf{A}_i} - \alpha)(\beta - \widehat{\mathbf{A}_i}) \\
&\leq \phi(\alpha) \frac{\beta - \sum_{i=1}^{n} w_i \widehat{\mathbf{A}_i}}{\beta - \alpha} + \phi(\beta) \frac{\sum_{i=1}^{n} w_i \widehat{\mathbf{A}_i} - \alpha}{\beta - \alpha} - \sum_{i=1}^{n} w_i \widehat{\boldsymbol{\phi}(\mathbf{A}_i)} \\
&\leq \frac{K}{2} \sum_{i=1}^{n} w_i (\widehat{\mathbf{A}_i} - \alpha)(\beta - \widehat{\mathbf{A}_i}).
\end{aligned}
\tag{4.51}
$$

4.5 Some examples

Consider the function $\phi(t) = t^p, p \in (-\infty, 0) \cup [1, \infty)$. We observe that $\phi''(t) = p(p-1)t^{p-2}$, $t > 0$. For $0 < m < M$, put

$$B_{m,M,p} := p(p-1) \times \begin{cases} M^{p-2} & \text{if } p \geq 2 \\ m^{p-2} & \text{if } (-\infty, 0) \cup [1, 2) \end{cases}$$

and

$$b_{m,M,p} := p(p-1) \times \begin{cases} m^{p-2} & \text{if } p \geq 2 \\ M^{p-2} & \text{if } (-\infty, 0) \cup [1, 2). \end{cases}$$

If $\operatorname{Sp}(A_i) \subset [m, M] \subset (0, \infty)$, $i \in \{1, \dots, n\}$ and $w_i \geq 0$, $i \in \{1, \dots, n\}$ with $\sum_{i=1}^n w_i = 1$, then by (4.43)

$$\begin{aligned} &\frac{1}{2} b_{m,M,p} \left[\sum_{i=1}^n w_i \widehat{\mathbf{A}_i^2} - \left(\sum_{i=1}^n w_i \widehat{\mathbf{A}_i} \right)^2 \right] \\ &\leq \sum_{i=1}^n w_i \widehat{\mathbf{A}_i^p} - \left(\sum_{i=1}^n w_i \widehat{\mathbf{A}_i} \right)^p \leq \frac{1}{2} B_{m,M,p} \left[\sum_{i=1}^n w_i \widehat{\mathbf{A}_i^2} - \left(\sum_{i=1}^n w_i \widehat{\mathbf{A}_i} \right)^2 \right], \end{aligned} \tag{4.52}$$

while from (4.51), we get

$$\begin{aligned} &\frac{1}{2} b_{m,M,p} \sum_{i=1}^n w_i (\widehat{\mathbf{A}_i} - m)(M - \widehat{\mathbf{A}_i}) \\ &\leq \frac{M - \sum_{i=1}^n w_i \widehat{\mathbf{A}_i}}{M - m} m^p + \frac{\sum_{i=1}^n w_i \widehat{\mathbf{A}_i} - m}{M - m} M^p - \sum_{i=1}^n w_i \widehat{\mathbf{A}_i^p} \\ &\leq \frac{1}{2} B_{m,M,p} \sum_{i=1}^n w_i (\widehat{\mathbf{A}_i} - m)(M - \widehat{\mathbf{A}_i}). \end{aligned} \tag{4.53}$$

For $p = -1$, we get

$$\begin{aligned} M^{-3} \left[\sum_{i=1}^n w_i \widehat{\mathbf{A}_i^2} - \left(\sum_{i=1}^n w_i \widehat{\mathbf{A}_i} \right)^2 \right] &\leq \sum_{i=1}^n w_i \widehat{\mathbf{A}_i^{-1}} - \left(\sum_{i=1}^n w_i \widehat{\mathbf{A}_i} \right)^{-1} \\ &\leq m^{-3} \left[\sum_{i=1}^n w_i \widehat{\mathbf{A}_i^2} - \left(\sum_{i=1}^n w_i \widehat{\mathbf{A}_i} \right)^2 \right] \end{aligned} \tag{4.54}$$

and

$$\begin{aligned} M^{-3} \sum_{i=1}^n w_i (\widehat{\mathbf{A}_i} - m)(M - \widehat{\mathbf{A}_i}) &\leq m^{-1} \frac{M - \sum_{i=1}^n w_i \widehat{\mathbf{A}_i}}{M - m} + M^{-1} \frac{\sum_{i=1}^n w_i \widehat{\mathbf{A}_i} - m}{M - m} - \sum_{i=1}^n w_i \widehat{\mathbf{A}_i^{-1}} \\ &\leq m^{-3} \sum_{i=1}^n w_i (\widehat{\mathbf{A}_i} - m)(M - \widehat{\mathbf{A}_i}). \end{aligned} \tag{4.55}$$

Consider the function $\phi(t) = -\ln t$, $t > 0$. Then $\phi''(t) = \frac{1}{t^2} \in [\frac{1}{M^2}, \frac{1}{m^2}]$ for $t \in [m, M]$. Then by (4.43), we get

$$\begin{aligned}\frac{1}{2M^2}\left[\sum_{i=1}^{n} w_i\widehat{\mathbf{A}_i^2} - \left(\sum_{i=1}^{n} w_i\widehat{\mathbf{A}_i}\right)^2\right] &\le \ln\left(\sum_{i=1}^{n} w_i\widehat{\mathbf{A}_i}\right) - \sum_{i=1}^{n} w_i\,\widehat{\ln(\mathbf{A}_i)} \\ &\le \frac{1}{2m^2}\left[\sum_{i=1}^{n} w_i\widehat{\mathbf{A}_i^2} - \left(\sum_{i=1}^{n} w_i\widehat{\mathbf{A}_i}\right)^2\right], \end{aligned}\tag{4.56}$$

while from (4.50)

$$\begin{aligned}&\frac{1}{2M^2}\sum_{i=1}^{n} w_i(\widehat{\mathbf{A}_i} - m)(M - \widehat{\mathbf{A}_i}) \\ &\le \sum_{i=1}^{n} w_i\,\widehat{\ln(\mathbf{A}_i)} - \ln M\frac{\sum_{i=1}^{n} w_i\widehat{\mathbf{A}_i} - m}{M-m} - \ln m\frac{M - \sum_{i=1}^{n} w_i\widehat{\mathbf{A}_i}}{M-m} \\ &\le \frac{1}{2m^2}\sum_{i=1}^{n} w_i(\widehat{\mathbf{A}_i} - m)(M - \widehat{\mathbf{A}_i}). \end{aligned}\tag{4.57}$$

Consider the function $\phi(t) = \exp t$, $t \in \mathbb{R}$. Then $\phi''(t) = \exp t \in [\exp m, \exp M]$ for $t \in [m, M]$. From (4.43), we then get

$$\begin{aligned}&\frac{1}{2}\exp(m)\left[\sum_{i=1}^{n} w_i\widehat{\mathbf{A}_i^2} - \left(\sum_{i=1}^{n} w_i\widehat{\mathbf{A}_i}\right)^2\right] \\ &\le \sum_{i=1}^{n} w_i\,\widehat{\exp(\mathbf{A}_i)} - \exp\left(\sum_{i=1}^{n} w_i\widehat{\mathbf{A}_i}\right) \\ &\le \frac{1}{2}\exp(M)\left[\sum_{i=1}^{n} w_i\widehat{\mathbf{A}_i^2} - \left(\sum_{i=1}^{n} w_i\widehat{\mathbf{A}_i}\right)^2\right], \end{aligned}\tag{4.58}$$

while from (4.51), we get

$$\begin{aligned}&\frac{1}{2}\exp(m)\sum_{i=1}^{n} w_i(\widehat{\mathbf{A}_i} - m)(M - \widehat{\mathbf{A}_i}) \\ &\le \exp(m)\frac{M - \sum_{i=1}^{n} w_i\widehat{\mathbf{A}_i}}{M-m} + \exp(M)\frac{\sum_{i=1}^{n} w_i\widehat{\mathbf{A}_i} - m}{M-m} - \sum_{i=1}^{n} w_i\,\widehat{\exp(\mathbf{A}_i)} \\ &\le \frac{1}{2}\exp(M)\sum_{i=1}^{n} w_i(\widehat{\mathbf{A}_i} - m)(M - \widehat{\mathbf{A}_i}). \end{aligned}\tag{4.59}$$

4.6 Integral inequalities

Let Ω be a locally compact Hausdorff space endowed with a Radon measure μ. A field $(A_t)_{t\in\Omega}$ of operators in $B(H)$ is called a continuous field of operators if the parametrization $t \longmapsto A_t$ is norm continuous on $B(H)$. If, in addition, the norm function $t \longmapsto \|A_t\|$ is Lebesgue integrable on Ω, we can form the Bochner integral $\int_\Omega A_t d\mu(t)$, which is the unique operator in $B(H)$ such that $\varphi(\int_\Omega A_t d\mu(t)) = \int_\Omega \varphi(A_t) d\mu(t)$ for every bounded linear functional φ on $B(H)$. Assume also that $\int_\Omega 1 d\mu(t) = 1$.

If $\gamma \le \phi''(t) \le \Gamma$, $t \in \mathring{I}$, then from (4.43) for $n = 2$, we get for A, B with $\operatorname{Sp}(A), \operatorname{Sp}(B) \subset I$ and $\nu \in [0,1]$ that

$$\begin{aligned}
&\frac{\gamma}{2}[(1-\nu)A^2 \otimes 1 + \nu 1 \otimes B^2 - ((1-\nu)A \otimes 1 + \nu 1 \otimes B)^2] \\
&\quad \le (1-\nu)\phi(A) \otimes 1 + \nu 1 \otimes \phi(B) - \phi((1-\nu)A \otimes 1 + \nu 1 \otimes B) \\
&\quad \le \frac{\Gamma}{2}[(1-\nu)A^2 \otimes 1 + \nu 1 \otimes B^2 - ((1-\nu)A \otimes 1 + \nu 1 \otimes B)^2].
\end{aligned} \tag{4.60}$$

Observe that, since $A \otimes 1$ commutes with $1 \otimes B$, then

$$\begin{aligned}
&(1-\nu)A^2 \otimes 1 + \nu 1 \otimes B^2 - ((1-\nu)A \otimes 1 + \nu 1 \otimes B)^2 \\
&\quad = (1-\nu)A^2 \otimes 1 + \nu 1 \otimes B^2 - (1-\nu)^2(A \otimes 1)^2 - 2(1-\nu)\nu(A \otimes 1)(1 \otimes B) + \nu^2(1 \otimes B)^2 \\
&\quad = (1-\nu)\nu(A \otimes 1)^2 + (1-\nu)\nu(1 \otimes B)^2 - 2(1-\nu)\nu(A \otimes 1)(1 \otimes B) \\
&\quad = (1-\nu)\nu(A \otimes 1 - 1 \otimes B)^2
\end{aligned}$$

and we get by (4.60) that

$$\begin{aligned}
&\frac{\gamma}{2}(1-\nu)\nu(A \otimes 1 - 1 \otimes B)^2 \\
&\quad \le (1-\nu)\phi(A) \otimes 1 + \nu 1 \otimes \phi(B) - \phi((1-\nu)A \otimes 1 + \nu 1 \otimes B) \\
&\quad \le \frac{\Gamma}{2}(1-\nu)\nu(A \otimes 1 - 1 \otimes B)^2.
\end{aligned} \tag{4.61}$$

We have:

Proposition 4.1. *Assume that ϕ is twice differentiable and $\gamma \le \phi''(t) \le \Gamma$, $t \in \mathring{I}$. Let $(A_t)_{t\in\Omega}$ and $(B_t)_{t\in\Omega}$ be continuous fields of operators in $B(H)$ with $\operatorname{Sp}(A_t), \operatorname{Sp}(B_t) \subset I$, $t \in \Omega$, then for $\nu \in [0,1]$,*

$$\begin{aligned}
&\frac{\gamma}{2}(1-\nu)\nu\left[\left(\int_\Omega A_t^2 d\mu(t)\right) \otimes 1 + 1 \otimes \left(\int_\Omega B_s^2 d\mu(s)\right) - 2\int_\Omega A_t d\mu(t) \otimes \int_\Omega B_s d\mu(s)\right] \\
&\quad \le (1-\nu)\int_\Omega \phi(A_t) d\mu(t) \otimes 1 + \nu 1 \otimes \int_\Omega \phi(B_s) - \int_\Omega\int_\Omega \phi((1-\nu)A_t \otimes 1 + \nu 1 \otimes B_s) d\mu(t) d\mu(s)
\end{aligned}$$

$$\leq \frac{\Gamma}{2}(1-\nu)\nu\left[\int_{\Omega} A_t^2 d\mu(t) \otimes 1 + 1 \otimes \int_{\Omega} B_s^2 d\mu(s) - 2\int_{\Omega} A_t d\mu(t) \otimes \int_{\Omega} B_s d\mu(s)\right]. \tag{4.62}$$

Proof. From (4.61), we get

$$\begin{aligned}
&\frac{\gamma}{2}(1-\nu)\nu(A_t^2 \otimes 1 + 1 \otimes B_s^2 - 2A_t \otimes B_s) \\
&\quad\leq (1-\nu)\phi(A_t) \otimes 1 + \nu 1 \otimes \phi(B_s) - \phi((1-\nu)A_t \otimes 1 + \nu 1 \otimes B_s) \\
&\quad\leq \frac{\Gamma}{2}(1-\nu)\nu(A_t^2 \otimes 1 + 1 \otimes B_s^2 - 2A_t \otimes B_s)
\end{aligned} \tag{4.63}$$

for all $t, s \in \Omega$.

Fix $s \in \Omega$. If we take the $\int_{\Omega}$ over $d\mu(t)$ in (4.63), then we get

$$\begin{aligned}
&\frac{\gamma}{2}(1-\nu)\nu\left(\left(\int_{\Omega} A_t^2 d\mu(t)\right) \otimes 1 + 1 \otimes B_s^2 - 2\int_{\Omega} A_t d\mu(t) \otimes B_s\right) \\
&\quad\leq (1-\nu)\int_{\Omega} \phi(A_t) d\mu(t) \otimes 1 + \nu 1 \otimes \phi(B_s) - \int_{\Omega} \phi((1-\nu)A_t \otimes 1 + \nu 1 \otimes B_s) d\mu(t) \\
&\quad\leq \frac{\Gamma}{2}(1-\nu)\nu\left(\int_{\Omega} A_t^2 d\mu(t) \otimes 1 + 1 \otimes B_s^2 - 2\int_{\Omega} A_t d\mu(t) \otimes B_s\right)
\end{aligned} \tag{4.64}$$

for all $s \in \Omega$.

Further, if we take the $\int_{\Omega}$ over $d\mu(s)$ in (4.64), then we get the desired result (4.62). □

We have the representation

$$X \circ Y = \mathcal{U}^*(X \otimes Y)\mathcal{U},$$

where $\mathcal{U} : H \to H \otimes H$ is the isometry defined by $\mathcal{U}e_j = e_j \otimes e_j$ for all $j \in \mathbb{N}$.

Corollary 4.17. *With the assumptions of Proposition* 4.1, *we have the inequalities*

$$\begin{aligned}
&\gamma(1-\nu)\nu\left[\int_{\Omega}\left(\frac{A_t^2 + B_t^2}{2}\right) d\mu(t) \circ 1 - \int_{\Omega} A_t d\mu(t) \circ \int_{\Omega} B_s d\mu(s)\right] \\
&\quad\leq \int_{\Omega} [(1-\nu)\phi(A_t) + \nu\phi(B_s)] d\mu(t) \circ 1 - \int_{\Omega}\int_{\Omega} \mathcal{U}^* \phi((1-\nu)A_t \otimes 1 + \nu 1 \otimes B_s)\mathcal{U} d\mu(t) d\mu(s) \\
&\quad\leq \Gamma(1-\nu)\nu\left[\int_{\Omega}\left(\frac{A_t^2 + B_t^2}{2}\right) d\mu(t) \circ 1 - \int_{\Omega} A_t d\mu(t) \circ \int_{\Omega} B_s d\mu(s)\right].
\end{aligned} \tag{4.65}$$

The proof follows by (4.62) by taking $\mathcal{U}^*$ to the left and $\mathcal{U}$ to the right and performing the required calculation.

Remark 4.1. If we consider the exponential function $\phi(u) = \exp u$, then we get

$$\begin{aligned}&\exp((1-\nu)A_t \otimes 1 + \nu 1 \otimes B_s)\\ &\quad = \exp[(1-\nu)A_t \otimes 1]\exp[\nu 1 \otimes B_s]\\ &\quad = [(\exp(1-\nu)A_t) \otimes 1](1 \otimes \exp(\nu B_s)) = (\exp(1-\nu)A_t) \otimes \exp(\nu B_s).\end{aligned}$$

Therefore,

$$\begin{aligned}&\mathcal{U}^* \exp((1-\nu)A_t \otimes 1 + \nu 1 \otimes B_s)\mathcal{U}\\ &\quad = \mathcal{U}^*[(\exp(1-\nu)A_t) \otimes \exp(\nu B_s)]\mathcal{U} = (\exp(1-\nu)A_t) \circ \exp(\nu B_s)\end{aligned}$$

and

$$\begin{aligned}&\int_\Omega \int_\Omega \mathcal{U}^* \exp((1-\nu)A_t \otimes 1 + \nu 1 \otimes B_s)\mathcal{U} d\mu(t) d\mu(s)\\ &\quad = \int_\Omega (\exp(1-\nu)A_t) d\mu(t) \circ \int_\Omega \exp(\nu B_s) d\mu(s).\end{aligned}$$

In this case, the inequality (4.65) becomes

$$\begin{aligned}&(1-\nu)\nu \exp(m)\left[\int_\Omega \left(\frac{A_t^2 + B_t^2}{2}\right) d\mu(t) \circ 1 - \int_\Omega A_t d\mu(t) \circ \int_\Omega B_t d\mu(t)\right]\\ &\le \int_\Omega [(1-\nu)\exp(A_t) + \nu \exp(B_t)] d\mu(t) \circ 1 - \int_\Omega (\exp(1-\nu)A_t) d\mu(t) \circ \int_\Omega \exp(\nu B_t) d\mu(t)\\ &\le (1-\nu)\nu \exp(M)\left[\int_\Omega \left(\frac{A_t^2 + B_t^2}{2}\right) d\mu(t) \circ 1 - \int_\Omega A_t d\mu(t) \circ \int_\Omega B_t d\mu(s)\right], \end{aligned} \tag{4.66}$$

provided that $\mathrm{Sp}(A_t), \mathrm{Sp}(B_t) \subset [m, M]$, $t \in \Omega$.

Finally, if we take $B_t = A_t$, $t \in \Omega$, then we get the simpler inequality

$$\begin{aligned}&(1-\nu)\nu \exp(m)\left[\int_\Omega A_t^2 d\mu(t) \circ 1 - \int_\Omega A_t d\mu(t) \circ \int_\Omega A_t d\mu(t)\right]\\ &\quad \le \int_\Omega \exp(A_t) d\mu(t) \circ 1 - \int_\Omega (\exp(1-\nu)A_t) d\mu(t) \circ \int_\Omega \exp(\nu A_t) d\mu(t)\\ &\quad \le (1-\nu)\nu \exp(M)\left[\int_\Omega A_t^2 d\mu(t) \circ 1 - \int_\Omega A_t d\mu(t) \circ \int_\Omega A_t d\mu(t)\right], \end{aligned} \tag{4.67}$$

provided that $\mathrm{Sp}(A_t) \subset [m, M]$, $t \in \Omega$.

Bibliography

[1] Ando, T., *Concavity of certain maps on positive definite matrices and applications to Hadamard products*, Linear Algebra Appl. 26 (1979), 203–241.

[2] Araki, H. and Hansen, F., *Jensen's operator inequality for functions of several variables*, Proc. Am. Math. Soc. 128(7) (2000), 2075–2084.

[3] Aujila, J. S. and Vasudeva, H. L., *Inequalities involving Hadamard product and operator means*, Math. Jpn. 42 (1995), 265–272.

[4] Beesack, P. R. and Pečarić, J. E., *On Jessen's inequality for convex functions*, J. Math. Anal. Appl. 110 (1985), 536–552.

[5] Dragomir, S. S., Refinements and reverses of tensorial arithmetic mean-geometric mean inequalities for selfadjoint operators in Hilbert spaces, Preprint *RGMIA Res. Rep. Coll.* **25** (2022), Art.

[6] Dragomir, S. S., Pearce, C. E. M. and Pečarić, J. E., *On Jessen's and related inequalities for isotonic sublinear functionals*, Acta. Sci. Math. (Szeged) 61 (1995), 373–382.

[7] Fujii, J. I., *The Marcus-Khan theorem for Hilbert space operators*, Math. Jpn. 41 (1995), 531–535.

[8] Furuta, T., Mićić Hot, J., Pečarić, J. and Seo, Y., *Mond–Pečarić Method in Operator Inequalities. Inequalities for Bounded Selfadjoint Operators on a Hilbert Space*, Element, Zagreb, 2005.

[9] Kitamura, K. and Seo, Y., *Operator inequalities on Hadamard product associated with Kadison's Schwarz inequalities*, Sci. Math. 1(2) (1998), 237–241.

[10] Korányi, A., *On some classes of analytic functions of several variables*, Trans. Am. Math. Soc. 101 (1961), 520–554.

[11] Wada, S., *On some refinement of the Cauchy–Schwarz Inequality*, Linear Algebra Appl. 420 (2007), 433–440.

N.-L. Wang, T. Kuzumaki, and S. Kanemitsu

5 Summation formulas as modular relations

Abstract: In this note, we clarify the relationship between equivalents of the functional equation and various summation formulas. We refer to equivalent assertions to the functional equation as modular relations. Summation formulas include Abel–Plana, Euler–Maclaurin, Poisson. Also the approximate (or classical) sampling formula centers around the equivalent summation formulas.

Keywords: Abel–Plana, Euler–Maclaurin, Poisson summation formula, functional equation, partial fraction expansion for the cotangent function

MSC 2020: 11M99, 26A99

5.1 Introduction

Various summation formulas (SF) such as Poisson SF, Euler–Maclaurin SF, Abel–Plana SF, etc. have been rather widely applied in disciplines related to mathematics. However, very little attention has been paid on their interrelationship. In number theory, there are some studies on the subject as equivalents to the functional equation.

More recently, Butzer et al. [3] proved equivalence of PSF (Poisson summation formula), EMSF (Euler–Maclaurin summation formula), APSF (Abel–Plana summation formula) to approximate sampling formula (ASF). Prior to this, Butzer et al. [2] proved similar equivalence among PSFB (Poisson SF for band-limited functions), GPF (general Parseval formula), RKB (reproducing kernel formula), and CSF (classical sampling formula). In [3] it is proved that all these are equivalent, since (ASF) and (CSF) are proved equivalent. We therefore restrict ourselves to the latter paper and pursue equivalence among these and the modular relation (this aspect has also been studied in [1]).

It is convenient to state the results on the Riemann zeta-function by using the Hurwitz zeta-function defined by

$$\zeta(s,a) = \sum_{n=0}^{\infty} \frac{1}{(n+a)^s}, \tag{5.1}$$

where $s = \sigma + it$ is the complex variable and a is not a non-positive integer (in most cases $0 < a \le 1$). The Hurwitz zeta-function has the counterpart: the Lerch zeta-function

N.-L. Wang, School of Mathematics and Computer Science, Shangluo Univ., Shangluo, Shaanxi 726000, P.R. China, e-mail: wangnianliangshangluo@aliyun.com
T. Kuzumaki, Faculty of Engineering, Gifu Univ., Gifu 501-1193, Japan, e-mail: kobayashi.takako.w7@f.gifu-u.ac.jp
S. Kanemitsu, Sanmenxia SUDA New Energy Research Institute, No. 1, Taiyang Road, Economic Development Zone, Sanmenxia, Henan 472000, P.R. China, e-mail: omnikanemitsu@yahoo.com

https://doi.org/10.1515/9783112215760-005

or the polylogarithm function of order s of the complex exponential argument, which is defined by

$$l_s(z) = \sum_{n=1}^{\infty} \frac{e^{2\pi i n z}}{n^s} \tag{5.2}$$

for $\sigma > 1$ in the first instance. These are connected by the Hurwitz formula $(0 < a < 1)$

$$\zeta(1-s,a) = \frac{\Gamma(s)}{(2\pi)^s}\left(e^{-\frac{\pi i s}{2}}\, l_s(a) + e^{\frac{\pi i s}{2}}\, l_s(1-a)\right), \tag{5.3}$$

whose reciprocal formula reads

$$l_s(a) = \frac{\Gamma(1-s)}{(2\pi)^{1-s}}\left(e^{\frac{\pi i}{2}(1-s)}\zeta(1-s,a) + e^{-\frac{\pi i}{2}(1-s)}\zeta(1-s,1-a)\right). \tag{5.4}$$

Here $\Gamma(s)$ is the Euler gamma function.

Both the Hurwitz and the Lerch zeta-functions are generalizations of the Riemann zeta-function

$$\zeta(s,1) = l_s(\mathbb{Z}) = \zeta(s).$$

The functional equation for the Riemann zeta-function (FER) reads

$$\pi^{-\frac{\pi}{2}s}\Gamma\left(\frac{s}{2}\right)\zeta(s) = \pi^{-\frac{\pi}{2}(1-s)}\Gamma\left(\frac{1-s}{2}\right)\zeta(1-s) \tag{5.5}$$

as a special case of the Hurwitz formula. There are some variants and we quote them without further notice.

For the Riemann zeta-function, cf. [7] and [21] (where many proofs of (5.5) are given) and for the Lerch zeta-function, cf. [12].

5.2 Euler–Maclaurin summation formula

We state EMSF and its applications.

The lth Bernoulli polynomial $B_l(a)$ may be defined by the umbral calculus:

$$B_l(a) = (B+a)^l = \sum_{r=0}^{l}\binom{l}{r}B_r a^{l-r}, \quad B_l(0) = B_l, \tag{5.6}$$

where B_l is called the lth Bernoulli number. They can be computed from the recursion $\sum_{r=0}^{l-1}\binom{l}{r}B_r = 0$ (5.6) is the sequence generated by

$$\frac{ze^{az}}{e^z - 1} = \sum_{n=0}^{\infty} B_n(a)\frac{z^n}{n!}, \quad (|z| < 2\pi). \tag{5.7}$$

Let $[x]$ denote the largest integer not exceeding x. Then define the lth periodic Bernoulli polynomial $\bar{B}_l(a)$ by

$$\bar{B}_l(x) = \bar{B}_l x - [x], \quad \bar{B}_1(x) = x - [x] - \frac{1}{2}. \tag{5.8}$$

Since

$$\frac{\mathrm{d}}{\mathrm{d}t}\bar{B}_r(x) = r\bar{B}_{r-1}(x), \tag{5.9}$$

one can find the Bernoulli polynomial by integration; in particular, all can be found from

$$\bar{B}_1(x) = x - [x] - \frac{1}{2} = -\frac{1}{\pi}\sum_{n=1}^{\infty}\frac{\sin 2\pi n}{n}, \tag{5.10}$$

in which the third equality holds true for non-integral values of x.

We state the Euler–Maclaurin (EMSF) in the following handy form

Theorem 5.1 (Euler–Maclaurin summation formula). *For $f(t) \in C^l[a,x]$, we have*

$$\sum_{a<n\le x} f(n) = \int_a^x f(t)\,\mathrm{d}t + \sum_{r=1}^{l}\frac{(-1)^r}{r!}\{\bar{B}_r(x)f^{(r-1)}(x) - \bar{B}_r(a)f^{(r-1)}(a)\} + \frac{(-1)^{l+1}}{l!}\int_a^x \bar{B}_l(t)f^{(l)}(t)\,\mathrm{d}t. \tag{5.11}$$

Proof follows by repeated applications of integration by parts to Euler's summation formula, which in turn follows from partial summation formula. The latter eventually follows from the integration by parts for Stieltjes integrals.

As one of far-reaching results, we state the integral representation for the partial sum of the Hurwitz zeta-function [8, p. 55].

The following theorem and its corollaries [8, pp. 55–61] are far-reaching, but in literature, it went unnoticed. A prototype can be found in [15].

Theorem 5.2 (Integral representations). *For every $u \in \mathbb{C}$ and any $l \in \mathbb{N}$ with $l > \operatorname{Re} u + 1$, we have the integral representation (for any $x \ge 0$)*

$$\begin{aligned} L_u(x,a) &= \sum_{r=1}^{l}\frac{\Gamma(u+1)}{\Gamma(u+2-r)}\frac{(-1)^r}{r!}\bar{B}_r(x)\,(x+a)^{u-r+1} \\ &\quad + \frac{(-1)^l}{l!}\frac{\Gamma(u+1)}{\Gamma(u+1-l)}\int_x^{\infty}\bar{B}_l(t)\,(t+a)^{u-l}\,\mathrm{d}t \\ &\quad + \begin{cases} \frac{1}{u+1}(x+a)^{u+1} + \zeta(-u,a), & u \ne -1, \\ \log(x+a) - \psi(a), & u = -1. \end{cases} \end{aligned} \tag{5.12}$$

Equation (5.12) leads to the asymptotic formula for $L_u(x, a)$ and the integral representation for $\zeta(-u, a)$ according as $x \to \infty$ and $x = 0$.

Theorem 5.3. *EMSF ⇒ FER.*

Proof. We prove

$$\zeta(s) = \frac{1}{\pi}(2\pi)^s \sin\frac{\pi}{2}s\Gamma(1-s)\zeta(1-s), \tag{5.13}$$

which is [21, p. 13, (2.1.1)] by a corollary to Theorem 5.2: After transforming the special case of (5.12) with $a = 1$, $l = 2$ etc., we have

$$\zeta(s) = -\frac{s(s+1)}{2}\int_0^\infty \frac{\bar{B}_2(x)}{x^{s+2}}\,dx, \tag{5.14}$$

which is absolutely convergent for $-1 < \sigma < 0$ and gives an analytic continuation for ζ. The second periodic Bernoulli polynomial $\bar{B}_2(x) = \{x\}^2 - \{x\} + \frac{1}{6} = 2\int_1^s \bar{B}_1(t)\,dt$ has the absolutely convergent Fourier series

$$\bar{B}_2(x) = \frac{1}{\pi^2}\sum_{n=1}^\infty \frac{\cos 2\pi nx}{n^2}. \tag{5.15}$$

Substituting this in (5.14) and using the Mellin transform formula

$$\int_0^\infty (\cos 2\pi x)x^{-s-1}\frac{dx}{x} = -(2\pi)^{s+1}\cos\frac{\pi}{2}(s+1)\Gamma(-s-1), \tag{5.16}$$

we conclude that

$$\zeta(s) = \frac{s(s+1)}{2\pi^2}(2\pi)^{s+1}\cos\frac{\pi}{2}(s+1)\Gamma(-s-1)\zeta(2-s-1), \tag{5.17}$$

whence (5.13). Since Theorem 5.2 is a consequence of EMSF, this proves the assertion. □

Let I be a given interval and let F be a function analytic in a region containing I. Then the boundary function $f(x)$

$$f(x) = [F(z)] = F(x+i0) - F(x-i0) \tag{5.18}$$

is called the hyperfunction generated by F and denoted $f(x) = \mathrm{H.F.}F(z)$ and F is called the generating function and denoted $F(z) = \mathrm{G.F.}f(x)$. In fact, a hyperfunction is an equivalence class of the family $\{F\}$ modulo holomorphic functions on I. Cf., e. g., [6].

The derivative of a hyperfunction $f(x) = \mathrm{H.F.}F(z)$ is defined as

$$Df(x) = f'(x) = \mathrm{H.F.}F'(z). \tag{5.19}$$

and a hyperfunction may be differentiated infinitely many times (Goursat's theorem). The Leibniz rule holds true.

Suppose $f(x) = \mathrm{H.F.}F(z)$, $F_\pm(z)$ are integrable at $z = a$ and $z = b$ and that C is an arbitrary (positively-oriented) closed contour passing through $z = a$ and $z = b$. Then define

$$D^{-1}f = \int_a^b f(x)\,\mathrm{d}x = -\int_C F(z)\,\mathrm{d}z. \tag{5.20}$$

Lemma 5.1. *Suppose f resp. g is integrable on $[a,x]$ resp. of $C^l[a,x]$. We denote integration resp. differentiation by D^{-1} resp. D. Then we have*

$$D^{-l}(fg) = \sum_{k=1}^{l} (-1)^{k-1}[D^{-k}fD^{k-1}g]_a^x + (-1)^l D^{-1}(D^{-n}fD^n g). \tag{5.21}$$

With

$$g(t) = \bar{B}_1(t) \tag{5.22}$$

Lemma 5.1 gives EMSF, Theorem 5.1. The Heaviside function (or the unit step function) is given by

$$H(x) = \mathrm{H.F.}\left(\frac{-1}{2\pi i}\log(-z)\right), \tag{5.23}$$

where $\log z$ means the principal branch. As a hyperfunction

$$H'(x) = \mathrm{H.F.}\left(\frac{-1}{2\pi i}\frac{\mathrm{d}}{\mathrm{d}z}\log(-z)\right) = \mathrm{H.F.}\left(\frac{-1}{2\pi i}\frac{1}{z}\right), \tag{5.24}$$

which corresponds to

$$(\log z)' = \frac{1}{z} \quad (z \neq 0). \tag{5.25}$$

On the x-axis, we have

$$\log z = \log|x| \pm \pi i H(-x),$$

where + resp. – is to be taken according to approach from above or from below. Hence, we have

$$z^a = \begin{cases} |x|^a & x > 0, \\ e^{\pm\pi i a}|x|^a & x < 0. \end{cases}$$

We have the differentiation formulas.

$$(|x|^{\alpha}H(x))' = \alpha|x|^{\alpha-1}H(x). \tag{5.26}$$

Corollary 5.1 (Hadamard finite part). *Suppose α is non-integral. Then, we have*

$$\int_{-\epsilon}^{a} f(t)|t|^{\alpha}H(t)\,\mathrm{d}t = \sum_{k=1}^{l}(-1)^{k-1}\frac{a^{\alpha+k}}{(\alpha+1)\cdots(\alpha+k)}f^{(k-1)}(a) + \frac{(-1)^{l}}{(\alpha+1)\cdots(\alpha+l)}\int_{-\epsilon}^{a} t^{\alpha+l}f^{(l)}(t)\,\mathrm{d}t, \tag{5.27}$$

where the last integral converges absolutely for $l > -\operatorname{Re}\alpha - 1$ and gives the finite part interpretation (FP) of the divergent integral:

$$FP\int_{0}^{a} t^{\alpha}f(t)\,\mathrm{d}t = \int_{-\epsilon}^{a} f(t)|t|^{\alpha}H(t)\,\mathrm{d}t. \tag{5.28}$$

With

$$g(t) = |t|^{\alpha}H(t), \tag{5.29}$$

and with the interval $[-\epsilon, a]$ ($a > 0, \epsilon > 0$), Lemma 5.1 proves Corollary 5.1 on the basis of (5.26).

Theorem 5.1 and Corollary 5.1 are in remarkable contrast in that both are formulas which give absolutely convergent integrals, although they are in totally different spirit and both follow from Lemma 5.1, repeated application of integration by parts.

5.3 Poisson summation formula

The Poisson summation formula (PSF) is as fundamental as the functional equation p. 118, (1.446) for zeta-functions and so is very useful in many respects. Here main resources are [5, 16, 19, 18], which are unified in [4] on the basis of [11] and [14].

Referring to the main method of [14] as the Principle, we state the main result of [4] as

Theorem 5.4. *The Poisson summation formula* (5.34) *à la Hamburger is manifestation of the Principle, with the Fourier transform pair. The Soni–Oberhettinger formula of the form*

$$\sum_{n=1}^{\infty} a_n f(\lambda_n) = \mathrm{P}(\cdot) + \sum_{n=1}^{\infty} b_n \hat{f}(\mu_n), \quad \hat{f}(y) = \int_{0}^{\infty} f(x)K(xy)\,\mathrm{d}x \tag{5.30}$$

is a consequence of the Principle with $\hat{f}(z)$ replaced by the Mellin transform.

This implies Hamburger's theorem.

Corollary 5.2 ([5]). *FER (functional equation for the Riemann zeta-function)* (5.5) *is equivalent to the following equalities.*

The theta-transformation formula (the Bochner modular relation)

$$\sum_{n=1}^{\infty} e^{-\pi n^2 x} = x^{-\frac{1}{2}} \sum_{n=1}^{\infty} e^{-\frac{\pi n^2}{x}} + \frac{1}{2}(x^{-\frac{1}{2}} - 1), \quad \operatorname{Re} x > 0. \tag{5.31}$$

The partial fraction expansion for the hyperbolic cotangent function $i \cot \pi i z$

$$1 + 2\sum_{n=1}^{\infty} e^{-2\pi n z} = \frac{1}{\pi z} + \frac{2z}{\pi} \sum_{n=1}^{\infty} \frac{1}{z^2 + n^2}. \tag{5.32}$$

The Fourier expansion

$$\sum_{n=1}^{[x]} (x - n) = \frac{x(x-1)}{2} - \frac{1}{2\pi^2} \sum_{n=1}^{\infty} n^{-2}(\cos 2\pi n x - 1). \tag{5.33}$$

The Poisson summation formula ($T > 0$)

$$T \sum_{n=-\infty}^{\infty} \varphi(Tn) = \sum_{n=-\infty}^{\infty} \hat{\varphi}\left(\frac{2\pi}{T} n\right) = \sum_{n=-\infty}^{\infty} \int_{-\infty}^{\infty} e^{-\frac{2\pi}{T} in} \varphi(u)\, du \tag{5.34}$$

valid for a function $\varphi(u)$, *which is of bounded variation in any finite interval satisfying convergence conditions and such that*

$$f(z) := \int_{-\infty}^{\infty} e^{izu} \varphi(u)\, du \tag{5.35}$$

is analytic in a certain (vertical) strip $-d < \sigma < c$, *say, where* $c, d > 0$.

The convergence conditions are as stated on [5, p. 136] and

$$\int_{-\infty}^{\infty} |\varphi(\gamma + iy)|\, dy < \infty, \qquad \int_{-\infty}^{\infty} |\varphi(-\gamma + iy)|\, dy < \infty \tag{5.36}$$

for some γ, $-d < \gamma < c$.

(5.34) is proved for the analytic function $f(z)$ in (5.35) and then inverted by the Fourier integral theorem. (5.34) in the form

$$T \sum_{n=-\infty}^{\infty} \varphi(Tn) = \sum_{n=-\infty}^{\infty} \int_{\infty}^{\infty} e^{\frac{2\pi}{T} in} \varphi(u)\, du \tag{5.37}$$

with $T = 2\pi$ leading to [5, (V*)], while (5.34) with $T = 2\pi\lambda$ is a familiar form [10, pp. 141–142]. To be more in detail, let

$$\varphi_\lambda(x) = \lambda\varphi(\lambda x), \quad g_\lambda(t) = 2\pi \sum_{n=\infty}^{\infty} \varphi_\lambda(t + 2\pi n).$$

Then

$$\hat{g}_\lambda(n) = \hat{\varphi}\left(\frac{n}{\lambda}\right)$$

and (5.34) reads

$$g_\lambda(0) = \sum_{n=-\infty}^{\infty} \hat{g}_\lambda(n). \tag{5.38}$$

(5.31) and (5.32) have the very rich the background. The former is the main part of Hecke's correspondence, referred to as the Riemann–Hecke–Bochner (RHB) correspondence, which is rather intrinsic to holomorphic modular forms. The latter is formulated as the Fourier–Bessel expansion and its extension (Fourier–Whittaker expansion) allows one to treat the case of generalized Maass forms. Cf. [9].

$$\sum_{n=1}^{[x]} (x - n) = \frac{1}{2}(x^2 - x) - \frac{1}{2}\left(\bar{B}_2(x) - \frac{1}{6}\right), \tag{5.39}$$

so that (5.33) amounts to (5.15).

In [3], f is supposed to be AC (absolutely continuous), i. e., it can be expressed as

$$f(t) = \int_{-\infty}^{t} f'(t)\,dt, \quad f' \in L^1$$

in the similar spirit as (5.35).

In the literature, there are many variations of the conditions on the function φ. E. g., in [17, p. 73] one can find a theorem on the same conditions as on [5, p. 138, Footnote 20)]:

Theorem 5.5. *If $f \in C^2(\mathbb{R})$, $\int_{-\infty}^{\infty} f(t)\,dt$ exists and $f'' \in L^1$, then we have*

$$\sum_{n=-\infty}^{\infty} f(n + u) = \sum_{n=-\infty}^{\infty} \hat{f}(n) e^{2\pi i n u}, \tag{5.40}$$

where

$$\hat{f}(n) = \int_{-\infty}^{\infty} e^{-2\pi i n t} f(t)\,dt$$

is the Fourier transform of f.

Corollary 5.3. *PSF ⇒ LSF (Lipschitz summation formula) ⇒ reciprocal Hurwitz formula* (5.4) *⇒ FER*

In [21, § 2.3], there are two more proofs via the theta transformation formula and the partial fraction expansion, respectively.

Below, we briefly describe the source of the partial fraction expansion (5.32) as intrinsic properties of the 0th polylogarithm function (5.42), especially the one with the complex exponential argument $\mathrm{Li}_0(e^{2\pi iz})$, $\mathrm{Im}\, z > 0$.

The polylogarithm function is defined by

$$\mathrm{Li}_s(z) = \sum_{s=1}^{\infty} \frac{z^n}{n^s}, \tag{5.41}$$

where the series is absolutely convergent for $\sigma > 1$ if $|z| = 1$ or for $s \in \mathbb{C}$ if $|z| < 1$.

For $|z| < 1$, we have the geometric series

$$\mathrm{Li}_0(z) = \sum_{n=1}^{\infty} z^n = \frac{z}{1-z} \tag{5.42}$$

and the (mono-)logarithm function

$$\mathrm{Li}_1(z) = \sum_{n=1}^{\infty} \frac{z^n}{n} = -\log(1-z). \tag{5.43}$$

Writing

$$\mathcal{L}(z) = -\frac{1}{z}\mathrm{Li}_s(z) = \frac{1}{z-1}, \quad z \neq 1,$$

we have the Ramanujan kernel

$$\mathcal{L}(e^z) = \frac{1}{e^z - 1},$$

which appears in

$$\Gamma(s)\zeta(s) = \mathcal{M}(\mathcal{L}(e^z)) = \int_0^{\infty} x^{s-1}\frac{1}{e^x - 1}\,\mathrm{d}x.$$

The following is the most relevant to us.

$$(l_0(z) =)\mathrm{Li}_0(e^{2\pi iz}) = \frac{e^{2\pi iz}}{1 - e^{2\pi iz}} = \frac{1}{2}(-1 + i\cot \pi z), \quad z \notin \mathbb{Z} \tag{5.44}$$

or

$$\pi \cot \pi z = -\pi - 2\pi i \mathrm{Li}_0(e^{2\pi iz}), \quad z \notin \mathbb{Z}. \tag{5.45}$$

Hereafter, we shall use the familiar notation $\cot \pi z$ to state our results, but it must be born in mind that they are intrinsic properties of the 0th polylogarithm function expressed in another way.

Lemma 5.2. *The cotangent function $\pi \cot \pi z$ has the global expressions*

$$\pi \cot \pi z = \frac{1}{z} + 2z \sum_{n=1}^{\infty} \frac{1}{z^2 - n^2} = -\pi i - \frac{2\pi i}{e^{-2\pi i z} - 1} = \pi i + \frac{2\pi i}{e^{2\pi i z} - 1}, \quad z \notin \mathbb{Z}, \tag{5.46}$$

where the first member is the partial fraction expansion, and the second resp. third expression is of exponential decay for $y >$ resp. $y < 0$. The expressions of exponential decay have the rapidly converging expressions

$$\pi \cot \pi z = -\pi i - 2\pi i \sum_{n=1}^{\infty} e^{2\pi i n z} \quad (y > 0), \tag{5.47}$$

$$\pi \cot \pi z = \pi i + 2\pi i \sum_{n=1}^{\infty} e^{-2\pi i n z} \quad (y < 0), \tag{5.48}$$

respectively.

The lemma suggests that we should use (5.46) as the pick-up function and regard the Lambert series as respective expressions.

Lemma 5.3. *Suppose that $f(z)$ is meromorphic in $\mathbb{C}$, where it has poles at $z_1, z_2, \ldots$. Let $D = \{z = x + iy | \alpha \le x \le \beta, |y| \le \gamma\}$ $\alpha < \beta$, $\gamma > 0$, $[\alpha] = m \ne \alpha$, $[\beta] = n \ne \beta$ or a similar figure. Let $C = \partial D$. Then as $\alpha \to -\infty$, $\beta \to \infty$ and $\gamma \to \infty$, we have*

$$\lim \frac{1}{2\pi i} \int_C f(z) \pi \cot \pi z \, dx = \sum_{k=-\infty}^{\infty} f(k) + \sum_{k=1}^{\infty} \operatorname*{Res}_{z=z_k} f(z) \pi \cot \pi z, \tag{5.49}$$

provided that the limit exists and the sums are convergent.

The earliest appearance of this and relevant results is on [13, pp. 49–65], also partly reproduced on [17, pp. 211–223]. This has the effect of picking up the values of $f(\nu)$ at the poles $z = \nu \in \mathbb{Z}$. Here the partial fraction expansion (5.46) plays its role of picking-up.

Theorem 5.6. *Suppose $f(z)$ is analytic in the domain $\{z | |y| \le \gamma\}$, $\gamma > 0$ and*

$$\int_{-\infty}^{\infty} \max_{|y| \le \delta} |f(x + iy)| \, dy < \infty. \tag{5.50}$$

Further suppose that $f \in C^2(\mathbb{R})$, $\int_{-\infty}^{\infty} f(t) \, dt$ exists and $f'' \in L^1$. Then (5.40) holds true.

Proof. (5.40) follows from Lemma 5.3 with the rectangle $D = \{z = x + iy | -\nu \le x \le \nu, |y| \le \gamma\}$, ν being positive half-integer and using rapidly decreasing series in Lemma 5.2 in the upper resp. the lower half-plane. We let $\gamma \to 0$ first and then let $\nu \to \infty$. □

5.4 Plana's summation formula

Although it is known as Plana's summation formula, Butzer et al. [3] elucidated the history and call it Abel–Plana summation formula, cf. [13, pp. 49–65]. We refer to it as APSF or Plana's summation formula. It is noteworthy that Plana's summation formula has an extension in quadratic fields à la Koshklyakov [11]. It has been treated in [14], from which we quote some essential results.

Let Ω be a number field of degree $\chi = r_1 + 2r_2 \le 2$ with discriminant Δ and the Dedekind zeta-function $\zeta_\Omega(s)$. Then since the contents of Lemma 5.2 are essentially the ingredients of the following lemma in the case of the rational field ($r_1 = 1, r_2 = 0, \Omega = \mathbb{Q}$), we state it full in anticipation of further developments of [16] and [5].

Lemma 5.4 (Koshlyakov). *The functional equation* ($A = \frac{2^{r_2}\pi^{\frac{\chi}{2}}}{\sqrt{|\Delta|}}$)

$$A^{-s}\Gamma^{r_1}\left(\frac{s}{2}\right)\Gamma^{r_2}(s)\zeta_\Omega(s) = A^{-(1-s)}\Gamma^{r_1}\left(\frac{1-s}{2}\right)\Gamma^{r_2}(1-s)\zeta_\Omega(1-s), \tag{5.51}$$

is equivalent to each of the following. The Bochner modular relation

$$\sum_{n=1}^{\infty} a_n X_{r_1,r_2}(Axn) = \mathrm{P}_X(x) + x^{-1}\sum_{n=1}^{\infty} a_n X_{r_1,r_2}(Ax^{-1}n), \quad \operatorname{Re} x > 0. \tag{5.52}$$

The partial fraction expansion

$$\sigma(x) = \frac{A}{\pi}\sum_{n=1}^{\infty} a_n K_{r_1,r_2}(A^2xn) = \mathrm{P}_K(x) + \frac{x}{\pi}\sum_{n=1}^{\infty}\frac{a_n}{n^2+x^2}. \tag{5.53}$$

Here

$$X(x) = X_{r_1,r_2}(x) = \frac{1}{2\pi i}\int_{(c)} \Gamma^{r_1}\left(\frac{s}{2}\right)\Gamma^{r_2}(s)x^{-s}\,\mathrm{d}s, \quad c > 0, \tag{5.54}$$

for $x > 0$ ($\operatorname{Re} x > 0$), where and in what follows (c) indicates the Bromwich path $\sigma = c$, $-\infty < t < \infty$. We may find

$$X_{1,0}(x) = 2e^{-x^2}, \quad X_{0,1}(x) = e^{-x}, \quad X_{2,0}(x) = 4K_0(2x). \tag{5.55}$$

Also, $\mathrm{P}_X(x)$ is the residual function given by

$$\mathrm{P}_X(x) = 2^{r_1}\zeta_\Omega^{(r)}(0) - 2^{r_1}\zeta_\Omega^{(r)}(0)x^{-1}. \tag{5.56}$$

Similarly,

$$K_{r_1,r_2}(x) = \frac{1}{2\pi i}\int_{(c)} \frac{\pi}{2\cos\frac{\pi s}{2}}G(1-s)x^{-s}\,\mathrm{d}s \tag{5.57}$$

and

$$K_{1,0}(x) = \sqrt{\pi}\, e^{-2x},$$
$$K_{0,1}(x) = \frac{1}{i}(K_0(2\varepsilon\sqrt{x}) - K_0(2\bar{\varepsilon}\sqrt{x})), \tag{5.58}$$
$$K_{2,0}(x) = 2(K_0(4\varepsilon\sqrt{x}) + K_0(4\bar{\varepsilon}\sqrt{x})),$$

where K_0 is a modified Bessel function.

And

$$\mathrm{P}_K(x) = -\frac{1}{2}\rho - \frac{\zeta_\Omega(0)}{\pi}\frac{1}{x}, \tag{5.59}$$

where ρ is the residue of the Dedekind eta-function at $s = 1$ given in the case $\chi \le 2$ by

$$\rho = -\frac{2^{r+1}\pi^{r_2}\zeta_\Omega^{(r)}(0)}{\sqrt{|\Delta|}}. \tag{5.60}$$

The σ-function in (5.53) is defined by ($c > 1$)

$$\sigma(x) = \frac{A}{\pi}\sum_{n=1}^{\infty} a_n K_{r_1,r_2}(A^2 xn) = \frac{1}{2\pi i}\int_{(c)} \frac{1}{2\cos\frac{\pi s}{2}} A^{1-2s} G(1-s)\zeta_\Omega(s)x^{-s}\,\mathrm{d}s, \tag{5.61}$$

which reduces to the cotangent function in the rational case, cf. (5.44).

For details, cf. [11] and [14].

Theorem 5.7 (Plana's summation formula, [11, p. 217], [20, p. 90, (9)]). *Let $f(x)$ and $F(s)$ be the Mellin transform pair in Definition 1.12 and suppose that the Mellin transform $F(s)$*

$$F(s) = \int_0^\infty f(x)x^{s-1}\,\mathrm{d}x, \quad \sigma > 0 \tag{5.62}$$

satisfies the growth condition

$$F(s) = O(e^{-\frac{\pi}{2}|t|}|t|^{m\sigma-n}), \tag{5.63}$$

uniformly in $\alpha \le \sigma \le a$, where $-1 < \alpha < 0$, $1 < a$, and where $m > 0$, $n > 0$ are subject to the condition

$$\frac{m}{2} + n \ge 1, \tag{5.64}$$

and that $F(s)$ is regular in the strip $\beta \le \sigma \le a$ except for a simple pole at $s = 0$ with residue $f(0)$. Suppose that the domain of $f(x)$ is extended as

$$0 \ne x = |x|e^{i\theta}, \quad -\frac{\pi}{2} \le \theta \le \frac{\pi}{2}. \tag{5.65}$$

and that $f(x)$ is analytic there, as in [11]. *Then*

$$Z_f(s) = \sum_{n=1}^{\infty} a_n f(n) = J(s) + \mathrm{P}(s), \tag{5.66}$$

where

$$\mathrm{P}(s) = \zeta_\Omega(0)f(0) + \rho \int_0^\infty f(x)\mathrm{d}x, \quad J(s) = 2\int_0^\infty \frac{f(ix) - f(-ix)}{2i}^s \sigma(x)\,\mathrm{d}x, \tag{5.67}$$

and where $\sigma(s)$ is Koshlyakov's function (5.61) *and ρ is defined in* (5.60). (5.66) *entails Plana's summation formula* [20, *p. 90, (9)*]

$$\sum_{n=1}^{\infty} f(n) = -\frac{1}{2}f(0) + \int_0^\infty f(x)\mathrm{d}x - 2\int_0^\infty \frac{f(ix) - f(-ix)}{2i}\frac{1}{e^{2\pi x} - 1}\,\mathrm{d}x, \tag{5.68}$$

under the condition (5.65) *and the infinite series and the integral in* (5.68) *are (absolutely) convergent.*

For quadratic fields $\Omega = \mathbb{Q}(\sqrt{\Delta})$, we have

$$\begin{aligned} J(s) &= \int_0^\infty \frac{f(ix) - f(-ix)}{2i} I(x)\,\mathrm{d}x \\ &= \frac{2A}{\pi i}\sum_{n=1}^{\infty} a_n(K_0(2\bar{\varepsilon}A\sqrt{xn}) - K_0(2\varepsilon A\sqrt{xn})), \quad \Delta < 0 \\ &= \frac{4A}{\pi}\sum_{n=1}^{\infty} a_n(K_0(4\bar{\varepsilon}A\sqrt{xn}) + K_0(4\varepsilon A\sqrt{xn})), \quad \Delta > 0. \end{aligned} \tag{5.69}$$

Theorem 5.8. *With the aid of a version of the Plana summation formula, Theorem* 5.7, *the Poisson summation formula* (5.34) *of Hamburger follows from the partial fraction expansion* (5.53), *which in turn follows from* (5.52) *and leads to the functional equation* (5.51).

5.5 Equivalent summation formulas

Regarding interrelation between SFs and FER, Hamburger's theorem implies FER $\Leftrightarrow$ PSF and "Plana summation formula $\Rightarrow$ Poisson summation formula" is established in Theorem 5.8. There is another set of equivalent assertions to the FER. cf. [8, Chapter 5]. Therefore, all these are equivalent.

Regarding direct proofs of equivalence of SFs, the following theorem of Butzer et al. is the summit.

Theorem 5.9 ([3]).
(i) *Euler–Maclaurin summation formula ⇔ Poisson summation formula.*
(ii) *Abel–Plana summation formula ⇔ Euler–Maclaurin summation formula.*

(i) is proved in [3, § 5] and (ii) in [3, § 3].

We hope to return to revealing the relationship between ASF, CSF, FER, etc. and [22] elsewhere.

Bibliography

[1] Butzer, P. L. and Hauss, M., Applications of sampling theory to combinatorial analysis, Stirling numbers, special functions and the Riemann zeta function, in: J. R. Higgins and R. L. Stens (eds.), *Sampling Theory in Fourier and Signal Analysis. Vol. 2: Advanced Topics*, pp. 1–37, 266–268, Oxford University Press, 1999.

[2] Butzer, P. L., Ferreira, P. J. S. G., Higgins, J. R., Schmeisser, G. and Stens, R. L., *The sampling theorem, Poisson's summation formula, general Parseval formula, reproducing kernel formula and the Paley–Wiener theorem for bandlimited signals—their interconnections*, Appl. Anal. 90 (2011), 431–461, https://doi.org/10.1080/00036811003627567.

[3] Butzer, P. L., Ferreira, P. J. S. G., Schmeisser, G. and Stens, R. L., *The summation formulae of Euler–Maclaurin, Abel–Plana, Poisson, and their interconnections with the approximate sampling formula in signal analysis*, Res. Math. 59 (2011), 359–400.

[4] Chaktaborty, K., Kanemitsu, S. and Yu, L.-W., On the Hamburger–Oberhettinger–Soni modular relations, in: *Mathematical Analysis: Theory and Applications*, 2024.

[5] Hamburger, H., *Über einige Beziehungen, die mit der Funktionalgleichung der Riemannschen ζ-Funktion äquivalent sind*, Math. Ann. 85 (1922), 129–140.

[6] Imai, I., *Applied Hyperfunction Theory*, Springer, Dordrecht, 1992.

[7] Ivić, A., *The Riemann Zeta Function*, Dover, New York, 2003.

[8] Kanemitsu, S. and Tsukada, H., *Vistas of Special Functions*, World Scientific, Singapore, 2007.

[9] Kanemitsu, S. and Tsukada, H., *Contributions to the Theory of Zeta-Functions: The Modular Relation Supremacy*, World Scientific, Singapore, 2014.

[10] Katznelson, Y., *An Introduction to Harmonic Analysis*, 3rd ed., Cambridge UP, Cambridge, 2004.

[11] Koshlyakov, N. S., *Investigation of some questions of analytic theory of the rational and quadratic fields I, II (Russian)*, Izv. Akad. Nauk SSSR, Ser. Mat. 18 (1954), 113–144, 213–260, 307–326. Errata: 19 (1955), 277.

[12] Laurinchikas, A. and Garunkstis, R., *The Lerch Zeta-Function*, Kluwer Academic Publ., Dordrecht–Boston–London, 2002.

[13] Lindelöf, E., *Calcul des Résidus et Ses Applications a la Théorie des Fonctions*, Gauthier-Villars, Paris, 1905.

[14] Liu, G.-D., Chakraborty, K. and Kanemitsu, S., *A unifying principle in the theory of modular relations*, Mathematics 11(3) (2023), 535.

[15] Mellin, Hj., *Die Dirchletschen Reihen, die zahlentheoretischen Funktionen und die unendlichen Produkte von endlichem Geschlecht*, Acta Soc. Sci. Fenn. 31 (1902), 1–48.

[16] Oberhettinger, F. and Soni, K. L., *On some relations which are equivalent to the functional equations involving the Riemann zeta function*, Math. Z. 127 (1972), 17–34.

[17] Rademacher, H., *Topics in Analytic Number Theory*, Springer Verl., Berlin–Heidelberg, 1973.

[18] Soni, K. L., *On some relations which are equivalent to the functional equations of type of Riemann's zeta function and tables of divisor transforms*, Ph. D. Thesis, Oregon State University, Corvallis, 1964.

[19] Soni, K. L., *Some relations associated with an extension of Koshliakov's formula*, Proc. Am. Math. Soc. 17 (1966), 543–551.

[20] Srivastava, H. M. and Choi, J.-S., *Series Associated with the Zeta and Related Functions*, Kluwer Academic Publishers, Dordrecht, 2001.
[21] Titchmarsh, E. C., *The Theory of the Riemann Zeta-Function*, 2nd ed., OUP, Oxford, 1986.
[22] Zayed, A. I., *A proof of new summation formula by using sampling theorem*, Proc. Am. Math. Soc. 117 (1993), 699–710.

Halimeh Ardakani, Yeol Je Cho, and Madjid Eshaghi Gordji

6 Some geometric properties on Banach lattices

Abstract: This paper is devoted to some properties on Banach lattices related to positively limited sets and almost limited sets. At first, the concept of the strong limited p-Schur property ($1 \leq p \leq \infty$) is introduced and studied, i. e., spaces on which every weakly p-compact and almost limited set is relatively compact. Next, the weak DP* property of order p is defined and spaces with this property are characterized. As an application of these results, by the class of disjoint p-convergent operators, some characterizations of Banach lattices with the weak DP* property of order p are given. Moreover, Banach lattices with the strong Dunford–Pettis relatively compact property of order p ($1 \leq p < \infty$), that is, spaces in which every weakly p-compact and almost Dunford–Pettis set is relatively compact are studied. We also introduce the notion of the weak Dunford–Pettis property of order p and then characterize this property in terms of sequences. The last section is devoted to three properties on Banach lattices related to positively limited sets, which we call the positively limited Schur property of order p ($1 \leq p \leq \infty$), that is, spaces on which every weakly p-compact and positively limited set is relatively compact, the positive DP* property of order p and the weak positively limited Schur property of order p. In particular, it is proved that the weak positively limited Schur property of order p coincides with the order continuity of the norm.

Keywords: Limited set, Dunford–Pettis set, positively limited set, Gelfand–Phillips property, Schur property, weakly p-summable sequence

MSC 2020: 46B42, 46B50, 47B65

6.1 Introduction and preliminaries

Throughout this paper, let X denote a Banach space, E a Banach lattice, B_E the closed unit ball and E^+ the positive cone of E, respectively. The *solid hull* of a subset A of E is the set

$$Sol(A) = \{y \in E : |y| \leq |x| \quad \text{for some } x \in A\}.$$

Halimeh Ardakani, Department of Mathematics, Payame Noor University, P. O. Box 19395-4697, Tehran, Iran, e-mail: ardakani@pnu.ac.ir
Yeol Je Cho, Department of Mathematics Education, Gyeongsang National University, Jinju 52828, Korea, e-mail: yjchomath@gmail.com
Madjid Eshaghi Gordji, Department of Mathematics, Semnan University, P. O. Box 35195-363, Semnan, Iran, e-mail: meshaghi@semnan.ac.ir

https://doi.org/10.1515/9783112215760-006

The lattice operations in E are weakly sequentially continuous if, for every weakly null sequence (x_n) in E, $|x_n| \xrightarrow{w} 0$. Also, the lattice operations in E^* are weak* sequentially continuous if, for every weak* null sequence (x_n^*) in E^*, $|x_n^*| \xrightarrow{w^*} 0$.

Recall that a bounded subset A of X is called a *limited set* if every weak* null sequence (x_n^*) in X^* converges uniformly to zero on A, that is, $\sup_{x\in A}|x_n^*(x)| \to 0$. Banach spaces whose limited sets are relatively compact are called *Gelfand–Phillips spaces* (shortly, GP spaces). A Banach space X has the DP* *property* if each relatively weakly compact set in X is limited. Refer to the papers [12, 17, 20] for some results about limited sets, GP spaces and DP* properties.

If A is a subset of X such that, for each weakly null sequence (x_n^*) in X^*, $\lim_{n\to\infty}\sup_{a\in A}|\langle a, x_n^*\rangle| = 0$, then we say that A is Dunford–Pettis. Banach spaces whose Dunford–Pettis sets are relatively compact are called Dunford–Pettis relatively compact property or $DP_{rc}P$ spaces. A dual Banach space with the weak Radon-Nikodym property has the $DP_{rc}P$. A Banach space X has the Dunford–Pettis property if each weakly compact operator on X is Dunford–Pettis (i. e., it carries weakly null sequences to norm null ones) [17, 19, 20].

Later, the disjoint version of limited and Dunford–Pettis sets arises in Banach lattice theory in a natural way.

A bounded subset A of E is said to be an almost Dunford–Pettis set if every disjoint weakly null sequence (x_n^*) in E^* converges uniformly to zero on A [10]. Accordingly, the stronger version of the $DP_{rc}P$ is considered and the class of Banach lattices with the strong $DP_{rc}P$, which is shared by those Banach lattices whose almost Dunford–Pettis subsets are relatively compact is introduced. It was proved that discrete KB-spaces are exactly the same as the Banach lattices with the strong $DP_{rc}P$ [7].

A Banach lattice E has the Schur (positive Schur) property if each weakly null (positive weakly null) sequence in E is norm null [15, 32]. Also, E has the weak Dunford–Pettis property if each weakly compact operator from E into each Banach space is an almost Dunford–Pettis operator (i. e., it carries disjoint weakly null sequences to norm null ones) [35].

Following [14], a bounded subset A of a Banach lattice E is called an *almost limited set* if every disjoint weak* null sequence (x_n^*) in E^* converges uniformly to zero on A. Clearly, every limited set in a Banach lattice is almost limited. A Banach lattice E has the *weak* DP* *property* if each relatively weakly compact set in E is almost limited. Each Banach lattice with the DP* property has the weak DP* property, but the converse is not true. Indeed, $L^1[0,1]$ has the weak DP* property but it does not have the DP* property.

Recently, the concept of positively limited sets in Banach lattices in the paper [3]. A norm bounded subset A of a Banach lattice E is said to be *positively limited* when every positive weak* null sequence in E^* converges uniformly to zero on A. It is proved that every almost limited set in a Banach lattice is positively limited, but the converse is not true. In fact, the closed unit ball B_c is positively limited, but it is not almost limited.

They also considered the positive DP* property, a property which is in general weaker than the weak DP* property.

We recall some notation and notions of weakly p-summable sequences which will be needed. For each $1 \leq p < \infty$, a sequence (x_n) in X is called *weakly p-summable* if $(x^*(x_n)) \in \ell_p$ for each $x^* \in X^*$. We denote the set of all weakly p-summable sequences in X by $\ell_p^w(X)$. A sequence (x_n) in X is said to be *weakly p-convergent* to $x \in X$ if the sequence $(x_n - x) \in \ell_p^w(X)$. The weakly ∞-convergent sequences are simply the weakly convergent sequences. A norm bounded set A in a Banach space X is called *relatively weakly p-compact* if each sequence in A has a weakly p-convergent subsequence. If the limit point of each weakly p-convergent subsequence is in A, then we call A a weakly p-compact set. A Banach space X is *weakly p-compact* if the closed unit ball B_X is a weakly p-compact set.

Zeekoie introduced the notions of Schur property of order p (*p-Schur property*) and also the positive Schur property of order p (*p-positive Schur property*) and then characterized this property in terms of disjoint weakly p-summable sequences by proving that a Banach lattice E has the p-positive Schur property if and only if each disjoint sequence $(x_n) \in \ell_p^w(E)$ with positive terms is norm convergent to 0. Also, some conditions were investigated under which the p-Schur property and p-positive Schur property are coincide. The notion of the p-Schur property is a generalization of the Schur property. Following Zeekoei, X the 1-Schur property if and only if X does not contain a copy of c_0 [37].

Recently, in [16], the concept of the *limited p-Schur property*; that is, Banach spaces whose all limited weakly p-compact subsets are relatively compact ($1 \leq p \leq \infty$) is introduced and some conditions under which some operator spaces have the limited p-Schur property are established. Clearly, each Banach space with the GP property has the limited p-Schur property.

Using the concept "weakly p-compact set in a Banach space", the DP* property of order p (p-DP* *property*) is introduced in [21] and some characterizations of Banach spaces with this property are obtained.

We introduce a stronger form of the limited p-Schur property for Banach lattices (Definition 6.1) and characterize the class of spaces with this property. Then, using weakly p-summable sequence techniques, we consider the p-positive Schur property in Banach lattices and investigate spaces in which p-positive Schur property is equivalent to the positive Schur property. A Banach lattice E has the *positive Schur property* if each positive weakly null sequence in E is norm null [32]. Finally, we introduce the weak DP* property of order p (i. e. p-weak DP* property), which is a generalization of the classical weak DP* property and by a class of disjoint p-convergent operators, research Banach lattices with the p-weak DP* property. The reader should see [14, 32] for the weak DP* and positive Schur properties in Banach lattices.

Throughout the paper, p' denotes the conjugate number of p for $1 < p < \infty$; if $p = 1$, $\ell_{p'}$ plays the role of c_0.

The concept of the strong limited p-Schur property, that is, spaces on which every weakly p-compact and almost limited set is relatively compact, is introduced and studied. Also, some conditions under which two properties, p-Schur and p-positive Schur coincide are provided. Moreover, the p-weak DP* property is defined and spaces with this property are characterized.

Castillo mentioned the p-convergent operators for the first time. In fact, p-convergent operators are precisely those operators which transformed weakly p-compact subsets into relatively compact subsets. Equivalently, an operator $T : E \to X$ is *p-convergent* if, for every sequence $(x_n) \in \ell_p^w(E)$, $\|Tx_n\| \to 0$. Later disjoint p-convergent were introduced. An operator T is called *disjoint p-convergent* if, for every disjoint weakly p-summable sequence (x_n), $\|Tx_n\| \to 0$. A Banach space X has the *Dunford–Pettis property of order p (p-Dunford–Pettis property)* if each weakly compact operator T from X into each Banach space Y is p-convergent [13, 37].

A Banach space X has the *p-DP$_{rc}$P* if each weakly p-compact and Dunford–Pettis subset of X is relatively compact; or equivalently, every Dunford–Pettis sequence $(x_n) \in \ell_p^w(X)$ is norm null [25].

We introduce the concept of the strong p-DP$_{rc}$P for Banach lattices and consider the connection between this concept with other well-known properties. In particular, we provide Banach lattices in which two properties p-Schur and p-positive Schur coincide. Also, the p-weak Dunford–Pettis property is introduced and using disjoint p-convergent operators some characterizations are obtained. As an application, we provide the conditions under which two properties p-weak Dunford–Pettis and p-Dunford–Pettis will be the same. Finally, Banach lattices with the positive p-DP$_{rc}$P are studied and some of the results derived from this property are given.

Motivated by the above works and replacing positively limited sets by almost limited sets, in Section 6.7, we introduce the concept of the positively limited p-Schur property on Banach lattices, a property which is in general stronger than the strong limited p-Schur property, obtain some characterizations of this property and give some examples.

Section 6.8 is concerned with the notion of the positive DP* property of order p which is in general weaker than the positive DP* property. In this section, we characterize the positive DP* property of order p in terms of disjoint sequences and also give an operator characterization for this property.

In the last section of the present paper, motivated by the notions of positively limited sequences $(x_n) \in \ell_p^w(E)_+$, we introduce the concept of the weak positively limited p-Schur property on Banach lattices and obtain some characterizations of this property. In particular, it is proved that Banach lattices with the weak positively limited p-Schur property are exactly the same Banach lattices with order continuous norm.

We collect all the concepts mentioned above and other concepts which will be used in the present paper as follows:

(1) A Banach lattice E has the following:
   - *Positive Schur property* if each positive weakly null sequence in E is norm null (converges to zero) [32].

- *Dual positive Schur property* if each positive weak* null sequence in E^* is norm null [34].
- *Positive* DP* *property* if each relatively weakly compact set in E is positively limited [3].
- *Positive GP property* if each positively limited set in E is relatively compact [3].
- *p-Schur property* if every sequence $(x_n) \in \ell_p^w(E)$ is norm null [37].
- *p-Positive Schur property* if every sequence $(x_n) \in \ell_p^w(E)_+$ is norm null [37].
- *p*-DP* *property* if each relatively weakly p-compact set in E is limited [21].
- *p-Weak* DP* *property* if each relatively weakly p-compact set in E is almost limited [4].
- *Limited p-Schur property* if every limited weakly p-compact set in E is relatively compact [16].
- *Strong limited p-Schur property* if every almost limited weakly p-compact set in E is relatively compact [4].

(2) An operator $T : E \to X$ is called:
- *p-Convergent* if $\|Tx_n\| \to 0$ for every sequence $(x_n) \in \ell_p^w(E)$ [13].
- *Disjoint p-convergent* if $\|Tx_n\| \to 0$ for every disjoint sequence $(x_n) \in \ell_p^w(E)$ [37].

Throughout this paper, we assume that $1 \le p < \infty$, unless otherwise stated. The reader is referred to [1, 30] for information on positive operators and Banach lattices.

6.2 Strong limited *p*-Schur property

Recently, the notion of the p-Schur property ($1 \le p < \infty$) as a generalization of the Schur property is introduced. In fact, a Banach space X has the *p-Schur property* if every weakly p-summable sequence in X is norm null. Moreover, it has been shown that X has the 1-Schur property if and only if X contains no copy of c_0. As we said before, a Banach space X has the limited p-Schur property if each weakly p-compact limited set in X is relatively compact; or equivalently, every limited sequence $(x_n) \in \ell_p^w(X)$ is norm null. Every Banach space with the Schur property has the p-Schur and so the limited p-Schur property, but the converse is false. As an example, for $1 < p < \infty$ and $1 < q < p'$, all ℓ_q spaces have the p-Schur property, but none of them have the Schur property. Also, c_0 has the limited p-Schur property, but it does not have the p-Schur property [16, 37]. Accordingly, using almost limited sets instead of limited sets, the concept of strong limited p-Schur property is studied in [4].

Definition 6.1. A Banach lattice E has the *strong limited p-Schur property* if each almost limited weakly p-compact subset of E is relatively compact.

It can be easily shown that, a Banach lattice E has the strong limited p-Schur property if and only if each almost limited sequence $(x_n) \in \ell_p^w(E)$ is norm null. It is easy to see that, if $1 \le p < q$ and E has the strong limited q-Schur property, then E has the strong limited p-Schur property. Notice that, the strong limited p-Schur property implies the limited p-Schur property, but the following example shows that the converse is false, in general.

Example 6.1. $L^1[0,1]$ has the limited p-Schur property, but it does not have the strong limited p-Schur property, for $p \ge 2$.

Proof. Since $L^1[0,1]$ is separable, then it has the limited p-Schur property. But $L^1[0,1]$ does not have the strong limited p-Schur property, for $p \ge 2$. In fact, the *Rademacher sequences* (r_n) in $L^1[0,1]$ are weakly 2-summable (see [31, Proposition 3.6]) and almost limited by the weak DP* property, but $\|r_n\| = 1$ for all $n \ge 1$. □

Actually, $L^1[0,1]$ has the 1-Schur property since it contains no copy of c_0.

Each Banach lattice with the strong GP property, such as the classical Banach lattices c_0, ℓ_p, Schur spaces and more generally discrete Banach lattices with order continuous norm, have the strong limited p- Schur property (see [2, Theorem 2.1]). The following example shows that the converse is not true, in general.

Example 6.2. For each compact metric space K, $C(K)$ has the strong limited p-Schur property, but it does not have the strong GP property.

Proof. It is a well-known fact that every weakly p-summable sequence is weakly null. By [2, Theorem 2.12], every weakly null almost limited sequence in $C(K)$ is norm null, which implies that $C(K)$ has the strong limited p-Schur property. But $C(K)$ does not have the strong GP property. Indeed, there is an almost limited *Rademacher sequences* equivalent to the unit vector basis of ℓ_1 in $C(K)$, which is not relatively compact. □

Let us recall that an element $e \in E$ is a *weak unit* if $E = B_e$, where B_e is the band generated by e. For example, $C[0,1]$ has the weak unit $u(t) = t$, but $M[0,1]$ and ℓ_∞^* do not have any weak unit.

Proposition 6.1. *If E has order continuous norm or E^* has a weak unit, then the strong limited Schur (i. e., the strong limited ∞-Schur) and strong GP properties are equivalent.*

Proof. It is clear that the strong GP property implies the strong limited Schur property.

For the converse, by the strong limited Schur property, each almost limited weakly null sequence in E is norm null. Then, by hypothesis on E and [2, Theorem 2.15], E has the strong GP property. This completes the proof. □

Note that, each discrete Banach lattice with order continuous norm has the strong limited p-Schur property, but the converse is not true. Indeed, discrete Banach lattice c has the strong limited p-Schur property, but it does not have order continuous norm (see [2, Theorem 2.4]).

Now, for the proof of the Theorem 6.1, we need the following two lemmas.

Lemma 6.1. *Continuous linear image of an almost limited set in a Banach lattice E under a positive linear projection is almost limited.*

Proof. Assume that U is a closed sublattice of a Banach lattice E, P is a linear positive projection of E onto U and A is an almost limited set in E. To finish the proof, we have to show that $P(A)$ is an almost limited set in U; that is, for each disjoint weak* null sequence (f_n) in U^*, we have

$$\sup_{x\in A}|\langle f_n, Px\rangle| \to 0.$$

By [30, 1.4. E4], P^* is a lattice homomorphism and so (P^*f_n) is a disjoint weak* null sequence in E^*. Since A is an almost limited set in E, we have

$$\sup_{x\in A}|\langle f_n, Px\rangle| = \sup_{x\in A}|\langle P^*f_n, x\rangle| \to 0$$

as $n \to \infty$ and so $P(A)$ is an almost limited set. This completes the proof. □

Lemma 6.2. *Let Y be a complemented subspace of a Banach space X and $A \subseteq Y$ is a limited set in X. Then A is a limited set in Y.*

Proof. Since Y is a complemented subspace of X, then there is an onto positive projection $P : X \to Y$. If (x_n^*) is a weak* null sequence in Y^*, then we can also extend each x_n^* to the whole of X by putting $u_n^* = x_n^* \; o \; P$. It is clear that (u_n^*) is a weak* null sequence in X^*. Since A is a limited set in X, we have

$$\sup_{a\in A}|\langle x_n^*, a\rangle| = \sup_{a\in A}|\langle x_n^* \; o \; P, a\rangle| = \sup_{a\in A}|\langle u_n^*, a\rangle| \to 0.$$

This completes the proof. □

Note that, if F is a positively complemented sublattice of a Banach lattice E such that F has the property (d) and $A \subseteq F$ is an almost limited set in E, then A is an almost limited set in F. It is a well-known result that a Banach space X has the Schur (resp., p-Schur) property if and only if each closed separable linear subspace of X has the Schur (resp., p-Schur) property.

Theorem 6.1. *For each $1 \le p \le \infty$, the following are equivalent:*
(1) *E has the strong limited p-Schur property.*
(2) *Every closed separable sublattice of E is contained in a positively complemented sublattice Z with the property (d) of E with the strong limited p-Schur property.*
(3) *E is the direct sum of two spaces with the strong limited p-Schur property.*

Proof. (1) ⇒ (2) It is obvious. In fact, the strong limited p-Schur property is inherited by closed sublattices.

(2) $\Rightarrow$ (1) Suppose that A is a weakly p-compact almost limited subset of E and that $(x_n) \subseteq A$. Then there is a subsequence (x_{n_k}) of (x_n) that is almost limited weakly p-convergent to some $x \in A$. Consider the closed linear span of (x_n). By hypothesis, it is contained in a positively complemented sublattice Z with the property (d) of E with the strong limited p-Schur property. Hence, (x_n) is an almost limited sequence in Z. Since Z has the strong limited p-Schur property, then (x_{n_k}) is norm convergent to x.

(1) $\Rightarrow$ (3) Consider $E = E \oplus \{0\}$.

(3) $\Rightarrow$ (1) Let $E = Y \oplus Z$ such that Y and Z have the strong limited p-Schur property. Consider the positive linear projections $P_Y : E \to Y$ and $P_Z : E \to Z$ and assume that A is a weakly p-compact almost limited subset of E. So by Lemma 6.8, two sets $P_Y(A)$ and $P_Z(A)$ are weakly p-compact almost limited. Then for each sequence $(x_n) \subseteq A$ there is $y_n \in P_Y(A)$ and $z_n \in P_Z(A)$ such that $x_n = y_n + z_n$. Since $P_Y(A)$ and $P_Z(A)$ have the strong limited p-Schur property, the sequences (y_n) and (z_n) have convergent subsequences (y_{n_k}) and (z_{n_k}), that is, there are $y \in P_Y(A)$ and $z \in P_Z(A)$ such that $y_{n_k} \to y$ and $z_{n_k} \to z$. Since A is a weakly p-compact set, then $x_{n_k} \to y + z \in A$. Hence A is compact and so E has the strong limited p-Schur property. This completes the proof. □

Note that the complementedness of the Lemma 6.2 cannot be removed. In fact, the unit vector basis (e_n) of c_0, as a closed separable sublattice of ℓ_∞, is limited in ℓ_∞, but it is not limited (almost limited) in c_0.

It can be easily shown that, a Banach space X has the p-Schur property if and only if X has the limited p-Schur and DP* properties. So, one can conclude that ℓ_∞ does not have the limited p-Schur property. In fact, ℓ_∞ has the DP* property, but it does not have the p-Schur property. So, we can say more:

Corollary 6.1. *If a Banach space X contains a sublattice isomorphic to ℓ_∞, then X does not have the limited p-Schur property.*

For each σ-Dedekind complete Banach lattice, we have the following theorem.

Theorem 6.2. *If E is a σ-Dedekind complete Banach lattice, then the following are equivalent:*
(1) *E has the strong GP property.*
(2) *E is discrete with the strong limited p-Schur property.*
(3) *E is discrete with the limited p-Schur property.*

Proof. (1) $\Rightarrow$ (2) It is clear that E has the strong limited p-Schur property. Also, from [2, Theorem 2.3], E is discrete.

(2) $\Rightarrow$ (3) It is clear.

(3) $\Rightarrow$ (1) Since E has the limited p-Schur property, then by Corollary 6.1, E contains no sublattice isomorphic to ℓ_∞ and so E has order continuous norm [30, Corollary 2.4.3]. Also, E is discrete and then, by [2, Theorem 2.1], E has the strong GP property. This completes the proof. □

Each weakly p-compact set is weakly compact, but the converse is not true. In fact, non-discrete Banach lattice $L^1[0,1]$ has the 1-Schur property and so there is an order interval in $L^1[0,1]$ which is not weakly 1-compact (see [1, Proposition 3.6]). But $L^1[0,1]$ has order continuous norm and so each order interval in $L^1[0,1]$ is weakly compact.

Lemma 6.3. *Every order bounded disjoint sequence in a Banach lattice E is weakly p-summable for all $1 \leq p \leq \infty$.*

Proof. It is useful to know that every order bounded disjoint sequence in a Banach lattice E converges weakly to zero (see [1, p. 192]). Actually, it has been shown that, this sequence is weakly 1-summable. Since $\ell_p^w(X) \subset \ell_q^w(X)$ for all $1 \leq p \leq q$, every weakly 1-summable sequence is weakly p-summable. This completes the proof. □

It is proved in [15] that a σ-Dedekind complete Banach lattice has order continuous norm if and only if it has the GP property.

In the following theorem, we obtain the same result with the limited p-Schur property that is weaker than the GP property. Recall that a Banach lattice E has the *property* (d) if the sequence $(|f_n|)$ is weak* null for every disjoint weak* null sequence (f_n) in E^*. Clearly, each σ-Dedekind complete Banach lattice has the property (d), but the converse is false, in general. In fact, ℓ_∞/c_0 has the property (d), but it is not σ-Dedekind complete [29].

Theorem 6.3. *Let E be a Banach lattice. Then for the following assertions:*
(1) *E has the property (d) and the strong limited p-Schur property.*
(2) *E has order continuous norm.*
(3) *E is σ-Dedekind complete with the limited p-Schur property.*

Proof. (1) $\Rightarrow$ (2) $\Leftrightarrow$ (3) are valid.

(1) $\Rightarrow$ (2) Let (x_n) be an order-bounded disjoint sequence in E. Then (x_n) is weakly p-summable and almost limited by the property (d) (see [29, Proposition 2.1]). Also, E has the strongly limited p-Schur property and then (x_n) is norm null. Hence, by [1, Theorem 4.14], E has order continuous norm.

(2) $\Rightarrow$ (3) Each Banach lattice with order continuous norm has the GP (and so the limited p-Schur) property. It is clear that E is σ-Dedekind complete.

(3) $\Rightarrow$ (2) Let (x_n) be an order bounded disjoint sequence in E. Then (x_n) is weakly p-summable and limited in a σ-Dedekind complete Banach lattice E [3]. Also, E has the limited p-Schur property and then (x_n) is norm null. Hence, by [1, Theorem 4.14], the norm of E is order continuous. This completes the proof. □

It can be concluded that a σ-Dedekind complete Banach lattice E has the limited p-Schur property if and only if E has the GP property. If E is discrete, then all the statements of the Theorem 6.3 are equivalent.

From Corollary 6.1, we have the following two corollaries.

Corollary 6.2. *If a Banach lattice E contains a sublattice isomorphic to ℓ_1, then its dual E^* does not have the limited p-Schur property.*

Proof. If E contains a sublattice isomorphic to ℓ_1, then, by [30, Proposition 2.3.12], E^* contains a sublattice isomorphic to ℓ_∞ and so E^* does not have the limited p-Schur property. This completes the proof. □

Corollary 6.3. *If a Banach space X contains a complemented copy of ℓ_1, then X^* does not have the limited p-Schur property.*

Proof. If X contains a complemented copy of ℓ_1, then, by [17, Theorem 10], X^* contains an isomorphic copy of ℓ_∞ and so X^* does not have the limited p-Schur property. This completes the proof. □

Note that the converse of Corollary 6.3 is true for Banach lattices. In fact, By [17, Theorem 10], if a Banach lattice E contains no complemented copy of ℓ_1, then E^* contains no copy of c_0. By [30, Theorem 2.5.6], E^* is a KB-space and so it has order continuous norm. Hence E^* has the limited p-Schur property.

Proposition 6.2. *If a Banach space X contains no copy of ℓ_1, then X^* has the 1-Schur property.*

Proof. If X contains no copy of ℓ_1, then, by [17, Theorem 10], X^* contains no copy of c_0 and so, by [37, Proposition 3.2.3], X^* has the 1-Schur property. This completes the proof. □

Proposition 6.3. *For each Banach lattice E, the following assertions hold:*
(1) *If E has the p-Schur property, then it has the GP property.*
(2) *If E is discrete with the p-Schur property, then it has the strong GP property.*

Proof. To prove that (1), it suffice to note that, if E has the p-Schur property, then E contains no copy of c_0, that is, E is a KB-space. Hence, E has order continuous norm and so it has the GP property.

The second part (2) can be deduced from [2, Theorem 2.3]. In fact, in this case E is a discrete KB-space. This completes the proof. □

6.3 *p*-Positive Schur and *p*-weak DP* properties

A Banach lattice E has the positive Schur property if each positive weakly null sequence in E is norm null [15, 32]. It has been shown for discrete Banach lattices and also for Banach lattices with the strong GP property, the positive Schur and Schur properties are equivalent [2, 15].

Using weakly p-summable sequences, the notion of the p-positive Schur property has been introduced.

A Banach lattice E has the *p-positive Schur property* if every sequence $(x_n) \in \ell_p^w(E)_+$ is norm null, alternatively, E has the p-positive Schur property if and only if every disjoint sequence $(x_n) \in \ell_p^w(E)_+$ is norm null.

Clearly, the positive Schur property implies the p-positive Schur property, but the converse is not true. In fact, ℓ_p spaces $(1 < p < \infty)$, weakly sequentially complete (wsc) Banach lattices and reflexive spaces do not have the positive Schur property, but all of them have the 1-Schur property [37].

In the following theorem, we show that the positive Schur and p-positive Schur properties coincide in each Banach lattice with the finite type. In Chapter 16 of [18], one finds a discussion of type and cotype in Banach lattices. Moreover, we introduce and study the concept of the p-weak DP* property and then, by disjoint p-convergent operators, characterize some Banach lattices with the p-weak DP* property.

Theorem 6.4. *Let E be a Banach lattice with the type q (with $1 < q \leq 2$) and let $p \geq q'$. Then the following are equivalent:*
(1) *E has the p-positive Schur property.*
(2) *E has the positive Schur property.*

Proof. (1) $\Rightarrow$ (2) From [30, Corollary 3.6.8], it is enough to show that every relatively weakly compact set A in E is an L-weakly compact set (i. e., $\|x_n\| \to 0$ for every disjoint sequence (x_n) in the $sol(A)$). For this, let (x_n) be a disjoint sequence in $sol(A)$. Then, by [37, Lemma 4.2.1, Proposition 3.1.5], two sequences (x_n) and $(|x_n|)$ are weakly p-summable. Hence by the p-positive Schur property of E, the sequence $(|x_n|)$ (and hence (x_n) itself) is norm null. Therefore A is an L-weakly compact set, as desired.

(2) $\Rightarrow$ (1) It is evident. This completes the proof. □

A Banach lattice E is *weak p-consistent* if it follows from $(x_n) \in \ell_p^w(E)$ that $(|x_n|) \in \ell_p^w(E)$. If E is a weak p-consistent Banach lattice (for instance, an AM-space with unit), then E has the p-positive Schur property if and only if E has the p-Schur property [37].

In the following theorem, we show that the same result holds for some Banach lattices with the strong limited p-Schur property, too.

Theorem 6.5. *Let E be a Banach lattice with the type q (with $1 < q \leq 2$) and let $p \geq q'$. Then the following are equivalent:*
(1) *E has the p-Schur property.*
(2) *E has the strong limited p-Schur and p-positive Schur properties.*

Proof. (1) $\Rightarrow$ (2) It is obvious.

(2) $\Rightarrow$ (1) Let A be a weakly p-compact set in E. Clearly, A is a relatively weakly compact set. Since E has the p-positive Schur property, by Theorem 6.4, A is L-weakly compact. Then it follows from [14, Theorem 2.6] that A is an almost limited set in E. On the other hand, E has the strong limited p-Schur property and then A is relatively compact. Hence E has the p-Schur property. This completes the proof. □

Proposition 6.4. *Every Banach lattice E with the p-positive Schur property has the 1-Schur property.*

Proof. Note that c_0 does not have the p-positive Schur property. So, if a Banach lattice E has the p-positive Schur property, then E contains no copy of c_0 and, by [37, Proposition 3.2.3], E has the 1-Schur property. This completes the proof. □

Similar to the DP* property, the so-called *weak* DP* *property* of a Banach lattice is introduced in [14]. In fact, a Banach lattice E has the weak DP* property if all relatively weakly compact subsets are almost limited. In other words, E has the weak DP* property if and only if, for every weakly null sequence (x_n) in E and every disjoint weak* null sequence (x_n^*) in E^*, $x_n^*(x_n) \to 0$.

Theorem 6.6. *If E^* has the limited p-Schur property and a weak unit, Y is a Banach space and $T : E \to Y$ is an operator, then we have the following:*
1. *If Y is wsc, then T is a weakly compact operator.*
2. *If Y has the Schur property, then T is a compact operator.*
3. *If Y has the* DP* *property, then T is a limited operator.*
4. *If Y has the weak* DP* *property, then T is an almost limited operator.*

Proof. Since E^* has the limited p-Schur property, then E^* does not contain any sublattice isomorphic to ℓ_∞ and, by [30, Theorem 2.4.14], E^* is a *KB*-space. On the other hand, E^* has a weak unit and so, by [30, 2.5.E1], the closed unit ball B_E is weakly conditionally compact. Hence each of the statements of the theorem can be concluded easily. This completes the proof. □

From [21], a Banach space X is said to have the DP* *property of order p* or briefly the p-DP* *property* whenever every weakly p-compact set in X is limited. So, it is natural to study the Banach lattices E satisfying the p-weak DP* property. Thereby, we have a scale of properties in the sense that all Banach lattices have the 1-weak DP* property and, if $p < q$ and E has the q-weak DP* property, then it has the p-weak DP* property. The strongest property, the ∞-weak DP* property, coincides with the weak DP* property. We characterize some Banach lattices with the p-weak DP* property, discuss some examples and then consider some applications of disjoint p-convergent operators on Banach lattices.

Definition 6.2. A Banach lattice E has the *p-weak* DP* *property* if all weakly p-compact subsets are almost limited.

It can be easily shown that each Banach lattice with the positive Schur property is a *KB*-space with the p-weak DP* property. Also, the p-weak DP* property is equivalent to, for every sequence $(x_n) \in \ell_p^w(E)$ and every disjoint weak* null sequence (x_n^*) in E^*, $x_n^*(x_n) \to 0$. The weak DP* property implies the p-weak DP* property, but the converse is false.

Theorem 6.7. *Every discrete Banach lattice E with the p-Schur and without the Schur property has the p-weak* DP* *property, but it does not have the weak* DP* *property.*

Proof. If E has the p-Schur property, then it is clear that E has the p-weak DP* property. Moreover, E contains no copy of c_0, that is, E is a KB-space. Also, E is discrete and so E has the strong GP property. But E does not have the weak DP* property, since E does not have the Schur property [2]. This completes the proof. □

For example, for each $1 < p < \infty$ and $1 < q < p'$, all the spaces ℓ_q have the p-Schur (and so the p-weak DP*) and strong GP properties. But none of them have the weak DP* property (see [2, Corollary 2.7]). Of course, none of the spaces ℓ_q has the Schur property.

The following result is easily verified.

Proposition 6.5. *For a Banach lattice E, the following assertions are equivalent:*
(1) *E has the p-*DP* *and limited p-Schur properties.*
(2) *E has the p-weak* DP* *and strong limited p-Schur properties.*
(3) *E has the p-Schur property.*

Theorem 6.8. *Let E be a σ-Dedekind complete Banach lattice with the type q (with $1 < q \leq 2$) and $p \geq q'$. Then the following are equivalent:*
(1) *E has the p-weak* DP* *property.*
(2) *For every disjoint sequence $(x_n) \in \ell_p^w(E)$ and every disjoint* weak* *null sequence (x_n^*) in E^*, $x_n^*(x_n) \to 0$.*
(3) *For every disjoint sequence $(x_n) \in \ell_p^w(E)_+$ and every disjoint* weak* *null sequence (x_n^*) in E_+^*, $x_n^*(x_n) \to 0$.*
(4) *The solid hull of a weakly p-compact set in E is almost limited.*

Proof. We only prove (3) ⇒ (4) Let W be a weakly p-compact set in E and let $B := sol(W)$. Then each disjoint sequence in B is weakly p-summable. So, by hypothesis, for every disjoint sequence (x_n) in $B \cap E^+$ and every disjoint weak* null sequence (x_n^*) in E_+^*, $x_n^*(x_n) \to 0$. Hence, by [14, Theorem 2.5], B is almost limited. This completes the proof. □

Now, we define the p-bi-sequence property and then we characterize Banach lattices with this property. Also, we provide an example of a Banach lattice on which the p-weak DP* and p-bi-sequence properties are equivalent.

Definition 6.3. A Banach lattice E has the *p-bi-sequence property* if for every disjoint sequence $(x_n) \in \ell_p^w(E)_+$ and every weak* null sequence (x_n^*) in E_+^*, $x_n^*(x_n) \to 0$.

If a Banach lattice has the p-weak DP* property, then it has the p-bi-sequence property, but the converse is false in general. In fact, Banach lattice c of all convergent sequence of scalars has the bi-sequence property, but it does not have the p-weak DP* property. Note that, c is a Banach lattice with the strong limited p-Schur property and without the p-Schur property and so, by Theorem 6.5, it does not have the p-weak DP* property.

Theorem 6.9. *Let E be a Banach lattice with the type q (with $1 < q \leq 2$) and $p \geq q'$. Then these are equivalent:*

(1) *For each disjoint sequence $(x_n) \in \ell_p^w(E)$ and every disjoint* weak* *null sequence (x_n^*) in E^*, $x_n^*(x_n) \to 0$.*
(2) *E has the p-bi-sequence property.*
(3) *For every sequence $(x_n) \in \ell_p^w(E)_+$ and every disjoint* weak* *null sequence (x_n^*) in E^*, $x_n^*(x_n) \to 0$.*
(4) *For every sequence $(x_n) \in \ell_p^w(E)_+$ and every* weak* *null sequence (x_n^*) in E_+^*, $x_n^*(x_n) \to 0$.*

Proof. (1) ⇒ (2) It follows from [34, Proposition 2.4].

(2) ⇒ (4) Let $(x_n) \in \ell_p^w(E)$ be a positive sequence and let (x_n^*) be a positive weak* null sequence of E^*. Then the set $A = \{x_n : n \in \mathbb{N}\} \cup \{0\}$ is weakly p-compact and so it is relatively weakly compact. Hence every disjoint sequence (y_n) in $sol(A)$ is weakly p-summable. Since E has the p-bi-sequence property, $\sup_{k \in A} |\langle x_k^*, y_n\rangle| \to 0$ and so, for an operator $T : E \to c_0$ defined by

$$Tx = (\langle x, x_n^* \rangle), \quad \forall x \in E,$$

we have $\|Ty_n\| \to 0$. Now, by the same method in [34, Proposition 2.4], we conclude that $x_n^*(x_n) \to 0$.

(4) ⇒ (3) ⇒ (1) Obvious. This completes the proof. □

In the following theorem, we show that for some Banach lattices with the property (d), the p-weak DP* and p-bi-sequence properties are equivalent. We state the following result without proof and refer the reader to [29, Theorem 3.1]. Also, it describes another characterization for the p-weak DP* property.

Theorem 6.10. *Let E be a Banach lattice with the property (d) and the type q (with $1 < q \leq 2$) and $p \geq q'$. Then, the following are equivalent:*

(1) *E has the p-weak* DP* *property.*
(2) *For every disjoint sequence $(x_n) \in \ell_p^w(E)$ and every disjoint* weak* *null sequence (x_n^*) in E^*, $x_n^*(x_n) \to 0$.*
(3) *For every disjoint sequence $(x_n) \in \ell_p^w(E)_+$ and every disjoint* weak* *null sequence (x_n^*) in E_+^*, $x_n^*(x_n) \to 0$.*
(4) *The solid hull of a weakly p-compact set is almost limited.*
(5) *For every sequence $(x_n) \in \ell_p^w(E)_+$ and every* weak* *null sequence (x_n^*) in E_+^*, $x_n^*(x_n) \to 0$.*
(6) *E has the p-bi-sequence property.*

If a Banach lattice has the p-Schur or the p-DP* property, then it also has the p-weak DP* property, but the converse is false. In fact, Banach lattice $\ell_\infty \oplus L^1[0,1]$ has the p-weak DP* property, but it has neither the p-positive Schur nor the p-DP* property. Since ℓ_∞ has the DP* property, but it does not have the p-positive Schur property. Also, $L^1[0,1]$ has the positive Schur and weak DP* properties, but it does not have the p-DP* property. Note that each separable Banach space with the p-DP* property has the p-Schur property.

Castillo mentioned that *p-convergent* operators in [13]. In fact, p-convergent operators are precisely those operators which transformed weakly p-compact subsets into relatively compact subsets. Equivalently, an operator $T : X \to Y$ between two Banach spaces is called p-convergent if $\|Tx_n\| \to 0$ for every sequence $(x_n) \in \ell_p^w(X)$. Also, an operator $T : E \to Y$ is called *disjoint p-convergent* if it takes disjoint sequences $(x_n) \in \ell_p^w(E)$ to norm null ones [37]. It is clear that every p-convergent operator is disjoint p-convergent. For the converse, we prove the following practical lemma.

Lemma 6.4. *Let E be a Banach lattice with the type q (with $1 < q \leq 2$) and $p \geq q'$ and F be discrete with order continuous norm. Then every positive operator $T : E \to F$ is disjoint p-convergent if and only if it is p-convergent.*

Proof. Let W be a weakly p-compact subset of E and let $T : E \to F$ be a positive disjoint p-convergent operator. From [13], it is enough to show that $T(W)$ is relatively compact. Let A be the solid hull of W. Then every disjoint sequence (x_n) in A is weakly p-summable and so, by hypothesis, $\|Tx_n\| \to 0$. By a consequence of [1, Theorems 13.3 & 13.5], $T(A)$ is an almost order bounded set in F. Since F is discrete with order continuous norm, $T(A)$ is relatively compact. Hence T is a p-convergent operator. This completes the proof. □

An operator T on a Banach lattice E is said to be an *almost Dunford–Pettis operator* if the sequence $\|Tx_n\| \to 0$ for every disjoint weakly null sequence (x_n) in E. By [14, Theorem 3.5], E has the weak DP* property if and only if each operator $T : E \to c_0$ is an almost Dunford–Pettis operator. With the same techniques of [14, Theorem 3.5], we can characterize some σ-Dedekind complete Banach lattices with the p-weak DP* property.

Theorem 6.11. *Let E be a σ-Dedekind complete Banach lattice with the type q (with $1 < q \leq 2$) and $p \geq q'$. Then the following are equivalent:*
1. *E has the p-weak DP* property.*
2. *Every continuous operator $T : E \to c_0$ is disjoint p-convergent.*
3. *Every positive operator $T : E \to c_0$ is disjoint p-convergent.*
4. *Every positive operator $T : E \to c_0$ is p-convergent.*

Proof. (2) ⇒ (3) It is clear.

(3) ⇔ (4) Because c_0 is discrete with order continuous norm, it follows from Lemma 6.4.

(3) ⇒ (1) By Theorem 6.8, we have to show that for each disjoint sequence $(x_n) \in \ell_p^w(E)_+$ and each disjoint weak* null sequence (x_n^*) in E_+^*, $x_n^*(x_n) \to 0$. Consider the positive operator $T : E \to c_0$ defined by $Tx = (\langle x, x_n^* \rangle)$ for all $x \in E$. According to (3), T is a disjoint p-convergent operator. Therefore, $\|Tx_n\| \to 0$, and hence $x_n^*(x_n) \to 0$, as desired.

(1) $\Rightarrow$ (2) Consider the operator $T : E \to c_0$. We have to show that $\|Tx_n\| \to 0$ for each disjoint sequence $(x_n) \in \ell_p^w(E)$. Assume, by way of contradiction, that $\|Tx_n\| \nrightarrow 0$ for some disjoint sequence $(x_n) \in \ell_p^w(E)$. Then we can suppose that there would exist some $\epsilon > 0$ satisfying $\|Tx_n\| > \epsilon$ for all $n \in \mathbb{N}$. Since (x_n) is weakly p-summable, then it is weakly null and so, by [1, Ex. 22, p. 73] and the similar idea in Theorem 3.5 of [14], one can find a disjoint weak* null sequence (x_n^*) in E^*. Since E has the p-weak DP* property, by Theorem 6.8, $x_n^*(x_n) \to 0$. This is a contradiction. Hence $\|Tx_n\| \to 0$ for each disjoint sequence $(x_n) \in \ell_p^w(E)$ and so T is a disjoint p-convergent operator. This completes the proof. □

We can uniquely determine every bounded linear operator $T : E \to c_0$ by a weak* null sequence $(x_n^*) \subset E^*$ such that $Tx = (\langle x, x_n^* \rangle)$ for all $x \in E$. But, if $(x_n^*) \subset E^*$ is a disjoint weak* null sequence, then T is called a disjoint operator.

The final result deals the relationship between the weak DP* (resp. the p-weak DP*) property and disjoint completely continuous (resp. disjoint p-convergent) operators. Recall that an operator on a Banach lattice E is called *completely continuous* if it maps weakly null sequences in E to norm null ones.

Theorem 6.12. *If E is a Banach lattice with the property (d) with the type q (with $1 < q \leq 2$) and $p \geq q'$, then for the following assertions:*
(1) *Every disjoint operator $T : E \to c_0$ is completely continuous.*
(2) *E has the weak* DP* *property.*
(3) *E has the p-weak* DP* *property.*
(4) *Every disjoint operator $T : E \to c_0$ is p-convergent.*

Proof. (1) $\Leftrightarrow$ (2) $\Rightarrow$ (3) $\Leftrightarrow$ (4) are valid.

(1) $\Rightarrow$ (2) Assume, by way of contradiction, that there is a weakly null sequence (x_n) in E and a disjoint weak* null sequence (x_n^*) in E^* such that $|\langle x_n^*, x_n \rangle| > \epsilon$, for all n and some $\epsilon > 0$. Consider the operator $T : E \to c_0$ defined by

$$Tx = (\langle x, x_n^* \rangle), \quad \forall x \in E.$$

It is clear that T is a disjoint operator, but T is not completely continuous. In fact, the sequence (x_n) is weakly null, but $\|Tx_n\| \geq x_n^*(x_n) \geq \epsilon$ for all $n \geq 1$ and some $\epsilon > 0$.

(2) $\Rightarrow$ (1) Let $T : E \to c_0$ be a disjoint operator and let (x_n) be a weakly null sequence in E. Since E has the weak DP* property, then (x_n) is almost limited. By [27, Theorem 2.7], the operator T is order bounded. Moreover, E and c_0 have the property (d) and c_0 has the strong GP property. Then, by [27, Proposition 3.9] and [2, Definition 3.1], $\|Tx_n\| \to 0$ and so T is completely continuous.

(2) $\Rightarrow$ (3) It is clear.

(3) $\Leftrightarrow$ (4) It is enough to repeat the techniques of Theorem 6.11. This completes the proof. □

6.4 Strong p-DP$_{rc}$P and p-weak DP property

Recently, Banach spaces with the p-DP$_{rc}$P are studied and some results are obtained. Motivated by the notion of the p-DP$_{rc}$P, we define the so-called strong p-DP$_{rc}$P. We are then able to describe a relationship between strong p-DP$_{rc}$P and p-DP$_{rc}$P [5].

Definition 6.4. A Banach lattice E has the *strong p-DP$_{rc}$P* if each almost Dunford–Pettis weakly p-compact subset of E is relatively compact.

It is easily seen that E has the strong p-DP$_{rc}$P if and only if each almost Dunford–Pettis sequence $(x_n) \in \ell_p^w(E)$ is norm null. If $1 \leq p < q$ and E has the strong q-DP$_{rc}$P, then E has the strong p-DP$_{rc}$P.

Theorem 6.13. *Each Banach lattice with the strong p-DP$_{rc}$P is a KB-space, but the converse is false.*

Proof. If a Banach lattice has the strong p-DP$_{rc}$P, then it contains no copy of c_0 (since c_0 does not have the strong p-DP$_{rc}$P) and so it is a KB-space. For the converse, the space $L^1[0,1]$ is an example of a KB-space without the (strong) p-DP$_{rc}$P. In fact, Rademacher sequences (r_n) in are weakly 2-summable (see [31, Proposition 3.6]) and Dunford–Pettis in $L^1[0,1]$ (by the Dunford–Pettis property), but $\|r_n\| = 1$ for all n. Thus, $L^1[0,1]$ does not have the (strong) p-DP$_{rc}$P for all $2 \leq p \leq \infty$. This completes the proof. □

Note that the space $L^1[0,1]$ is a non-discrete Banach lattice with the 1-Schur property. We need to consider some preliminary results for Banach spaces with the 1-Schur property. If X is weakly sequentially complete, then it has no copy of c_0 (and so it has the 1-Schur property). The converse holds, if X has an unconditional basis [28, Theorem 1.c.13].

In fact, James space J is a Banach space without any copy of c_0, that is J has the 1-Schur property, while it is not weakly sequentially complete. But the situation is different in Banach lattices. A Banach lattice E does not contain a copy of c_0 if and only if E has the 1-Schur property. The converse of Theorem 6.13 holds when E is discrete.

Proposition 6.6. *If E is a discrete Banach lattice, then the following are equivalent:*
(1) *E has the strong DP$_{rc}$P.*
(2) *E has the strong p-DP$_{rc}$P.*
(3) *E has the p-DP$_{rc}$P.*
(4) *E has the DP$_{rc}$P.*

Proof. Only (4) ⇒ (1) needs a proof. To this end, it is enough to note that each Banach lattice with the DP$_{rc}$P is a KB-space and each discrete KB-space has the strong DP$_{rc}$P [7, Theorem 2.3]. This completes the proof. □

Each Banach lattice with the strong p-DP$_{rc}$P has the p-DP$_{rc}$P, but the converse is false.

Example 6.3. The Lorentz sequence space $d_{\omega,1}$ has the DP$_{rc}$P and so it has the p-DP$_{rc}$P. But $d_{\omega,1}$ cannot have the strong p-DP$_{rc}$P, for $1 < p < \infty$. In fact, $d_{\omega,1}$ is a Banach lattice with the weak Dunford–Pettis property and without the p-Schur property for $1 < p < \infty$ [35, Example 1, p. 231].

The following properties are easily verified.

Proposition 6.7. (1) *If a Banach space X contains a complemented copy of ℓ_1, then X^* cannot have the p-DP$_{rc}$P.*

(2) *If a Banach lattice E contains a sublattice isomorphic to ℓ_1, then E^* cannot have the p-DP$_{rc}$P.*

Proof. (1) If X has a complemented copy of ℓ_1, then by [17, Theorem 10 in Chapter V] X^* has an isomorphic copy of ℓ_∞ and so it cannot have the p-DP$_{rc}$P.

(2) If E contains a sublattice isomorphic to ℓ_1, then, by [30, Proposition 2.3.12], E^* contains a sublattice isomorphic to ℓ_∞ and so it cannot have the strong p-DP$_{rc}$P. This completes the proof. □

We give some sufficient conditions under which two properties strong ∞-DP$_{rc}$P and strong DP$_{rc}$P are the same. At first recall that the concept of (L)-weak* sequentially continuous lattice operations. A sequence (f_n) in E^* is called an L-sequence if, for each weakly null sequence (x_n) in $E, f_n(x_n) \to 0$. Also, E^* has (L)-weak* sequentially continuous lattice operations if for every L-sequence (f_n) of E^* satisfying $f_n \xrightarrow{w^*} 0$, $|f_n| \xrightarrow{w^*} 0$ and, in this case, each disjoint weakly compact set (that is, set whose disjoint sequences in the solid hull is weakly null) is weakly conditionally compact. Note that almost Dunford–Pettis sets are disjoint weakly compact [36].

Theorem 6.14. *If E^* has (L)-weak* sequentially continuous lattice operations, then the strong ∞-DP$_{rc}$P and the strong DP$_{rc}$P are equivalent.*

Proof. Clearly, the strong DP$_{rc}$P implies that the strong ∞-DP$_{rc}$P. For the converse, suppose that E has the strong ∞-DP$_{rc}$P. Then each almost Dunford–Pettis weakly null sequence in E is norm null. Let $A \subset E$ be an almost DP set. Since E^* has (L)-weak* sequentially continuous lattice operations, A is weakly conditionally compact. On the other hand, the difference set $A-A$ is almost DP. Thus we can prove that A is relatively compact and so E has the strong DP$_{rc}$P. This completes the proof. □

Remark 6.1. Continuous linear image of an almost Dunford–Pettis set under a positive linear projection is almost Dunford–Pettis. Also, if Y is a complemented subspace of X and $A \subseteq Y$ is a Dunford–Pettis set in X, then A is a Dunford–Pettis set in Y.

Note that a Banach space X has the Schur (resp., p-Schur) property if and only if each closed linear separable subspace of X has the Schur (resp., p-Schur) property. From the above remark, we observe the following interesting result.

Theorem 6.15. *For each $1 \le p \le \infty$, the following are equivalent:*

(1) *E has the strong p-DP$_{rc}$P.*
(2) *Every closed separable sublattice of E is contained in a positively complemented sublattice Z of E with the strong p-DP$_{rc}$P.*
(3) *E is the direct sum of two spaces with the strong p-DP$_{rc}$P.*

Proof. (1) $\Rightarrow$ (2) It is obvious. In fact, strong p-DP$_{rc}$P is inherited by closed sublattices.

(2) $\Rightarrow$ (1) Suppose that A is a weakly p-compact almost Dunford–Pettis subset of E and that $(x_n) \subseteq A$. Then there is a subsequence (x_{n_k}) of (x_n) that is almost Dunford–Pettis weakly p-convergent to some $x \in A$. Consider the closed linear span of (x_n). By hypothesis, it is contained in a complemented sublattice Z of E with the strong p-DP$_{rc}$P. So (x_n) is almost Dunford–Pettis in Z. Since Z has the strong p-DP$_{rc}$P, (x_{n_k}) is norm convergent to x.

(1) $\Rightarrow$ (3) Consider $E = E \oplus \{0\}$.

(3) $\Rightarrow$ (1) Let $E = Y \oplus Z$ such that Y and Z have the strong p-DP$_{rc}$P. Consider the positive linear projections $P_Y : E \to Y$ and $P_Z : E \to Z$. Assume that A is a weakly p-compact almost Dunford–Pettis subset of E. Then $P_Y(A)$ and $P_Z(A)$ are weakly p-compact and they are almost Dunford–Pettis. For each $(x_n) \subseteq A$ there is $y_n \in P_Y(A)$ and $z_n \in P_Z(A)$ such that $x_n = y_n + z_n$. Since $P_Y(A)$ and $P_Z(A)$ have the strong p-DP$_{rc}$P and so the sequences (y_n) and (z_n) have convergent subsequences (y_{n_k}) and (z_{n_k}), that is, there are $y \in P_Y(A)$ and $z \in P_Z(A)$ such that $y_{n_k} \to y$ and $z_{n_k} \to z$. So $x_{n_k} \to y + z \in A$, since A is weakly p-compact. Hence A is compact and so E has the strong p-DP$_{rc}$P. This completes the proof. ☐

Now, we introduce and study the notion of p-weak Dunford–Pettis property, which plays an important role in the present paper and provide some characterizations of Banach lattices with this property.

Definition 6.5. A Banach lattice E has the p-weak Dunford–Pettis property if each weakly compact operator T from E into each Banach space Y is disjoint p-convergent.

Clearly, the weak Dunford–Pettis property implies the p-weak Dunford–Pettis property, but the converse is false. In fact, every discrete Banach lattice with the p-Schur and without the Schur property has the p-weak Dunford–Pettis property, but it does not have the weak Dunford–Pettis property. It is important to note that E has the Schur property if and only if E has the weak Dunford–Pettis property and strong DP$_{rc}$P.

In the following proposition, a similar characterization of the p-Schur property is discussed:

Proposition 6.8. *For each Banach lattice E, the following assertions are equivalent:*
(1) *E has p-Dunford–Pettis property and p-DP$_{rc}$P.*
(2) *E has p-weak Dunford–Pettis property and strong p-DP$_{rc}$P.*
(3) *E has the p-Schur property.*

We have the following example of a space with the p-weak Dunford–Pettis property and without the weak Dunford–Pettis property.

Example 6.4. All the spaces ℓ_q have the p-Schur and so p-weak Dunford–Pettis property for $1 < p < \infty$ and $1 < q < p'$. On the other hand, these spaces have the strong $\mathrm{DP}_{rc}\mathrm{P}$ property and so non of them can have the weak Dunford–Pettis property.

Each Banach lattice with the p-Dunford–Pettis property has the p-weak Dunford–Pettis property, but the converse is false.

Example 6.5. Lorentz sequence space $d_{\omega,1}$ has the weak Dunford–Pettis (and so p-weak Dunford–Pettis) property. But it does not have p-Dunford–Pettis property. In fact, $d_{\omega,1}$ is a Banach lattice with the $\mathrm{DP}_{rc}\mathrm{P}$ and without the p-Schur property for $1 < p < \infty$.

Using Proposition 1 of [35], we obtain a similar characterization for the p-weak Dunford–Pettis property in a Banach lattice in terms of disjoint, weakly p-summable sequences.

Proposition 6.9. *If E is a Banach lattice. Then the following are equivalent:*
(1) *E has the p-weak Dunford–Pettis property.*
(2) *Every weakly compact operator $T : E \to c_0$ is disjoint p-convergent.*
(3) *For every disjoint sequence $(x_n) \in \ell_p^w(E)_+$ and every weakly null sequence (x_n^*) in E^*, $x_n^*(x_n) \to 0$.*

The following proposition characterizes some Banach lattices with the p-weak Dunford–Pettis property.

Proposition 6.10. *For a Banach lattice E with the type q (with $1 < q \leq 2$) and $p \geq q'$. Then the following are equivalent:*
(1) *E has the p-weak Dunford–Pettis property.*
(2) *For every disjoint sequence $(x_n) \in \ell_p^w(E)$ and every disjoint weakly null sequence (x_n^*) in E^*, $x_n^*(x_n) \to 0$.*
(3) *The solid hull of a weakly p-compact set is almost Dunford–Pettis.*
(4) *Each weakly p-compact set is almost Dunford–Pettis.*
(5) *For every sequence $(x_n) \in \ell_p^w(E)$ and every disjoint weakly null sequence (x_n^*) in E_+^*, $x_n^*(x_n) \to 0$.*

Proof. The reader should note that our arguments are similar to arguments in the proof of Theorem 2.7 of [10]. Only (2) $\Rightarrow$ (3) needs to be proved.

Let W be a weakly p-compact set in E. Each disjoint sequence in $B := Sol(W)$ is weakly p-summable. So by hypothesis for every disjoint weakly null sequence (x_n^*) in E^*, $x_n^*(x_n) \to 0$. Hence B is almost Dunford–Pettis. This completes the proof. □

The following example shows that the p-weak Dunford–Pettis property is not preserved under linear homeomorphisms, unlike the p-Dunford–Pettis property. Indeed, each complemented sublattice (closed ideal) of a Banach lattice with the weak Dunford–Pettis property have this property too [35, Proposition 3] and the same will be true for the p-weak Dunford–Pettis property.

Example 6.6. The Orlicz space $L^{\phi^*}(0,1)$, where $\phi(r) = (e-1)^{-1}(e^{r^2} - 1)$ has the weak Dunford–Pettis property (and so it has the p-weak Dunford–Pettis property) and also it has a complemented copy of ℓ_2 spanned by the Rademacher functions. Hence $L^{\phi^*}(0,1)$ is isomorphic to $L^{\phi^*}(0,1) \oplus \ell_2$ [15]. Since ℓ_2 does not have the p-weak Dunford–Pettis property for all $p \geq 2$, $L^{\phi^*}(0,1) \oplus \ell_2$ does not have the p-weak Dunford–Pettis property for all $p \geq 2$.

We now formulate a characterization of the p-Schur property.

Theorem 6.16. *For a Banach lattice E with the type q (with $1 < q \leq 2$) and $p \geq q'$. Then the following are equivalent:*
(1) *E has the p-Schur property.*
(2) *E has the strong p-$DP_{rc}P$ and p-positive Schur property.*

Proof. (1) $\Rightarrow$ (2) It is obvious.

(2) $\Rightarrow$ (1) Let A be a weakly p-compact set in E. Since E has the p-positive Schur property, by [37, Proposition 3.1.21], each operator on E is disjoint p-convergent. Then E has the p-weak Dunford–Pettis property and Proposition 6.10 implies that A is almost Dunford–Pettis and so relatively compact, by the strong p-$\mathrm{DP}_{rc}\mathrm{P}$ of E. Thus E has the p-Schur property. This completes the proof. ☐

6.5 p-Weak Dunford–Pettis property and some classes of operators

In this section, we provide some results on disjoint weakly compact operators and their applications in the study of geometrical properties of Banach lattices. In terms of disjoint weakly compact operators into c_0, we give an operator characterization of those Banach lattices with the p-weak Dunford–Pettis property. Every disjoint weakly compact operator $T : E \to c_0$ can be uniquely determined by a disjoint weakly null sequence $(x_n^*) \subset E^*$ such that $Tx = (\langle x, x_n^* \rangle)$, for all $x \in E$.

Similar [24, Proposition 2.3], we can provide a characterization of almost Dunford–Pettis sets:

Proposition 6.11. *A subset $A \subset E$ is almost Dunford–Pettis if and only if, for each disjoint weakly compact operator $T : E \to c_0$, $T(A)$ is relatively compact.*

Chen et al. [14] characterized the weak DP* property (i. e., each relatively weakly compact set is almost limited) of a Banach lattice: a σ-Dedekind complete Banach lattice E has the weak DP* property if and only if each continuous operator $T : E \to c_0$ is almost Dunford–Pettis. Comparing this with Proposition 6.10 in the present paper, we naturally posed the following theorem.

Theorem 6.17. *Let E be a Banach lattice with the type q (with $1 < q \leq 2$) and $p \geq q'$. Then the following are equivalent:*
(1) *E has the p-weak Dunford–Pettis property.*
(2) *Every disjoint weakly compact operator $T : E \to c_0$ is p-convergent.*
(3) *Every disjoint weakly compact operator $T : E \to c_0$ is disjoint p-convergent.*

Proof. (1) ⇒ (2) Let A be a weakly p-compact set in E. Then by the p-weak Dunford–Pettis property of E, A is almost Dunford–Pettis. From Proposition 6.11 for every disjoint weakly compact operator $T : E \to c_0$, $T(A)$ is a relatively compact set in c_0, that is, T is p-convergent.

(2) ⇒ (3) It is clear.

(3) ⇒ (1) We have to show that, for each disjoint sequence $(x_n) \in \ell_p^w(E)$ and each disjoint weakly null sequence (x_n^*) in E^*, $x_n^*(x_n) \to 0$. Consider the disjoint weakly compact operator $T : E \to c_0$ defined by $Tx = (\langle x, x_n^* \rangle)$ for all $x \in E$. According to (3), T is a disjoint p-convergent operator. Therefore, $\|Tx_n\| \to 0$, and hence $x_n^*(x_n) \to 0$, as desired. This completes the proof. □

In order to study the connection between p-weak Dunford–Pettis property of E and E^*, we need the following lemma. In fact, the relationship between disjoint p-convergent and Dunford–Pettis operators is described.

Lemma 6.5. *Let E be a Banach lattice with the type q (with $1 < q \leq 2$) and $p \geq q'$ and F be discrete with order continuous norm. Then every positive operator $T : E \to F$ is disjoint p-convergent if and only if T is Dunford–Pettis.*

Proof. Let W be a relatively weakly compact subset of E and $T : E \to F$ be a positive disjoint p-convergent operator. It is enough to show that $T(W)$ is relatively compact. From [37, Lemma 4.2.1] every disjoint sequence (x_n) in $A := Sol(W)$ is weakly p-summable and so $\|Tx_n\| \to 0$. By a consequence of two theorems 13.3 and 13.5 of [1], $T(A)$ is an almost order-bounded set in F. Since F is discrete with order-continuous norm, $T(A)$ is relatively compact. Hence, T is a Dunford–Pettis operator. This completes the proof. □

Theorem 6.18. *Let E be a Banach lattice with the type q (with $1 < q \leq 2$) and $p \geq q'$. If E^* has the p-weak Dunford–Pettis property, then E has the p-weak Dunford–Pettis property.*

Proof. Let $(x_n) \in \ell_p^w(E)_+$ be a disjoint sequence and (x_n^*) be a weakly null sequence in E^*. Consider the disjoint positive weakly compact operator $T : E^* \to c_0$ defined by $Tf = (\langle f, x_n \rangle)$, for all $f \in E^*$. According to the p-weak Dunford–Pettis property of E^*, T is a disjoint p-convergent operator. Therefore, by Lemma 6.5, T is a Dunford–Pettis operator. Hence $\|Tx_n^*\| \to 0$ and so $x_n^*(x_n) \to 0$. This completes the proof. □

From [35, Theorem 4], E^* has the positive Schur property if and only if E has the weak Dunford–Pettis property and E contains no complemented copy of ℓ_1. For the p-positive Schur and p-weak Dunford–Pettis properties, this theorem is also expressed.

At first, we formulate the following lemma, which describes the coincidences of disjoint p-convergent and compact operators.

Lemma 6.6. *Let E be a Banach lattice with the type q (with $1 < q \leq 2$) and $p \geq q'$ such that E^* be a KB-space and F be a discrete Banach lattice with order continuous norm. Then each positive operator $T : E \to F$ is disjoint p-convergent if and only if it is compact.*

Proof. Suppose that $T : E \to F$ is a positive disjoint p-convergent operator. Since F is discrete with order continuous norm, then by Lemma 6.5 T is a Dunford–Pettis operator. On the other hand, E^* is a KB-space and so by [30, Theorem 2.4.14] each disjoint sequence (x_n) in B_E is weakly null. Then by [1, Theorem 13.5] an operator T carries unit ball of E into almost order bounded subsets of F, which are relatively compact. Hence T is compact. This completes the proof. □

Using Lemma 6.6, we obtain a characterization for the p-positive Schur property as follows.

Theorem 6.19. *Let E be a Banach lattice with the type q (with $1 < q \leq 2$) and $p \geq q'$. Then these are equivalent:*
(1) *E^* has the p-positive Schur property.*
(2) *E has the p-weak Dunford–Pettis property and E^* is a KB-space.*

Proof. (1) ⇒ (2) If E^* has the p-positive Schur property, then the identity operator on E^* is disjoint p-convergent and so E^* has the p-weak Dunford–Pettis property. Then by Theorem 6.18, E has the p-weak Dunford–Pettis property. Also E^* has order continuous norm which is equivalent E^* is a KB-space.

(2) ⇒ (1) We have to show that every sequence $(x_n^*) \in \ell_p^w(E^*)_+$, $\|x_n^*\| \to 0$. Consider the weakly compact positive operator $T : E \to c_0$ defined by $Tx = (\langle x, x_n^* \rangle)$, for all $x \in E$. According to Proposition 6.9, T is disjoint p-convergent. By Lemma 6.6, T is a compact operator and so $\|x_n^*\| \to 0$, which implies that E^* has the p-positive Schur property. This completes the proof. □

The following condition on the underlying Banach lattices ensures that disjoint p-convergent operators are compact.

Theorem 6.20. *Let E be an AM-space with unit and F be a discrete Banach lattice with order continuous norm. Then each operator $T : E \to F$ is disjoint p-convergent if and only if it is compact.*

Proof. Since the closed unit ball B_E is order bounded, each disjoint sequence (x_n) in B_E is order bounded and so it is weakly p-summable. Moreover, $T : E \to F$ is disjoint p-convergent and then $\|T(x_n)\| \to 0$. Hence, by [1, Theorem 13.5], $T(B_E)$ is an almost order bounded in F. Since F is discrete with order continuous norm, $T(B_E)$ is relatively compact and so T is a compact operator. This completes the proof. □

The following theorem leads to an improvement of proposition 1 of [35]. Recall that E has the weak Dunford–Pettis property if and only if every weakly compact operator $T : E \to c_0$ is almost Dunford–Pettis.

In the following theorem, we show that the same result can be obtained with disjoint weakly compact operators.

Theorem 6.21. *If E is a Banach lattice, then the following are equivalent:*
(1) *E has the weak Dunford–Pettis property.*
(2) *Every disjoint weakly compact operator $T : E \to c_0$ is Dunford–Pettis.*
(3) *Every disjoint weakly compact operator $T : E \to c_0$ is almost Dunford–Pettis.*

Proof. (1) $\Rightarrow$ (2) Let A be a relatively weakly compact set in E. Then by the weak Dunford–Pettis property of E, A is an almost DP set [10, Theorem 2.7]. From Proposition 6.11, for every disjoint weakly compact operator $T : E \to c_0$, $T(A)$ is a relatively compact set in c_0 and so T is Dunford–Pettis.

(2) $\Rightarrow$ (3) It is clear.

(3) $\Rightarrow$ (1) From [9, Corollary 2.6], assume, by way of contradiction, that there is a weakly null sequence (x_n) in E^+ and a disjoint weakly null sequence (x_n^*) in E^* such that $|\langle x_n^*, x_n\rangle| > \epsilon$, for all n and some $\epsilon > 0$. Consider the operator $T : E \to c_0$ defined by

$$Tx = (\langle x, x_n^*\rangle), \quad \forall x \in E.$$

It is clear that T is a disjoint weakly compact operator. But T is not almost Dunford–Pettis, since (x_n) is a positive weakly null sequence and $\|Tx_n\| \geq x_n^*(x_n) \geq \epsilon$ for all n [11, Theorem 2.2]. This completes the proof. □

An operator $T : E \to X$ is order weakly compact if $T[-x, x]$ is relatively weakly compact, for all $x \in E^+$. Alternatively, T is order weakly compact if and only if $\|T(x_n)\| \to 0$ for each order bounded disjoint sequence in E [1, Theorem 18.6].

Theorem 6.22. *Each disjoint p-convergent operator $T : E \to X$ is order weakly compact.*

Proof. Let (x_n) be an order bounded disjoint sequence in a Banach lattice E. Then (x_n) is weakly null (see [1, p. 186]). Actually, it has been shown that, this sequence is weakly 1-summable. So, by the known fact $\ell_p^w(X) \subset \ell_q^w(X)$ for all $1 \leq p \leq q$ and $\frac{1}{p} + \frac{1}{q} = 1$, the sequence (x_n) is weakly p-summable. Since $T : E \to X$ is a disjoint p-convergent operator, $\|T(x_n)\| \to 0$ and so T is order weakly compact. This completes the proof. □

The converse of Theorem 6.22 is false. In fact, identity operator on c_0 is order weakly compact, but it is not disjoint p-convergent.

A Banach lattice E is weak p-consistent if for each sequence $(x_n) \in \ell_p^w(E)$, we have $(|x_n|) \in \ell_p^w(E)$. In each weak p-consistent Banach lattice, two properties p-positive Schur and p-Schur are the same [37, Proposition 3.3.6].

From [37, Lemma 4.2.1] and with the same techniques used to prove [11, Theorem 2.2], we formulate the following lemma.

Lemma 6.7. *Let E be a weak p-consistent Banach lattice with the type q (with $1 < q \leq 2$), $p \geq q'$ and T be an operator from E into a Banach space Y. Then the following are equivalent:*
(1) *T is a disjoint p-convergent operator.*
(2) *T is order weakly compact and each weakly p-compact set in E is approximately order bounded with respect to the lattice semi-norm q_T.*
(3) *$\|Tx_n\| \to 0$ for each sequence $(x_n) \in \ell_p^w(E)_+$.*
(4) *$\|Tx_n\| \to 0$ for each disjoint sequence $(x_n) \in \ell_p^w(E)_+$.*

Theorem 6.23. *Let E be a weak p-consistent Banach lattice with the type q (with $1 < q \leq 2$), $p \geq q'$ and T be an operator from E into a Banach space Y. Then T is disjoint p-convergent if and only if it is a p-convergent operator.*

Proof. Let $T : E \to Y$ be a disjoint p-convergent operator. We show that T is p-convergent, that is, $\|Tx_n\| \to 0$ for each sequence $(x_n) \in \ell_p^w(E)$. Since E is a weak p-consistent Banach lattice, $(x_n^+) \in \ell_p^w(E)$ and $(x_n^-) \in \ell_p^w(E)$. On the other hand, T is disjoint p-convergent and so, by Lemma 6.7, $\|Tx_n\| \leq \|Tx_n^+\| + \|Tx_n^-\| \to 0$. Hence T is p-convergent. This completes the proof. □

Corollary 6.4. *Let E be a weak p-consistent Banach lattice with the type q (with $1 < q \leq 2$) and $p \geq q'$. Then E has the p-weak Dunford–Pettis property if and only if E has the p-Dunford–Pettis property.*

Proof. Assume that E has the p-weak Dunford–Pettis property and T is a weakly compact operator from E into a Banach space Y, then T is disjoint p-convergent. By Theorem 6.23, T is p-convergent. Hence E has the p-Dunford–Pettis property. This completes the proof. □

It is clear that each Dunford–Pettis operator is disjoint p-convergent, but the converse is false. In fact, each operator on an AL-space (for instance, $Id_{L^1[0,1]}$) is disjoint p-convergent, while it is not necessarily Dunford–Pettis. But on an AM-space with unit, these two operators are the same.

Corollary 6.5. *For each compact space K, each operator T on $C(K)$ is Dunford–Pettis if and only if it is disjoint p-convergent.*

Proof. Each disjoint p-convergent operator T is order weakly compact. Since the closed unit ball of $C(K)$ is order bounded, then T is weakly compact. Also $C(K)$ has the Dunford–Pettis property and so each weakly compact operator on $C(K)$ is Dunford–Pettis. This completes the proof. □

Theorem 6.24. *Suppose that $T : E \to X$ admits a factorization through a Banach lattice F with the p-positive Schur property*

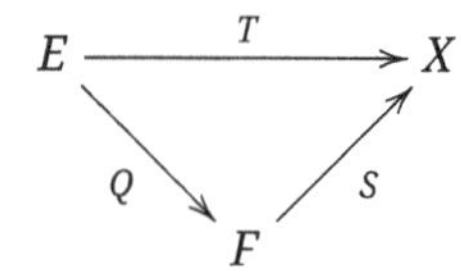

such that $Q : E \to F$ is a lattice homomorphism. Then T is disjoint p-convergent.

Proof. Let (x_n) be a disjoint weakly p-summable sequence in E. Since $Q : E \to F$ is a lattice homomorphism, then $(Q(x_n))$ is a disjoint weakly p-summable sequence in F and so it is norm null, by the p-positive Schur property of F. Hence $\|T(x_n)\| = \|S(Q(x_n))\| \to 0$, which implies that T is disjoint p-convergent. This completes the proof. □

6.6 Positive p-DP$_{rc}$P

This section focuses on the so-called positive p-DP$_{rc}$P for Banach lattices as well as several results concerning some well-known properties of order p. The concept of positive p-DP$_{rc}$P is defined as follows.

Definition 6.6. A Banach lattice E has the positive p-DP$_{rc}$P if each Dunford–Pettis sequence $(x_n) \in \ell_p^w(E)_+$ is norm null.

The positive p-DP$_{rc}$P is characterized in terms of the sequences as follows.

Theorem 6.25. *For a Banach lattice E, these are equivalent:*
(1) *E has the positive p-DP$_{rc}$P.*
(2) *Each disjoint Dunford–Pettis sequence $(x_n) \in \ell_p^w(E)$ is norm null.*

Proof. (1) ⇒ (2) Let $(x_n) \in \ell_p^w(E)$ be a disjoint Dunford–Pettis sequence. Then from [37, 23], the sequence $(|x_n|)$ is weakly p-summable positive and Dunford–Pettis. By hypothesis (1), $\|x_n\| = \||x_n|\| \to 0$.

(2) ⇒ (1) Suppose that $(x_n) \in \ell_p^w(E)_+$ is a Dunford–Pettis sequence, but $\|x_n\| \nrightarrow 0$. Then (x_n) is a weakly null positive sequence in E which is not norm null. By [26, Corollary 5], there is a disjoint positive subsequence (x_{n_k}) of (x_n) such that $\|x_{n_k}\| \nrightarrow 0$, which is a contradiction. Note that (x_{n_k}) is a disjoint Dunford–Pettis weakly p-summable sequence and by hypothesis it must be norm null. This completes the proof. □

As a consequence of Theorem 6.25, we obtain the following characterization.

Corollary 6.6. *If E is a Banach lattice. Then the following are equivalent:*
(1) *E has the p-positive Schur property.*
(2) *E has the p-weak Dunford–Pettis property and positive p-DP$_{rc}$P.*

Proof. (1) ⇒ (2) It is clear.

(2) ⇒ (1) Let $(x_n) \in \ell_p^w(E)_+$ be a disjoint sequence. We have to show that $\|x_n\| \to 0$. Since E has the p-weak Dunford–Pettis property, then, by Theorem 6.9, the sequence (x_n) is Dunford–Pettis and so by the positive p-DP$_{rc}$P, $\|x_n\| \to 0$. This completes the proof. □

If E has the strong p-DP$_{rc}$P, then E has the positive p-DP$_{rc}$P. The converse is false, in general. In fact, $L^1[0,1]$ has the positive Schur property and so it has the positive p-DP$_{rc}$P. But it is clear that $L^1[0,1]$ does not have the strong p-DP$_{rc}$P for $2 \le p \le \infty$.

Proposition 6.12. *Each Banach lattice E with the positive p-DP$_{rc}$P is a KB-space.*

Proof. Note that c_0 does not have the positive p-DP$_{rc}$P. In fact, c_0 is a Banach lattice with the p-weak Dunford–Pettis property and without the p-positive Schur property. So, if a Banach lattice E has the positive p-DP$_{rc}$P, then E does not contain a copy of c_0 and so E is a KB-space. This completes the proof. □

Since discrete KB-spaces have the strong DP$_{rc}$P, then we conclude in each discrete Banach lattice positive p-DP$_{rc}$P and p-DP$_{rc}$P are the same. According to the result obtained in the previous section, the following result can be generally summarized.

Corollary 6.7. *If E is a discrete Banach lattice, then the following are equivalent:*
(1) *E has the strong DP$_{rc}$P.*
(2) *E has the strong p-DP$_{rc}$P.*
(3) *E has the p-DP$_{rc}$P.*
(4) *E has the DP$_{rc}$P.*
(5) *E has the positive DP$_{rc}$P.*
(6) *E has the positive p-DP$_{rc}$P.*

Naturally, one may think we can define the so-called positive strong p-DP$_{rc}$P in a Banach lattice E to be Banach lattices on which every almost Dunford–Pettis sequence $(x_n) \in \ell_p^w(E)_+$ is norm null. Indeed, the introduction of this notion is superfluous because we have the following theorem, which gives another characterization for the positive p-DP$_{rc}$P.

Theorem 6.26. *For a Banach lattice E, these are equivalent:*
(1) *Each almost Dunford–Pettis sequence $(x_n) \in \ell_p^w(E)_+$ is norm null.*
(2) *Each disjoint almost Dunford–Pettis sequence $(x_n) \in \ell_p^w(E)$ is norm null.*
(3) *Each disjoint Dunford–Pettis sequence $(x_n) \in \ell_p^w(E)$ is norm null.*

Proof. (1) ⇒ (2) It follows from [37, Proposition 3.1.5] and [23, Lemma 3.7]. In fact, for each disjoint sequence $(x_n) \in \ell_p^w(E)$, we have sequence $(|x_n|) \in \ell_p^w(E)$ and also it is Dunford–Pettis.

(2) ⇒ (3) It is clear.

(3) ⇒ (1) Suppose that $(x_n) \in \ell_p^w(E)_+$ is an almost Dunford–Pettis sequence, but $\|x_n\| \nrightarrow 0$. Then (x_n) is a weakly null positive sequence in E which is not norm null. By [26, Corollary 5], there is a disjoint positive subsequence (x_{n_k}) of (x_n) such that $\|x_{n_k}\| \nrightarrow 0$,

which is a contradiction. Note that (x_{n_k}) is a disjoint Dunford–Pettis (by [23, Lemma 3.7]) weakly p-summable sequence and by hypothesis it must be norm null. This completes the proof. □

In [23, Theorem 3.15], the authors derive a characterization of the positive $\mathrm{DP}_{rc}\mathrm{P}$ of a Banach lattice. In fact, a Banach lattice E has the positive $\mathrm{DP}_{rc}\mathrm{P}$ if and only if each almost Dunford–Pettis set in E is an L-weakly compact set if and only if E is a KB-space and each almost Dunford–Pettis set in E is approximately order bounded.

According to this result, we state the following proposition in which we characterize positive p-$\mathrm{DP}_{rc}\mathrm{P}$ in discrete Banach lattices.

Theorem 6.27. *If E is a discrete Banach lattice, then the following are equivalent:*
(1) *E has the positive p-DP_{rc}P.*
(2) *Each almost Dunford–Pettis set in E is an L-weakly compact set.*
(3) *E is a KB-space.*

Proof. (1) ⇒ (2) If E has the positive p-$\mathrm{DP}_{rc}\mathrm{P}$ and E is discrete, then, by Corollary 6.7, E has the positive $\mathrm{DP}_{rc}\mathrm{P}$ and so each almost Dunford–Pettis set in E is an L-weakly compact set.

(2) ⇒ (3) It follows from [23, Theorem 3.15].

(3) ⇒ (1) Each discrete KB-space has the strong $\mathrm{DP}_{rc}\mathrm{P}$ and so, by Corollary 6.7, it has the positive p-$\mathrm{DP}_{rc}\mathrm{P}$. This completes the proof. □

6.7 Positively limited *p*-Schur property

In this section, we study the concept of the positively limited p-Schur property and discuss Banach lattices with this property [6]. The main goal is to investigate:“ what is the relationship between positively limited p-Schur property and other known properties such as limited p-Schur, strong limited p-Schur, p-Schur, p-positive Schur and positive GP?”.

Definition 6.7. A Banach lattice E has the *positively limited Schur property of order p* (*positively limited p-Schur property*) if each positively limited weakly p-compact set in E is relatively compact.

It can be easily shown that E has the positively limited p-Schur property if and only if each positively limited sequence $(x_n) \in \ell_p^w(E)$ is norm null. If E has the positively limited q-Schur property and $1 \le p < q$, then E has the positively limited p-Schur property. The first result gives a necessary condition for this property.

Theorem 6.28. *Each Banach lattice E with the positively limited p-Schur property has order continuous norm.*

Proof. Let (x_n) be an order-bounded disjoint sequence in E. It follows from [1, p. 192] and the fact that $\ell_p^w(E) \subset \ell_q^w(E)$ (for all $1 \le p \le q$), (x_n) is a weakly 1-summable and so it is a weakly p-summable sequence in E. Also, each order-bounded sequence in E is positively limited [3]. Therefore, the positively limited p-Schur property of E implies that the sequence (x_n) is norm null and so E has order-continuous norm. This completes the proof. □

The converse is not true in general.

Example 6.7. $L^1[0,1]$ is a Banach lattice with order continuous norm and without the positively limited p-Schur property, for all $p \ge 2$. In fact, the *Rademacher sequences* (r_n) in $L^1[0,1]$ are weakly 2-summable (cf., [31, Proposition 3.6]) and positively limited (by the positive DP* property), but $\|r_n\| = 1$ for all $n \ge 1$.

The following theorem provides a sufficient condition under which the converse of Theorem 6.28 also holds.

Theorem 6.29. *Let E be a Banach lattice with the weakly sequentially continuous lattice operations. Then the following assertions are equivalent:*
(1) *E has the positively limited p-Schur property.*
(2) *E has order continuous norm.*
(3) *E has the positive GP property.*

Proof. Only (2) ⇒ (3) needs a proof. To this end, it is enough to note that, by [15, Corollary 2.3], each Banach lattice with the weakly sequentially continuous lattice operations and order continuous norm is discrete and so it has the positive GP property (cf., [3, Theorem 3.6]). This completes the proof. □

Since every almost limited set in a Banach lattice is positively limited, the positively limited p-Schur property implies the strong limited p-Schur and so the limited p-Schur property, but the converse is not true in general. If E^* has weak* sequentially continuous lattice operations, then three properties positively limited p-Schur, strong limited p-Schur and limited p-Schur coincide in E (cf., [3]).

Example 6.8. $C[0,1]$ and c are Banach lattices with the strong limited p-Schur property [2] and without the positively limited p-Schur property (Theorem 6.28). Also, $L^1[0,1]$ is a Banach lattice with the limited p-Schur property (by the GP property) and without the positively limited p-Schur property, for $p \ge 2$.

The reader should note that the limited p-Schur property is inherited by closed subspaces. Thus, if a Banach lattice has the limited p-Schur property, then it does not contain any sublattice isomorphic to ℓ_∞. Hence by [30, Corollary 2.4.3], we can conclude the following proposition (cf., [33, Theorem 4.5]).

Proposition 6.13. *Let E be a σ-Dedekind complete Banach lattice. Then the following assertions are equivalent:*

(1) *E has GP property.*
(2) *E has the limited p-Schur property.*
(3) *E has order continuous norm.*

We give a Banach lattice of measurable vector-valued functions with the positively limited p-Schur property. Let $(S, \sum, \mu)$ be an arbitrary positive measure space. For $1 \leq p \leq \infty$; let $L_p(\mu, X) = L_p(S, \sum, \mu, E)$ denote the Banach space of X-valued p-Bochner μ-integrable functions with its usual norm. It is proved that $L_p(\mu, X)$ has the GP property whenever X has the GP property (cf. [19, Theorem 3.1]).

Using the same arguments, we can show that $L_p(\mu, E)$ has the positively limited p-Schur property, if E has this property.

To obtain an interesting characterization of the positively limited p-Schur property, we need to consider the following lemma for positively limited sets in Banach lattices. Let us say that a closed subspace X of a Banach lattice E is positively complemented whenever there exists a positive projection on E whose range is X.

Lemma 6.8. *Let E be a Banach lattice. Then*
(1) *Continuous linear image of each positively limited set in E under a positive operator is positively limited.*
(2) *If F is a positively complemented sublattice of E and $A \subseteq F$ is a positively limited set in E. Then A is a positively limited set in F.*

Proof. (1) It is remarked in the paper [3].

(2) Let F be a positively complemented sublattice of E. Then there is an onto positive projection $P : E \to E$ with range F. If (x_n^*) is a positive weak* null sequence in F^*, then we can also extend each x_n^* to the whole of E by putting $u_n^* = x_n^* \; o \; P$. It is clear that (u_n^*) is a positive weak* null sequence in E^*. Since A is a positively limited set in E,

$$\sup_{a \in A} |\langle x_n^*, a \rangle| = \sup_{a \in A} |\langle x_n^* \; o \; P, a \rangle| = \sup_{a \in A} |\langle u_n^*, a \rangle| \to 0.$$

Therefore, A is a positively limited set in F.

As an alternative proof, remember that positive operators preserve positively limited sets (cf. [3, Theorem 2.11]). Therefore, if $A \subset F$ is a positively limited set in E, then $P(A) = A$ is a positively limited set in F. This completes the proof. □

Note that a Banach space X has the Schur (resp. p-Schur) property if and only if each closed separable linear subspace of X has the Schur (resp. p-Schur) property.

Theorem 6.30. *For each $1 \leq p \leq \infty$, the following are equivalent:*
(1) *E has the positively limited p-Schur property.*
(2) *Every closed separable sublattice of E is contained in a positively complemented sublattice Z of E with the positively limited p-Schur property.*
(3) *E is the direct sum of two positively complemented sublattices with the positively limited p-Schur property.*

Proof. (1) ⇒ (2) It is obvious. In fact, the positively limited p-Schur property is inherited by closed sublattices.

(2) ⇒ (1) Suppose that A is a weakly p-compact positively limited set in E. We have to show that A is relatively compact. For this, let $(x_n) \subseteq A$ be a sequence. Then there is a subsequence (x_{n_k}) of (x_n) that is positively limited weakly p-convergent to some $x \in A$. Consider the closed linear span of (x_n). By hypothesis, it is contained in a positively complemented sublattice Z of E with the positively limited p-Schur property. By Lemma 6.8, (x_n) is a positively limited sequence in Z. Since Z has the positively limited p-Schur property, (x_{n_k}) is norm convergent to x and so A is relatively compact.

(1) ⇒ (3) Consider $E = E \oplus \{0\}$.

(3) ⇒ (1) Consider $E = Y \oplus Z$ such that Y and Z are positively complemented sublattices of E with the positively limited p-Schur property. Assume that A is a weakly p-compact positively limited set in E. Consider the positive linear projections $P_Y : E \to Y$ and $P_Z : E \to Z$. By Lemma 6.8, two sets $P_Y(A)$ and $P_Z(A)$ are weakly p-compact and positively limited. Then for each sequence $(x_n) \subseteq A$ there is $y_n \in P_Y(A)$ and $z_n \in P_Z(A)$ such that $x_n = y_n + z_n$. Since $P_Y(A)$ and $P_Z(A)$ have the positively limited p-Schur property, the sequences (y_n) and (z_n) have convergent subsequences (y_{n_k}) and (z_{n_k}); that is, there are $y \in P_Y(A)$ and $z \in P_Z(A)$ such that $y_{n_k} \to y$ and $z_{n_k} \to z$. By the weakly p-compactness of A, $x_{n_k} \to y + z \in A$. This implies that A is a relatively compact set in E and so E has the positively limited p-Schur property. This completes the proof. □

It is not true that a Banach lattice E has the positively limited p-Schur property provided it has a closed ideal F such that both F and E/F have this property.

Theorem 6.31. *Let E be a Banach lattice and F be a closed ideal of E. If E/F has the positively limited p-Schur property, then E has the same property if (and only if) every weakly p-summable sequence $(x_n) \subset F$ which is positively limited in E is norm null.*

Proof. The result can be proved similar to [19, Proposition 5.1]. It is enough to show that the continuous linear image of a positively limited set under a quotient map is positively limited. Consider the quotient map $Q : E \to E/F$. If $A \subset E$ is a positively limited set and (f_n) is a positive weak* null sequence in $(E/F)^* = F^\perp$, then, for each $x \in E$, $x + F \in E/F$. Hence $\langle f_n, x\rangle = \langle f_n, x + F\rangle \to 0$ as $n \to \infty$, that is, (f_n) is a positive weak* null sequence in E^*. Since A is a positively limited set in E, $\sup_{x\in A} |\langle f_n, Q(x)\rangle| = \sup_{x\in A} |\langle f_n, x\rangle| \to 0$. This implies that $Q(A)$ is a positively limited set in E/F. This completes the proof. □

As a consequence, we obtain the following.

Corollary 6.8. *Let E be a Banach lattice and F be a closed ideal of E with the p-Schur property. If E/F has the positively limited p-Schur property, then E has the same property too.*

Using the disjointness of (x_n), follow the argument of the proof of Theorem 4.34 of [1, p. 209] and Corollary 11.17 of [18, p. 224], we can prove that the following lemma which improves Lemma 3.4 of [22]. The reader is referred to the Chapter 16 of [18] to recall the discussion of type and cotype.

Lemma 6.9. *Let E be a Banach lattice with the type q (with $1 < q \leq 2$) and let $p \geq q'$. Then each disjoint sequence (x_n) in the solid hull of each norm-bounded subset W of E, belongs to $\ell_p^w(E)$.*

Proof. Let W be a norm-bounded subset of E and $(x_n) \subset E$ be a disjoint sequence lying in the solid hull of W. Pick a sequence $(y_n) \subset W$ satisfying $|x_n| \leq |y_n|$ for all n and fix $0 \leq x^* \in E^*$. From [1, Theorem 4.34], there is a sequence $(y_n^*) \in \ell_1^w(E^*)$ such that $|x^*(x_n)| \leq y_n^*(y_n)$ for all n. Since E^* has the cotype q', the sequence (y_n^*) is norm p-summable (cf. [18, Corollary 11.17]). Hence, we can conclude that $(x^*(x_n)) \in \ell_p$ for each $x^* \in E^*$ and so $(x_n) \in \ell_p^w(E)$. This completes the proof. □

In several papers, results concerning the coincidence of two properties positive Schur and Schur are discussed. Most of these results can be carried over to the setting of two properties of order p. As an application of the positively limited p-Schur property, we discuss one of them.

Theorem 6.32. *Let E be a Banach lattice with the type q (with $1 < q \leq 2$) and let $p \geq q'$. Then the following assertions are equivalent:*
(1) *E has the p-Schur property.*
(2) *E has the positively limited p-Schur and p-positive Schur properties.*

Proof. (1) ⇒ (2) It is obvious.

(2) ⇒ (1) Let A be a weakly p-compact set in E. Clearly A is relatively weakly compact. Since by [4, Theorem 3.1], E has the positive Schur property, A is an L-weakly compact and so it is a positively limited set in E (cf. [3, Theorem 2.6] & [14, Theorem 2.6]). Hence the positively limited p-Schur property of E implies that A is a relatively compact set in E. Thus E has the p-Schur property. This completes the proof. □

6.8 Positive DP* property of order *p*

This section focus on the positive DP* property of order p which is weaker than the p-weak DP* property. When a Banach lattice has the type q ($1 < q \leq 2$) and $p \geq q'$, we characterize the positive DP* property of order p in terms of sequences. In particular, in terms of disjoint p-convergent operators into c_0, we give an operator characterization of those Banach lattices with this property.

Definition 6.8. A Banach lattice E has the *positive* DP* *property of order p (p-positive* DP* *property)* if each weakly p-compact set in E is positively limited.

In other words, E has the p-positive DP* property if and only if, for every weakly p-summable sequence (x_n) in E and every positive weak* null sequence (x_n^*) in E^*, $x_n^*(x_n) \to 0$.

Each Banach lattice with the positive DP* property has the p-positive DP* property too, but the converse is false. Using the fact that a Banach lattice has the Schur property if and only if it has the two properties positive DP* and positive GP, we have the following theorem.

Theorem 6.33. *Every discrete Banach lattice E with the p-Schur and without the Schur property has the p-positive* DP* *property, but it does not have the positive* DP* *property.*

Proof. Since E has the p-Schur property, it has the p-positive DP* property and also it contains no copy of c_0; that is, E is a KB-space. On the other hand, E is discrete and so it has the positive GP property. But E does not have the positive DP* property, since it does not have the Schur property. This completes the proof. □

All the spaces ℓ_q, where $1 < p < \infty$ and $1 < q < p'$ have the p-positive DP* property (by the p-Schur property). But none of them have the positive DP* property.

Theorem 6.34. *Let E be a Banach lattice. Then the following assertions are equivalent:*
(1) *E has the p-positive* DP* *property.*
(2) *Every positive operator $T : E \to c_0$ is p-convergent.*

Proof. (1) ⇒ (2) Let A be a weakly p-compact set in E and $T : E \to c_0$ be a positive operator. It suffices to show that $T(A)$ is a relatively compact set in c_0. Since E has the p-positive DP* property, A is a positively limited set in E. By [3, Theorem 2.11], $T(A)$ is a positively limited set in c_0. Hence by the positive GP property of c_0, $T(A)$ is relatively compact.

(2) ⇒ (1) We show that for every weakly p-summable sequence (x_n) in E and every positive weak* null sequence (x_n^*) in E^*, $x_n^*(x_n) \to 0$. We can uniquely determine every positive operator $T : E \to c_0$ by a positive weak* null sequence $(x_n^*) \subset E^*$ such that $Tx = (\langle x, x_n^* \rangle)$, for all $x \in E$. By hypothesis on T, $\|Tx_n\| \to 0$. Consequently, $x_n^*(x_n) \to 0$ and so E has the p-positive DP* property. This completes the proof. □

The separable injectivity of c_0 immediately yields the next result.

Corollary 6.9. *If E is a separable Banach lattice such that E^* has* weak* *sequentially continuous lattice operations, then the following assertions are equivalent:*
(1) *E has the p-Schur property.*
(2) *Every positive operator $T : E \to c_0$ is p-convergent.*

Now, we have the following easy result and omit its proof.

Proposition 6.14. *For a Banach lattice E, the following assertions are equivalent:*
(1) *E has the p-*DP* *and limited p-Schur properties.*

(2) *E has the p-positive* DP* *and positively limited p-Schur properties.*
(3) *E has the p-weak* DP* *and strong limited p-Schur properties.*
(4) *E has the p-Schur property.*

Each Banach lattice with the p-weak DP* property has the p-positive DP* property, but the converse is not true. In fact, Banach lattice c has the p-positive DP* property (the closed unit ball B_c is positively limited), but it does not have the p-weak DP* property (since c has the strong limited p-Schur property, but it does not have the p-Schur property).

Here we say a Banach lattice E has the property (d) if $|x_n^*| \xrightarrow{w^*} 0$ for every disjoint weak* null sequence (x_n^*) in E^* (cf., [14]).

Each Banach lattice with the property (d) has the p-positive DP* property if and only if it has the p-weak DP* property. Also, in each Banach lattice E such that E^* has the weak* sequentially continuous lattice operations, three properties p-DP*, p-weak DP* and p-positive DP* coincide.

Theorem 6.35. *Let E be a Banach lattice with the type q $(1 < q \leq 2)$ and $p \geq q'$. Then the following assertions are equivalent:*
(1) *E has the p-positive* DP* *property.*
(2) *For every sequence $(x_n) \in \ell_p^w(E)_+$ and every positive* weak* *null sequence (x_n^*) in E^*, $x_n^*(x_n) \to 0$.*
(3) *For every disjoint sequence $(x_n) \in \ell_p^w(E)$ and every positive* weak* *null sequence (x_n^*) in E^*, $x_n^*(x_n) \to 0$.*
(4) *For every disjoint sequence $(x_n) \in \ell_p^w(E)_+$ and every positive* weak* *null sequence (x_n^*) in E^*, $x_n^*(x_n) \to 0$.*
(5) *The solid hull of a weakly p-compact set in E is positively limited.*

Proof. We only prove (4) ⇒ (5). Let W be a weakly p-compact set in E and $B := Sol(W)$ be the solid hull of W. By Lemma 6.9, each disjoint sequence in B is weakly p-summable. So, by hypothesis, for every disjoint sequence (x_n) in $B \cap E^+$ and every positive weak* null sequence (x_n^*) in E^*, $x_n^*(x_n) \to 0$. Hence, by [3, Theorem 2.2], $B := Sol(W)$ is positively limited. This completes the proof. □

Chen et al. [14, Theorem 3.5] characterized the weak DP* property of a Banach lattice: a σ-Dedekind complete Banach lattice E has the weak DP* property if and only if each continuous operator $T : E \to c_0$ is almost Dunford–Pettis.

In terms of positive operators into c_0, we give an operator characterization of the p-positive DP* property. For this, we need the following lemma of [4].

Lemma 6.10. *Let E be a Banach lattice with the type q $(1 < q \leq 2)$, $p \geq q'$ and F be a discrete Banach lattice with order continuous norm. Then every positive operator $T : E \to F$ is disjoint p-convergent if and only if it is p-convergent.*

Using this lemma and Theorem 6.35, we have the following theorem.

Theorem 6.36. *Let E be a Banach lattice with the type q $(1 < q \leq 2)$ and $p \geq q'$. Then the following assertions are equivalent:*
1. *E has the p-positive* DP* *property.*
2. *Every continuous operator $T : E \to c_0$ is disjoint p-convergent.*
3. *Every positive operator $T : E \to c_0$ is disjoint p-convergent.*
4. *Every positive operator $T : E \to c_0$ is p-convergent.*

Proof. (1) ⇔ (4) It follows from Theorem 6.34.

(3) ⇔ (4) Because c_0 is discrete with order continuous norm, it follows from Lemma 6.4.

(2) ⇒ (3) Obvious.

(1) ⇒ (2) Consider the operator $T : E \to c_0$. We have to show that $\|Tx_n\| \to 0$ for each disjoint sequence $(x_n) \in \ell_p^w(E)$. Assume, by way of contradiction, that $\|Tx_n\| \nrightarrow 0$ for some disjoint sequence $(x_n) \in \ell_p^w(E)$. Then we can suppose that there would exist some $\epsilon > 0$ satisfying $\|Tx_n\| > \epsilon$ for all $n \in \mathbb{N}$. Since (x_n) is weakly null, by [1, Ex. 22, p. 73] and by the similar idea in Theorem 3.5 of [14], one can find a (disjoint) positive weak* null sequence (x_n^*) in E^*. Since E has the p-positive DP* property, by Theorem 6.35, $x_n^*(x_n) \to 0$. This is a contradiction. Hence, $\|Tx_n\| \to 0$ for each disjoint sequence $(x_n) \in \ell_p^w(E)$ and so T is a disjoint p-convergent operator. This completes the proof. □

It should be noted that the positive DP* property (and also p-positive DP* property) is not inherited by a closed ideal of a Banach lattice E. Consider c_0 as a closed ideal of ℓ_∞.

Theorem 6.37. *Positively complemented sublattices of a Banach lattice E with the positive* DP* *property have this property too.*

Proof. Let F be a positively complemented sublattice of a Banach lattice E with the positive DP* property. To prove that F has the positive DP* property, it suffices to show that every positive operator $T : F \to c_0$ is almost Dunford–Pettis (cf. [3, Definition 3.9] & [8, Theorem 4.2]). Let $P : E \to E$ be the positive projection whose range is F. Since E has the positive DP* property, it follows that $TP : E \to c_0$ is an almost Dunford–Pettis operator. If (x_n) is a disjoint weakly null sequence in F, then it is a disjoint weakly null sequence in E. Hence, $\|Tx_n\| = \|TPx_n\| \to 0$ and so T is an almost Dunford–Pettis operator.

As an alternative proof, if $A \subset F$ is a relatively weakly compact set in F, then it is a relatively weakly compact set in E. By the positive DP* property of E, the set A is positively limited in E. Since F is a positively complemented sublattice of E, the set A is positively limited in F (cf. Lemma 6.8). Thus F has the positive DP* property. This completes the proof. □

With the same techniques used in Theorem 6.37, it can be shown that each positively complemented sublattice F of a Banach lattice E with the p-positive DP* property have this property too. To this end, it suffices to show that every positive operator $T : F \to c_0$ is p-convergent (cf. Theorem 6.34).

Also, it can be shown that a complemented sublattice F of a Banach lattice E with the type q $(1 < q \leq 2)$ and $p \geq q'$ and the p-positive DP* property have this property too. To prove that F has the p-positive DP* property, in view of Theorem 6.11 it is enough to show that every positive operator $T : F \to c_0$ is p-convergent.

6.9 Weak positively limited Schur property of order p

In this section, we study the weak positively limited Schur property of order p on Banach lattices. We show that this property coincides with the order continuity of the norm.

Definition 6.9. A Banach lattice E has the *weak positively limited Schur property of order p (weak positively limited p-Schur property)* if each weakly p-summable positively limited sequence with the positive terms in E is norm null.

Each Banach lattice with the positively limited p-Schur property has the weak positively limited p-Schur property, but the converse is false. In fact, $L^1[0,1]$ has the weak positively limited p-Schur property (since it has the positive Schur property), but it does not have the positively limited p-Schur property, for $p \geq 2$.

We need the following lemma, which can be proved by the same techniques of [23, Lemma 3.7] (and also [14, Lemma 2.2]), to give a characterization of Banach lattices with the weak positively limited p-Schur property.

Lemma 6.11. *Let E be a σ-Dedekind complete Banach lattice. If (x_n) is a positively limited disjoint weakly null sequence in E, then the positive sequences $(|x_n|)$, (x_n^+) and (x_n^-) are all limited weakly null.*

Theorem 6.38. *For a σ-Dedekind complete Banach lattice E, the following assertions are equivalent:*
(1) *E has the weak positively limited p-Schur property.*
(2) *Each disjoint positively limited sequence $(x_n) \in \ell_p^w(E)$ is norm null.*
(3) *Each disjoint almost limited sequence $(x_n) \in \ell_p^w(E)$ is norm null.*
(4) *Each disjoint limited sequence $(x_n) \in \ell_p^w(E)$ is norm null.*

Proof. (1) $\Rightarrow$ (2) Let $(x_n) \in \ell_p^w(E)$ be a disjoint positively limited sequence. Then by [37, Corollary 3.1.6] and [3, Theorem 2.3], the sequence $(|x_n|)$ is weakly p-summable and positively limited in E. Hence by the weak positively limited p-Schur property of E, $\|x_n\| = \||x_n|\| \to 0$.

(2) $\Rightarrow$ (3) $\Rightarrow$ (4) Obvious.

(4) $\Rightarrow$ (1) Suppose that $(x_n) \in \ell_p^w(E)_+$ be a positively limited sequence, but $\|x_n\| \nrightarrow 0$. By [30, Corollary 2.3.5], there is a disjoint positive subsequence (x_{n_k}) of (x_n) such that $\|x_{n_k}\| \nrightarrow 0$, which is a contradiction. In fact, since (x_{n_k}) is a positive and disjoint positively

limited weakly null sequence in E, it must be a limited sequence in E (cf., Lemma 6.11) and so, by hypothesis (4), $\|x_{n_k}\| \to 0$. This completes the proof. □

In the following theorem, we give an interesting characterization of the weak positively limited p-Schur property:

Theorem 6.39. *Let E be a Banach lattice. Then the following assertions are equivalent:*
(1) *E has the weak positively limited p-Schur property.*
(2) *E has order continuous norm.*

Proof. (1) ⇒ (2) Let (x_n) be an order-bounded disjoint sequence in E. Then (x_n) is a weakly p-summable (cf. [4, Lemma 2.10]) and positively limited (cf. [3, p. 2]) sequence in E. Then, by [37, Corollary 3.1.6] and [3, Theorem 2.3], the positive sequence $(|x_n|)$ is weakly p-summable and positively limited in E. By the weak positively limited p-Schur property of E, $\|x_n\| = \||x_n|\| \to 0$ and so E has order continuous norm.

(2) ⇒ (1) Since E has order continuous norm, it is a σ-Dedekind complete Banach lattice and also it has the limited p-Schur property. Hence, each limited sequence $(x_n) \in \ell_p^w(E)$ is norm null. Thus by Theorem 6.38, E has the weak positively limited p-Schur property. This completes the proof. □

Note that, in Theorem 6.38, the σ-Dedekind completeness of a Banach lattice E can not be removed. Actually, each weakly null and almost limited sequence in $C[0,1]$ is norm null (cf., [2, Theorem 2.12]) and so the assertion (3) of Theorem 6.38 holds. But $C[0,1]$ does not have order continuous norm and so by Theorem 6.39, it cannot have the weak positively limited p-Schur property.

As a consequence, we have the following corollary:

Corollary 6.10. *Let E be a Banach lattice such that E^* has the* weak* *sequentially continuous lattice operations. Then the following assertions are equivalent:*
(1) *E has the positively limited p-Schur property.*
(2) *E has the weak positive limited p-Schur property.*

Proof. (1) ⇒ (2) It is clear.

(2) ⇒ (1) If E has the weak positive limited p-Schur property, then by Theorem 6.39, it has order continuous norm. Hence E has the limited p-Schur property; that is, each limited weakly p-summable sequence (x_n) in E is norm null. From [3, Theorem 2.5] each positively limited weakly p-summable sequence (x_n) in E is norm null and so E has the positively limited p-Schur property. This completes the proof. □

Recall from [8] that a Banach lattice E has the *bi-sequence property* if for every disjoint weakly null sequence (x_n) in E and every weak* null sequence (x_n^*) in E_+^*, $x_n^*(x_n) \to 0$. The p-bi-sequence property is introduced as follows:

Definition 6.10. A Banach lattice E has the *p-bi-sequence property* if, for every weakly p-summable disjoint sequence (x_n) in E and every positive weak* null sequence (x_n^*) in E^*, $x_n^*(x_n) \to 0$.

In other words, E has the p-bi-sequence property if and only if each disjoint weakly p-summable sequence in E is positively limited. In the following theorem, we give an operator characterization for this property.

Theorem 6.40. *Let E be a Banach lattice. Then the following assertions are equivalent:*
(1) *E has the p-bi-sequence property.*
(2) *Every positive operator $T : E \to c_0$ is disjoint p-convergent.*

Proof. (1) $\Rightarrow$ (2) Obvious.

(2) $\Rightarrow$ (1) We show that, for every weakly p-summable disjoint sequence (x_n) in E and every positive weak* null sequence (x_n^*) in E^*, $x_n^*(x_n) \to 0$. Consider the positive operator $T : E \to c_0$ defined by positive weak* null sequences $(x_n^*) \subset E^*$ such that $Tx = (\langle x, x_n^* \rangle)$, for all $x \in E$. By hypothesis on T, $\|Tx_n\| \to 0$. Consequently, $x_n^*(x_n) \to 0$ and so E has the p-bi-sequence property. This completes the proof. □

Using above theorem, in every Banach lattice with the type q (with $1 < q \leq 2$) and $p \geq q'$, these two properties p-bi-sequence and p-positive DP* are the same.

Also, note that the p-bi-sequence property is not inherited by a closed ideal of a Banach lattice E. Consider c_0 as a closed ideal of ℓ_∞.

Similar to Theorem 6.37, we can prove that positively complemented sublattices of a Banach lattice E with the p-bi-sequence property have this property too.

Corollary 6.11. *If E is a separable Banach lattice such that E^* has* weak* *sequentially continuous lattice operations, then the following assertions are equivalent:*
(1) *E has the p-positive Schur property.*
(2) *Every positive operator $T : E \to c_0$ is disjoint p-convergent.*

In papers [3, 14], some results concerning the positive Schur property are discussed. A Banach lattice has the positive Schur property if and only if it has order continuous norm and the weak DP* property if and only if it has order continuous norm and the positive DP* property. We conclude this section by stating a similar characterization for the p-positive Schur property.

Theorem 6.41. *Let E be a Banach lattice. Then the following assertions are equivalent:*
(1) *E has the p-positive Schur property.*
(2) *E has order continuous norm and the p-positive* DP* *property.*
(3) *E has order continuous norm and the p-bi-sequence property.*

Proof. (1) $\Rightarrow$ (2) $\Rightarrow$ (3) Obvious.

(3) $\Rightarrow$ (1) By [37, Proposition 3.1.21], it is enough to show that for each disjoint sequence $(x_n) \in \ell_p^w(E)_+$, $\|x_n\| \to 0$. To this end, note that since E has the p-bi-sequence property, the sequence (x_n) is positively limited. On the other hand, E has the weak positive limited p-Schur property and so $\|x_n\| \to 0$. Thus E has the p-positive Schur property. This completes the proof. □

Bibliography

[1] Aliprantis, C. D. and Burkishaw, O., *Positive Operators*, Pure and Applied Mathematics Series, Academic Press, New York and London, 1985.

[2] Ardakani, H., Moshtaghioun, S. M., Modarres Mosadegh, S. M. S. and Salimi, M., *The strong Gelfand–Phillips property in Banach lattices*, Banach J. Math. Anal. 10 (2016), 15–26.

[3] Ardakani, H. and Chen, J. X., *Positively limited sets in Banach lattices*, J. Math. Anal. Appl. 526(1) (2023), 127220, https://doi.org/10.1016/j.jmaa.2023.127220.

[4] Ardakani, H. and Taghavinejad, Kh., *The strong limited p-Schur property in Banach lattices*, Oper. Matrices 16 (2022), 811–825.

[5] Ardakani, H., Taghavinejad, Kh. and Rezagholi, Sh., *The strong Dunford–Pettis relatively compact property of order p*, Filomat 37 (2023), 7339–7349.

[6] Ardakani, H. and Amjadi, Kh., *The positively limited p-Schur property in Banach lattices*, Ann. Funct. Anal. 14 (2023), 68, https://doi.org/10.1007/s43034-023-00292-y.

[7] Ardakani, H. and Modarres Mosadegh, S. M. S., *The strong relatively compact Dunford–Pettis property in Banach lattices*, Ann. Funct. Anal. 7 (2016), 270–280.

[8] Aqzzouz, B., Elbour, A. and Wickstead, A. W., *Positive almost Dunford–Pettis operators and their duality*, Positivity 15 (2011), 185–197.

[9] Bouras, K. and Moussa, M., *Banach lattices with weak Dunford–Pettis property*, Int. J. Math. Sci. 6 (2010), 203–207.

[10] Bouras, K., *Almost Dunford–Pettis sets in Banach lattices*, Rend. Circ. Mat. Palermo 62 (2013), 227–236.

[11] Aqzzouz, B. and Elbour, A., *Some characterizations of almost Dunford-–Pettis operators and applications*, Positivity 15 (2011), 369–380.

[12] Borwein, J., Fabian, M. and Vanderwerff, J., *Characterizations of Banach spaces via convex and other locally Lipschitz functions*, Acta Math. Vietnam. 22 (1997), 53–69.

[13] Castillo, J. and Sanchez, F., *Dunford–Pettis-like properties of continuous vector function spaces*, Rev. Mat 6 (1993), 43–59.

[14] Chen, J. X., Chen, Z. L. and Ji, G. X., *Almost limited sets in Banach lattices*, J. Math. Anal. Appl. 412 (2014), 547–563.

[15] Chen, Z. L. and Wickstead, A. W., *Relative weak compactness of solid hulls in Banach lattices*, Indag. Math. N. S. 9 (1995), 187–196.

[16] Dehghani, M. B., Moshtaghioun, S. M. and Dehghani, M., *On the limited p-Schur property of some operator spaces*, Int. J. Anal. Appl. 16 (2018), 50–61.

[17] Diestel, J., *Sequences and Series in Banach Spaces*, Graduate Texts in Math., vol. 92, Springer-Verlag, Berlin, 1984.

[18] Diestel, J., *Absolutely Summing Operators*, Cambridge University Press, 1995.

[19] Drewnowski, L. and Emmanuele, G., *On Banach spaces with the Gelfand–Phillips property II*, Rend. Circ. Mat. 38 (1989), 377–391.

[20] Emmanuele, G., *On Banach spaces with the Gelfand–Phillips property, III*, J. Math. Pures Appl. 72 (1993), 327–333.

[21] Fourie, J. H. and Zeekoei, E. D., DP**-properties of order p on Banach spaces*, Quaest. Math. 37 (2014), 349–358.

[22] Fourie, J. H. and Zeekoei, E. D., *On weak-star p-convergent operators*, Quaest. Math. 40 (2017), 1–17.

[23] Fahri, K. E., Machrafi, N. and Moussa, M., *Banach lattices with the positive Dunford–Pettis relatively compact property*, Extr. Math. 30 (2015), 161–179.

[24] Galindo, P. and Miranda, V. C. C., *Grothendieck-type subsets of Banach lattices*, J. Math. Anal. Appl. 506 (2022), 1–14.

[25] Ghenciu, I., *Some classes of Banach spaces and complemented subspaces of operators*, Adv. Oper. Theory 4 (2019), 369–387.

[26] Groenewegen, G. and Meyer-Nieberg, P., *An elementary and unified approach to disjoint sequence theorems*, Indag. Math. 48 (1986), 313–317.

[27] Lourenço, M. L. and Miranda, V. C. C., *The property (d) and almost limited completely continuous operators*, arXiv:2011.02890v1.

[28] Lindenstrauss, J. and Tzafriri, L., *Classical Banach Spaces, I: Sequence Spaces*, Ergeb. Math. Grenzgeb. (3), vol. 92, Springer, Berlin, 1977.

[29] Machrafi, N., Elbour, A. and Moussa, M., *Some characterizations of almost limited sets and applications*, arXiv:1312.2770v1.

[30] Meyer-Nieberg, P., *Banach Lattices*, Universitext, Springer-Verlag, Berlin, 1991.

[31] Palazuelos, C., Sanchez Perez, E. A. and Tradacete, P., *Maurey–Rosenthal factorization for p-summing operators and Dodds–Fremlin domination*, J. Oper. Theory 68 (2012), 205–222.

[32] Sanchez, J. A., *Positive Schur property in Banach lattices*, Extr. Math. 7 (1992), 161–163.

[33] Wnuk, W., *Banach Lattices with Order Continuous Norms*, Polish Scientific Publishers PWN, Warsaw, 1999.

[34] Wnuk, W., *On the dual positive Schur property in Banach lattices*, Positivity 2 (2012), 759–773.

[35] Wnuk, W., *Banach lattices with the weak Dunford–Pettis property*, Atti Semin. Mat. Fis. Univ. Modena 42(1) (1994), 227–236.

[36] Xiang, B., Chen, J. X. and Li, L., *Weak precompactness in Banach lattices*, Positivity 26 (2022), 1–13.

[37] Zeekoei, E. and Fourie, J., *Classes of Dunford–Pettis-type operators with applications to Banach spaces and Banach lattices*, Ph. D. Thesis, 2017.

George Xianzhi Yuan, Yuanlei Luo, and Yeol Je Cho

7 The nonlinear principles and fixed-point theorems for non-self $\mathfrak{F}$ contraction and non-expansive set-valued mappings by applying Caristi fixed-point theorem in locally complete convex spaces

Abstract: The goal of this paper is to establish fixed-point theorems for non-self $\mathfrak{F}$ contraction and non-self non-expansive set-valued mappings by applying Caristi fixed point as a tool in locally complete spaces. In particular, we first derive a few of key principles widely used in nonlinear analysis and optimization by using Caristi fixed-point theorem as a tool in both complete metric spaces and locally complete convex spaces, respectively. Secondly, as applications of Caristi fixed-point theorem being a starting point, we establish fixed-point theorems for non-self $\mathfrak{F}$ contraction and non-self non-expansive set-valued mappings. These results improve or unify corresponding results in the literature.

Keywords: Nonlinear analysis, fixed point, Caristi fixed-point theorem, Ekeland variational principle, Takahashi nonconvex minimization theorem, Oettli–Théra theorem, non-self mapping, $\mathfrak{F}$ contraction set-valued mapping, non-expansive set-valued mapping, boundary condition (α), P-Opial condition, weakly inward mapping, demiclosed, complete metric space, Banach disc (disk), locally complete convex space, Cantor intersection theorem

MSC 2020: 47H04, 47H10, 47J29, 49J53, 55M20, 90C33, 91B50

Funding: This work was supported in part by National Nature Science Foundation of China (71971031).

Availability of data and materials: Not applicable.

Declarations: The authors declare no competing interests.

George Xianzhi Yuan, School of Economics and Finance and College of Science, Chongqing University of Technology, Chongqing 400054, China; and Business School, Chengdu University, Chengdu 610601, China; and Business School, Sun Yat-Sen University, Guangzhou 510275, China, e-mail: yuanleluo2-c@my.cityu.edu.hk
Yuanlei Luo, College of Business, City University of Hong Kong, Hong Kong 999077, Hong Kong SAR, e-mail: george_yuan99@yahoo.com
Yeol Je Cho, Department of Mathematics Education, Gyeongsang National University, Jinju 52828, Republic of Korea, e-mail: yjchomath@gmail.com

https://doi.org/10.1515/9783112215760-007

7.1 Introduction

It is well know that Caristi fixed point, established by Caristi [12] in 1976, plays a very important role in optimization, nonlinear analysis and related topics, in particular for the study of fixed point theory and related applications, see Agarwal et al. [1], Almezel et al. [2], Ansatri [3], Ansatri and Sahu [4]; and also see Caristi and Kirk [13], Cho [15], Kirk [31–33], Kirk and Caristi [34], Qiu [55–58], Qiu and Rolewicz [61], Zeidler [79], Zeidler [79], Zhong et al. [82, 83], and references wherein for more in details.

The goal of this paper is to establish fixed-point theorems for non-self $\mathfrak{F}$ contraction and non-self non-expansive set-valued mappings by applying Caristi fixed point as a tool in locally complete spaces. In particular, we first derive a few of key principles widely used in nonlinear analysis and optimization by using Caristi fixed-point theorem as a tool in both complete metric spaces and locally complete convex spaces, respectively. Secondly, as applications of Caristi fixed-point theorem being a starting point, we establish fixed-point theorems for non-self $\mathfrak{F}$ contraction and non-self non-expansive set-valued mappings. These results improve or unify corresponding results in the literature.

The goal of this paper is first to discuss the famous and very important tool in nonlinear functional analysis called *Caristi fixed-point theorem* established by Caristi [12] in complete metric spaces and locally complete convex spaces. Then we show that a few of key principles established in nonlinear analysis are equivalent to Caristi fixed-point theorem. In particular, by following the idea given by Ansatri [3], Ansatri and Sahu [4], we derive Ekeland variational principle [20–22], Takahashi non-convex minimization theorem [68], and Oettli–Théra theorem in complete metric spaces (and indeed, also equivalent to Daneš drop theorem, and flower petal theorem as discussed by Penot [53]), and see also related discussion by Agarwal et al. [1], Almezel et al. [2], Qiu [55–58], Qiu and Rolewicz [61], Park [46–50], Yuan [73–77], Zeidler [79], Zhong et al. [82, 83] and comprehensive references wherein.

This paper consists of 4 sections: Section one is for a introduction with literature review briefly; Section two is preliminary by including some notions and basic results; Section three is for the discussion of Caristi fixed fixed-point theorems in complete metric spaces and locally complete convex spaces; Section four is to establish fixed-point theorems for non-self $\mathfrak{F}$ contraction and non-expansive set-valued mappings by applying Caristi fixed-point theorem in locally complete convex spaces, and ended by some remarks.

7.2 Preliminary

For the convenience of our discussion, throughout this paper, for a given set X, we denotes by 2^X the family of all subsets of X. If X is a metric space (X, d), we denote by $CB(X)$ the family of all non-empty closed bounded subsets of (X, d), where d is the metric of X.

In addition, all topological vector spaces are assumed to be Hausdorff, and we denote by $\mathbb{R}$ and $\mathbb{N}$ the real line, and the all positive integers, respectively. For any set $D \subset X$, the diameter of D is defined by $\operatorname{diam}(D) := \sup\{d(x,y), x, y \in D\}$.

We now recall some standard notions and concepts that are used or helpful for the related discussion or topic in this paper.

Definition 7.1. A set A in a vector space X is said to be convex if for any $x, y \in A$, $s, t \geq 0$, we have $(1-t)x + ty \in A$, whenever $t \in [0,1]$.

Definition 7.2. If A is a subset of a topological vector space X, the closure of A is denoted by $\overline{A}$, then the convex hull of A and its closed convex hull denoted by $C(A)$, and $\overline{C}(A)$, respectively, which is the smallest convex set containing A, and the smallest closed convex set containing A, respectively.

Definition 7.3. Let A be convex and $x_1, \ldots, x_n \in A$, and $t_i \geq 0$, $\sum_1^n t_i = 1$. Then $\sum_1^n t_i x_i$ is called a convex combination of $\{x_i\}$ for $i = 1, 2, \ldots, n$. If $\sum_1^n |t_i| \leq 1$, then $\sum_1^n t_i x_i$ is called an absolutely convex combination. It is easy to see that $\sum_1^n t_i x_i \in A$ for a convex set A.

Definition 7.4. A subset A of a vector space X is called circled (or balanced) if $\lambda A \subset A$ holds for all scalars λ satisfying $|\lambda| \leq 1$. We say that A is absorbing if for each $x \in X$, there is a real number $\rho_x > 0$ such that $\lambda x \in A$ for all $\lambda > 0$ with $|\lambda| \leq \rho_x$.

By Definition 7.4, it is easy to see that the system of all circled subsets of X is easily seen to be closed under the formation of linear combinations, arbitrary unions, and arbitrary intersections. In particular, every set $A \subset X$ determines a smallest circled subset $\hat{A}$ of X in which it is contained: $\hat{A}$ is called the circled hull of A. It is clear that $\hat{A} = \bigcup_{|\lambda| \leq 1} \lambda A$ holds, so that A is circled if and only if (in short, iff) $\hat{A} = A$. We use $\overline{\hat{A}}$ to denote for the closed circled hull of $A \subset X$.

In addition, if X is a topological vector space, we use the $\operatorname{int}(A)$ to denote the (topological relatively) interior of set $A \subset X$ and if $0 \in \operatorname{int}(A)$, then $\operatorname{int}(A)$ is also circled; and use the ∂A to denote the boundary of A in X.

Definition 7.5. A topological vector space is said to be locally convex if the origin (denoted by θ) has a fundamental set of (absolutely) convex θ-neighborhoods. This topology can be determined by seminorms which are defined in the obvious way (see Bayoumi [6, p. 52], Jarchow [29], Rolewicz [64], or Yuan [74–77]).

Definition 7.6. Let X is a vector space and $\mathbb{R}^+$ is a non-negative part of a real line $\mathbb{R}$. Then a mapping $P : X \longrightarrow \mathbb{R}^+$ is said to be a seminorm if it satisfies the following requirements:

(i) $P(x) \geq 0$ for all $x \in X$;
(ii) $P(\lambda x) = |\lambda| P(x)$ for all $x \in X$ and $\lambda \in \mathbb{R}$;
(iii) $P(x + y) \leq P(x) + P(y)$ for all $x, y \in X$.

A seminorm P is called a norm if $x = 0$ whenever $P(x) = 0$. A vector space X with a specific norm (denoted by $\|\cdot\|$) is called a normed space, which is always denoted by $(X, \|\cdot\|)$ throughout this paper.

Definition 7.7. A topological vector space X is said to be locally convex if it has a θ-basis consisting of convex neighborhoods.

We also recall following notion for the so-called gauge (see Balachandra [5]).

Definition 7.8. Let A be an absorbing subset of a vector space X. For $x \in X$, set $P_A = \inf\{\alpha > 0 : x \in \alpha A\}$, then the non-negative real-valued function P_A is called the gauge. The gauge of A is also known as the Minkowski functional.

Now by Pérez Carreras and Bonet [54, Chapter 5] and Horváth [27] (see also Qiu and Rolewicz [62]), we have the following definition and notions.

Definition 7.9. Let X be a real locally convex Hausdorff (i. e., separated) space (in short, LCS) and X^* be its topological dual. By following Pérez Carreras and Bonet [54, Chapter 3] and also Horváth [27], we call a bounded absolutely convex set B in X a disc, and denoted by "span[B]" the linear subspace spanned by B and denote p_B the Minkowski functional of B. Then $E_B := (\text{span}[B], p_B)$ is a normed space. If E_B is a Banach space, then B is called a Banach disc (or, Banach disk used by some existing literature).

Definition 7.10. A locally convex space X is said to be sequently complete if every Cauchy sequence in X is convergent. A locally convex space X is said to be a locally complete convex space if every locally Cauchy sequence is locally convergent. This is equivalent to saying that each bounded subset of X is contained in a certain Banach disc (Banach disk). Let $A \subset X$ be non-empty. Then A is said to be locally closed if for any locally convergent sequence in A, its local limit point belongs to A.

For more on the notion of locally complete convex spaces and its applications, see Qiu [57–59], Qiu and Rolewicz [62] and references wherein.

Finally, we also recall the following Cantor intersection theorem, which will be used in this paper.

Cantor intersection theorem. *Let (X, d) be a complete metric space and let $\{D_n\}$ be a decreasing sequence (that is, $D_{n+1} \subset D_n$) of non-empty closed subsets of X such that the diameter of D_n for $n \in \mathbb{N}$, with $\lim_{n\to\infty} \operatorname{diam}(D) = 0$. Then, the intersection $\bigcap_{n=1}^{\infty} D_n$ contains exactly one point.*

7.3 Non-linear principles related to Caristi fixed-point theorem in non-linear analysis

The goal of this section is first to discuss the famous and very important tool in nonlinear functional analysis called *Caristi fixed-point theorem* established by Caristi [12] in complete metric spaces and locally complete locally convex spaces. Then, as applications, by following the idea given by Ansatri [3], Ansatri and Sahu [4], we derive a few (of its equivalent) nonlinear principles in nonlinear analysis and optimization by including Ekeland variational principle [20], Takahashi nonconvex minimization theorem [68] and Oettli–Théra theorem [45].

By following Goebel and Kirk [25], we recall that for a given complete metric space (M, d), we denote by $CB(X)$ the family of all non-empty closed bounded subsets of X. Then a mapping $T : M \to M$ is said to be a contraction mapping with the Lipschitz constant $k \in (0, 1)$, if

$$d(T(x), T(y)) \leq kd(x, y)$$

for each $x, y \in M$. If we define a mapping $\phi : M \to [0, +\infty)$ by

$$\phi(x) := (1 - k)^{-1} d(x, Tx)$$

for each $x \in M$, as T is k contractive, where $k \in (0, 1)$, then we have

$$d(x, Tx) - kd(x, Tx) \leq d(x, Tx) - d(Tx, T^2x),$$

then it follows

$$d(x, Tx) \leq \phi(x) - \phi(Tx)$$

for each $x \in M$, and this implies that T has a unique fixed point in M as T is a contraction mapping with the Lipschitz constant $k \in (0, 1)$.

We know if the mapping ϕ above is only lower semicontinuous bounded below, for any mapping $T : M \to M$ which may not be a contraction mapping, but satisfies above inequality with ϕ, it then has at least one fixed point, as shown by Caristi fixed-point theorem established by Caristi [12] in 1976, for which we list below as Theorem 7.1 by applying a smart proof given by Du [18, 19] (see also related different proofs given by Almezel et al. [2], Ansatri [3], Ansatri and Sahu [4] for related discussion with more in details).

Theorem 7.1 (Caristi fixed-point theorem for single-valued mappings). *Let (X, d) be a complete metric space. Assume $T : X \to X$ is a mapping and $\phi : X \to \mathbb{R}$ is a lower semicontinuous mapping, which is bounded below such that for any $x \in X$,*

$$d(x, T(x)) \le \phi(x) - \phi(T(x)).$$

Then T has a fixed point in X.

Proof. It is indeed Theorem 2.1′ of Caristi [12] (see different proof given by Siegel [66], Brézis and Browder [7], Goebel and Kik [25], Wong [69], and another direct proof given by Kozlowski [38]). But here, we provide a directly proof given by Du [18, 19] without using Zorn's lemma, transfinite induction, or any other well-known axiom as follows. We first define a set-valued mapping $F : X \to 2^X$ by

$$F(x) := \{y \in X : d(x,y) \le \phi(x) - \phi(y)\}$$

for each $x \in X$. Clearly, we have that $Tx \in F(x)$ and hence $F(x) \ne \emptyset$ for each $x \in X$.

We now claim that for each $y \in F(x)$, $\phi(y) \le \phi(x)$ and $F(y) \subset F(x)$. Let $y \in F(x)$ be any given point. Then, $y \ne x$ and $d(x,y) \le \phi(x) - \phi(y)$, thus $\phi(y) \le \phi(x)$. As $F(y) \ne \emptyset$, let $z \in F(y)$. Then, $d(y,z) \le \phi(y) - \phi(x)$. It follows that $\phi(z) \le \phi(y) \le \phi(x)$, and hence

$$d(x,z) \le d(x,y) + d(y,z) \le \phi(x) - \phi(z).$$

Thus $z \in F(x)$, therefore we prove that $F(y) \subset F(x)$ for each $y \in F(x)$.

We now construct a sequence $\{x_n\}$ in X by induction, starting with any point $x_1 \in X$ as follows: Suppose $x_n \in X$ is known for $n \in \mathbb{N}$, then choose $x_{n+1} \in F(x_n)$ such that

$$\phi(x_{n+1}) \le \inf_{z \in F(x_n)} \phi(z) + \frac{1}{n}.$$

For each $n \in \mathbb{N}$, since $x_{n+1} \in F(x_n)$, we obtain

$$d(x_n, x_{n+1}) \le \phi(x_n) - \phi(x_{n+1}).$$

Thus $\phi(x_{n+1}) \le \phi(x_n)$ for each $n \in \mathbb{N}$.

Since ϕ is bounded below, the number $\lambda := \lim_{n \in \mathbb{N}} \phi(x_n)$ exists. Now for $m > n$, with $m, n \in \mathbb{N}$, by inequalities above, we obtain

$$d(x_n, d_m) \le \sum_{i=n}^{m-1} d(x_i, x_{i+1}) \le \phi(x_n) - \lambda.$$

Since $\lim_{n \to \infty} \phi(x_n) = \lambda$, we have

$$\sup_{n \to \infty} \{d(x_x, x_m) : m > n\} = 0.$$

Hence $\{x_n\}$ is a Cauchy sequence in X. By the completeness of X, there exists some $v \in X$ such that $\lim_{n \to \infty} x_n = v$. As ϕ is lower semicontinuous, it follows that for all $i \in \mathbb{N}$,

$$\phi(v) \le \liminf_{n \to \infty} \phi(x_n) = \inf_{n \in \mathbb{N}} \phi(x_n) \le \phi(x_i).$$

Now, we claim that $\bigcap_{n=1}^{\infty} F(x_n) = \{v\}$. Indeed, for $m > n$ with $m, n \in \mathbb{N}$, by inequalities above, we have

$$d(x_n, x_m) \leq \sum_{i=n}^{m-1} d(x_i, x_{i+1}) \leq \phi(x_n) - \phi(v).$$

Since $\lim_{m\to\infty} x_m = v$, the inequality above implies for each $n \in \mathbb{N}$,

$$d(x_n, v) \leq \phi(x_n) - \phi(v).$$

It implies that $v \in \bigcap_{n=1}^{\infty} F(x_n)$ and thus $\bigcap_{n=1}^{\infty} F(x_n) \neq \emptyset$. Moreover, we can see that

$$F(v) \subset \bigcap_{n=1}^{\infty} F(x_n).$$

Now, for any $c \in \bigcap_{n=1}^{\infty} F(x_n)$, by the inequality in the beginning above, we have

$$d(x_n, c) \leq \phi(x_n) - \phi(c) \leq \phi(x_n) - \inf_{z \in F(x_n)} \phi(z) \leq \phi(x_n) - \phi(x_{n+1}) + \frac{1}{n},$$

for all $n, m \in \mathbb{N}$. Therefore, we have $\lim_{n\to\infty} d(x_n, c) = 0$, or, equivalently to say, $\lim_{n\to\infty} x_n = v$. By the uniqueness of the limit of a convergent sequence, we have $c = v$. This means $\bigcap_{n=1}^{\infty} F(x_n) = \{v\}$.

On the other hand, since $F(v) \neq \emptyset$, and $F(v) \subset \bigcap_{n=1}^{\infty} F(x_n) = \{v\}$. Thus, we have $F(v) = \{v\}$. For the element $v \in X$, as $Tv \in F(v)$, it implies $v = Tv$, which is a fixed point of T. Therefore, T must have a fixed point in X, and this completes the proof. □

Here, we like to remark that, as mentioned above, a different proof without using transfinite induction skill has been given by Brézis and Browder [7], and also see Park [49, 50] gave a comprehensive review for the study of Caristi fixed-point theorem and related metric fixed pint theory under the general framework of quasi-metric spaces. In addition, see also Feng and Liu [24], Lau and Yao [40] and references wherein for some recent results in some other new directions, which are related to Caristi type fixed-point theorem and related content, not covered by this paper.

Now as a consequence of Theorem 7.1, we have the following conclusion, which is Proposition of Wong [69].

Theorem 7.2. *Let (X, d) be a non-empty complete metric space. Let $T : X \to X$ be a single-valued mapping with non-empty values. Suppose there exists a lower semicontinuous function $\phi : X \to [0, +\infty)$ such that for any $x \in X$ with $x \neq Tx$, there exists $y \in X \setminus \{x\}$ satisfying*

$$d(x, y) \leq \phi(x) - \phi(y).$$

Then, T has a fixed point.

Proof. Suppose T has not fixed point, then for each $x \in X$, $T(x) \neq x$, but there exists $y \in X$ with $y \neq x$ such that $d(x,y) \leq \phi(x) - \phi(y)$. We now define a new mapping $T^* : X \to X$ with $T^*(x) := y$ for each $x \in X$, where y is a fixed in $X \setminus \{x\}$ satisfying $d(x,y) \leq \phi(x) - \phi(y)$, as mentioned in the assumption. It is then clear that T^* is well defined, and we have $d(x, T^*(x)) \leq \phi(x) - \phi(T^*(x))$ for each $x \in X$. By Theorem 7.1, T^* has a fixed point $x^* \in X$ such that $T^*(x^*) = x^*$. But by the definition of T^*, $x^* \neq T^*(x^*)$, this is a contradiction. Thus, T mush have a fixed point in X. This completes the proof. □

Now by following the idea used by Ansatri [3] and Ansari and Sahu [4], we show that how Caristi fixed-point theorem plays a key role to its equivalent nonlinear principles in nonlinear analysis and optimization in complete metric spaces.

Theorem 7.3 (Caristi fixed-point theorem for set-valued mappings). *Let (X, d) be a non-empty complete metric space. Let $T : X \to 2^X$ be a set-valued mappings with non-empty values. Suppose there exists a lower semicontinuous function $\phi : X \to [0, +\infty)$ such that for any $x \in X$, there exists $y \in T(x)$ such that*

$$d(x,y) \leq \phi(x) - \phi(y).$$

Then T has a fixed point.

Proof. For each $x \in X$, we define a single-valued mapping $f : X \to X$ by $f(x) := y$, where choosing one $y \in X$ such that $y \in T(x)$ and $d(x,y) + \phi(y) \leq \phi(x)$.

Then f is well defined and $f(x) \in T(x)$ for each $x \in X$, and we also have

$$d(x,f(x)) + \phi(f(x)) \leq \phi(x).$$

Now by Theorem 7.1 for single-valued mapping $f : X \to X$, it follows that there exists $x \in X$ such that $x = f(x)$. By the definition of f, we have $x = f(x) \in T(x)$, thus x is a fixed point of the set-valued mapping T. This completes the proof. □

As an application of Caristi fixed-point theorem, we can prove the following Nadler's fixed point for contraction set-valued mappings.

Theorem 7.4 (Nadler's fixed-point theorem for contraction set-valued mappings).
Let (X, d) be a complete metric space. Then, each contraction set-valued mapping $T : X \to CB(X)$ has a fixed point, where $CB(X)$ denotes the family of all non-empty closed bounded subsets of X.

Proof. Let $T : X \to CB(X)$ be a Nadler contraction set-valued mapping with contraction constant h. Choose real number $\alpha \in \mathbb{R}$ such that $h < \alpha < 1$. Let $x \in X$, then the set $\{y \in T(x) : \alpha d(x,y) \leq d(x, T(x))\} \neq \emptyset$.

By the axiom of choice, there is a single-valued mapping $f : X \to X$ such that $f(x) \in T(x)$ for each $x \in X$. and $\alpha d(x,f(x)) \leq d(x, T(x))$. Thus, we have $d(f(x), T(f(x))) \leq d_H(T(x), T(f(X))) \leq h \cdot d(x,f(x))$.

Note that

$$\begin{aligned} d(x,f(x)) &= \frac{1}{a-h}(ad(x,f(x)) - h \cdot d(x,f(x))) \\ &\le \frac{1}{a-h}(d(x,T(x)) - d(f(x),T(f(x)))). \end{aligned}$$

Now set $\phi(x) := \frac{1}{a-h}d(x,T(x))$ for each $x \in X$. Then, we know that ϕ is continuous from X to $[0,\infty)$ and

$$d(x,f(x)) \le \phi(x) - \phi(f(x))$$

for each $x \in X$. Now by Theorem 7.3 above (Caristi fixed-point theorem), there exists $x \in X$ such that $x = f(x)$, which is a fixed point of f, so $x = f(x) \in T(x)$ is a fixed point of T. This completes the proof. □

Now, we prove Ekeland's variational principle by using Theorem 7.3, which is also called Caristi and Kirk Theorem by Caristi an Kirk [13].

Theorem 7.5 (Strong form of Ekeland's variational principle). *Let (X,d) be a complete metric space and $f : X \to \mathbb{R} \cup \{+\infty\}$ be a proper, bounded below and lower semicontinuous functional. Let $\epsilon > 0$ and $\hat{x} \in X$ be given such that $f(\hat{x}) \le \inf_{x\in X} f(x) + \epsilon$. Then, for a given $\lambda > 0$, there exists $\bar{x} \in X$ such that*
(1) *$f(\bar{x}) \le f(\hat{x})$;*
(2) *$d(\hat{x},\bar{x}) \le \lambda$; and*
(3) *$f(\bar{x}) < f(x) + \frac{\epsilon}{\lambda}d(x,\bar{x})$, for all $x \in X \setminus \{\bar{x}\}$.*

Proof. Without loss of generality, we may assume that $\lambda = 1$. Let $\epsilon > 0$ be given and choose $\hat{x} \in X$ such that $f(\hat{x}) \le \inf_{x\in X} f(x) + \epsilon$. Now let $X' = \{x \in X : f(x) \le f(\hat{x}) - \epsilon d(\hat{x},x)\}$. Then, X' is non-empty. By lower semicontinuity of f, it follows that X' is closed subset of a complete metric space, and hence, a complete metric space. For each $x \in X'$,

$$S(x) := \{y \in X : y \ne x, f(y) \le f(x) - \epsilon d(x,y)\},$$

for each $x \in X$. Now, we define a set-valued mapping $T : X \to 2^X$ by

$$T(x) = \begin{cases} x, & \text{if } S(x) = \emptyset, \\ S(x), & \text{if } S(x) \ne \emptyset. \end{cases} \tag{7.1}$$

Then, T is a set-valued mapping from X' to itself with non-empty values. Indeed, $T(x) = x \in X'$ if $S(x) = \emptyset$. Since $T(x) = S(x)$, if not, we have for all $y \in T(x)$,

$$\begin{aligned} \epsilon d(\hat{x},y) &\le \epsilon d(\hat{x},x) + \epsilon(x,y) \le f(\hat{x}) - f(x) + f(x) - f(y) \\ &\le f(\hat{x}) - f(y), \end{aligned}$$

and hence, $y \in X'$. Moreover, for all $x \in X$ and $y \in T(x)$, we have

$$\frac{1}{\epsilon}f(y) + d(x,y) \le \frac{1}{\epsilon}f(x).$$

Now, by Theorem 7.3, T has a fixed point $\bar{x} \in X'$. Consequently, $S(\bar{x}) = \emptyset$, that is, we have

$$f(x) > f(\bar{x}) - \epsilon d(\bar{x}, x)$$

for all $x \in X$ with $x \neq \bar{x}$. Since $\bar{x} \in X'$, we have

$$f(\bar{x}) \le f(\hat{x}) - \epsilon(\hat{x}, \bar{x}) \le f(\hat{x}).$$

Further, we have

$$\epsilon d(\hat{x}, \bar{x}) \le f(\hat{x}) - f(\bar{x}) \le f(\hat{x}) - \inf_{x\in X} f(x) \le \epsilon,$$

and hence, we have $d(\hat{x}, \bar{x}) \le 1$. This completes the proof. □

Now as an application of Theorem 7.5, we have the following Takahashi minimization theorem established by Takahashi in [67, 68].

Theorem 7.6 (Takahashi minimization theorem). *Let (X, d) a complete metric space and $f : X \to \mathbb{R} \cup \{\infty\}$ be a proper, bounded below and lower semicontinuous functional. Suppose that, for each $\hat{x} \in X$ with $\inf_{x\in X} f(x) < f(\hat{x})$, there exists $z \in X$ such that $z \neq \hat{x}$ and $f(z) + d(\hat{x}, z) \le f(\hat{x})$. Then, there exists $\bar{x} \in X$ such that $f(\bar{x}) = \inf_{x\in X} f(x)$.*

Proof. We prove the conclusion by using the Strong form of Ekeland's variational principle as a tool. By Theorem 7.5 above, for any given $\epsilon > 0$, there exists $\bar{x} \in X$ such that $f(\bar{x}) < f(x) + \epsilon d(x, \bar{x})$ for all $x \in X$ with $x \neq \bar{x}$. We now claim that $f(\bar{x}) = \inf_{x\in X} f(x)$ as follows.

Assume to the contrary that there exists $w \in X$ such that $f(w) > \inf_{x\in X} f(x)$. Now by the hypothesis, there exists $z \in X$ such that $z \neq w$ and

$$f(z) + d(w, z) \le f(w),$$

which contradicts inequality above. Hence, we have some $\bar{x} \in X$ such that $f(\bar{x}) = \inf_{x\in X} f(x)$. This completes the proof. □

In what follows, we also give the following theorem, which characterizes the completeness of the underlying metric space, which is as an application of Takahashi minimization theorem due to Takahashi [67].

Theorem 7.7. *A metric space (X, d) is complete if for every uniformly continuous function $f : X \to \mathbb{R} \cup \{+\infty\}$ and every $\hat{x} \in X$ with $\inf_{x\in X} f < f(\hat{x})$, there exists $z \in X$ such that $z \neq \hat{x}$ and*

$$f(z) + d(\hat{x}, z) \le f(\hat{x}),$$

then, there exists $\bar{x} \in X$ *such that* $f(\bar{x}) = \inf_{x\in X} f(x)$.

Proof. Let $\{x_n\}$ be a Cauchy sequence in X. Consider the function $f : X \to \mathbb{R} \cup \{+\infty\}$ defined by $f(x) := \lim_{n\to\infty} d(x_n, x)$ for each $x \in X$. The f is uniformly continuous and $\inf_{x\in X} f(x) = 0$. Let $f(\hat{x}) = 0$. Then, there exists an $x_m \in X$ such that

$$x_m \ne \hat{x}, \quad f(x_m) < \frac{1}{3} f(\hat{x}) \quad \text{and} \quad d(x_m, \hat{x}) - f(\hat{x}) < f(\hat{x}).$$

Thus, we have

$$3f(x_m) + d(x_m, \hat{x}) < f(\hat{x}) + 2f(\hat{x}) = 3f(\hat{x}).$$

Therefore, there exists an $\bar{x} \in X$ such that $f(\bar{x}) = \inf_{x\in X} f(x) = 0$, and so, $0 = f(\bar{x}) = \lim_{n\to\infty} d(x_n, \bar{x})$. Thus, $\{x_n\}$ converges to $\bar{x}$, and hence, X is complete. This completes the proof. □

Here we like to refer interested readers to related study for the characterization of complete metrics given by Cobzaş [16], Jachymski et al. [28], Park [49], Park and Rhoades [51] and related references therein and related comprehensive review on the discussion of various circumstances in which fixed point results imply completeness for metric spaces by involving Ekeland variational principle and of its equivalent, Caristi fixed-point theorem; and other fixed point results having this property in metric spaces, including quasi-metric spaces and in partial metric spaces; and in addition for the discussion on topology and order and on fixed points in ordered structures and their completeness properties.

Now by using Cantor intersection theorem in complete metric spaces as a tool, we have following so-called extended version for Ekeland's variational principle, which is equivalent to the Caristi–Kirk-type fixed-point theorem [13], as shown below.

Theorem 7.8 (Extended Ekeland's variational principle). *Let K be a non-empty closed subset of a complete metric space (X, d) and $F : K \times K \to \mathbb{R}$ be a bifunction mapping. Assume that $\epsilon > 0$ and the following assumptions are satisfied:*
(i) *For all $x \in K$, the set $L := \{y \in K : F(x, y) + \epsilon d(x, y) \le 0\}$ is closed;*
(ii) *$F(x, x) = 0$ for all $x \in K$;*
(iii) *$F(x, y) \le F(x, z) + F(z, y)$ for all $x, y, z \in K$.*

Now if $\inf_{y\in K} F(x_0, y) > -\infty$ for some $x_0 \in K$, then there exists $\bar{x} \in K$ such that
(a) *$F(x_0, \bar{x}) + \epsilon d(x_0, \bar{x}) \le 0$; and*
(b) *$F(\bar{x}.x) + \epsilon d(\bar{x}, x) > 0$ for all $x \in K$ with $x \ne \bar{x}$.*

Proof. For the sake of convenience, we set $d_\epsilon(u, v) := \epsilon d(u, v)$. Then d_ϵ is equivalent to d and (X, d_ϵ) is complete. For each $x \in X$, we define a set

$$S(x) := \{y \in K : F(x,y) + d_\epsilon(x,y) \le 0\}.$$

By condition (i), $S(x)$ is closed for each $x \in K$. From condition (ii), $x \in S(x)$ for all $x \in K$, thus $S(x)$ is non-empty for each $x \in X$. So, there exists $y \in S(x)$, that is

$$F(x,y) + d_\epsilon(x,y) \le 0,$$

and, also there exists $z \in S(y)$, such that

$$F(y,z) + d_\epsilon(z,y) \le 0.$$

By adding inequalities above together, and using condition (iii), we obtain

$$0 \ge F(x,y) + d_\epsilon(x,y) + F(y,z) + d_\epsilon(z,y) \ge F(x,z) + d_\epsilon(x,z).$$

Therefore, $z \in S(x)$, which implies that $S(y) \subset S(x)$. Now, we show that there is a sequence $\{x_n\}$ of K such that for each $n \in \mathbb{N}$,

$$x_{n+1} \in S(x_n), F(x_n, x_{n+1}) < \inf_{z \in S(x_n)} F(x_n, z) + \frac{1}{n+1}.$$

Let $v(x_0) := \inf_{z \in S(x_0)} F(x_0, z) > -\infty$, and construct a sequence in the following manner:

For x_1, let $x_1 \in S(x_0)$ such that $F(x_0, x_1) < v(x_0) + \frac{1}{1}$.

Since $x_1 \in S(x_0)$, we have $S(x_1) \subset S(x_0)$. By condition (iii), we have

$$v(x_1) = \inf_{z \in S(x_1)} F(x_1, z) \ge \inf_{z \in S(x_1)} F(x_0, z) - F(x_0, x_1) \ge v(x_0) - F(x_0, x_1) > -\infty.$$

Then, there exists $x_2 \in S(x_1)$ such that

$$F(x_1, x_2) \le v(x_1) + \frac{1}{2}.$$

Continuing in this way, we obtain a sequence $\{x_n\}$ such that for each $n \in \mathbb{N}$,

$$x_{n+1} \in S(x_n), F(x_n, x_{n+1}) < v(x_n) + \frac{1}{n+1},$$

and

$$v(x_{n+1}) \ge v(x_n) - F(x_n, x_{n+1}),$$

which implies that

$$-v(x_n) \le -F(x_n, x_{n+1}) + \frac{1}{n+1} \le v(x_{n+1}) - v(x_n) + \frac{1}{n+1}.$$

Thus, we have that for $n \in \mathbb{N}$,

$$v(x_{n+1}) + \frac{1}{n+1} \geq 0. \tag{7.2}$$

Now, if $z_1, z_2 \in S(x_n)$, then

$$d(z_1, z_2) \leq d(x_n, z_1) + d(x_n, z_2) \leq -F(x_n, z_1) - F(x_n, z_2) \leq -2v(x_n).$$

Thus it implies that the diameter of $S(x_n)$, denoted by $\operatorname{diam}(S(x_n))$, we have $\operatorname{diam}(S(x_n)) \leq -2v(x_n)$. Thus, by the inequality (2) above, we have $\lim_{n\to\infty} \operatorname{diam}(S(x_n)) = 0$.

By the fact that $\{S(x_n)\}$ is a family of closed sets such that $S(x_{n+1}) \subset S(x_n)$ for all $n \in \mathbb{N}$, and $\lim_{n\to\infty} \operatorname{diam}(S(x_n)) = 0$, now by Cantor intersection theorem in complete metric spaces, there exists exactly one point $\bar{x} \in X$ such that $\bigcap_{n=1}^{\infty} S(x_n) = \{\bar{x}\}$.

This implies that $\bar{x} \in S(x_0)$, that is, we have

$$F(x_0, \bar{x}) + d_\epsilon(x_0, \bar{x}) \leq 0,$$

which shows that the conclusion (a) holds.

Secondly, $\bar{x} \in \bigcap_{n=1}^{\infty} S(x_n)$ and, since $S(\bar{x}) \subset S(x_n)$ for all $n \in \mathbb{N}$, we thus have $S(\bar{x}) = \{\bar{x}\}$. It then follows that any $x \notin S(\bar{x})$ whenever $x \neq \bar{x}$, implying that

$$F(\bar{x}, x) + d_\epsilon(\bar{x}, x) > 0,$$

that is, the conclusion (b) holds. This completes the proof. □

As mentioned above, the following result shows the equivalences among extended Ekelend's variational principle, extended Takahashi's minimization theorem, Caristi–Kirk fixed-point theorem for set-valued mappings, and Oettli–Théra theorem in complete metric spaces, which is Theorem 3.13 of Ansatri [3] and Theorem 5.16 of Ansatri and Sahu [4], thus we omit its proof here by saving the space.

Theorem 7.9. *Let K be a non-empty closed subset of a complete metric space (X, d) and $F : K \times K \to \mathbb{R}$ be a (bifunction) mapping such that it is lower semicontinuous in the second argument and the following conditions hold:*

(i) *$F(x, x) = 0$ for all $x \in X$; and*

(ii) *$F(x, y) \leq F(x, z) + F(z, y)$ for all $x, y, z \in X$.*

Assume that there exists $\hat{x} \in X$ such that $\inf_{x\in X} F(\hat{x}, x) > -\infty$. Let

$$\hat{S} := \{x \in X : F(\hat{x}, x) + d(\hat{x}, x) \leq 0\}. \tag{7.3}$$

Then (by (i) it follows that $\hat{x} \in \hat{S} \neq \emptyset$), the following statements are equivalent:

(a) *(Extended Ekeland's variational principle):*
There exists $\bar{x} \in \hat{S}$ such that

$$F(\bar{x}, x) + d(\bar{x}, x) > 0, \quad \textit{for all } x \neq \bar{x}. \tag{7.4}$$

(b) *(Extended Takahashi's minimization theorem):*
Assume that

$$\begin{cases} \text{for every } \hat{x} \in \hat{S} \text{ with } \inf_{x \in X} F(\hat{x}, x) < 0 \text{ there exist} \\ x \in X \text{ such that } F(\hat{x}, x) + d(\hat{x}, x) \leq 0 \text{ for all } x \neq \hat{x} \end{cases} \tag{7.5}$$

Then, there exists $\bar{x} \in \hat{S}$ such that $F(\bar{x}, x) \geq 0$ for all $x \in X$.

(c) *(Caristi–Kirk fixed-point theorem):*
Let $T : X \to 2^X$ be a set-valued mapping such that

$$\begin{cases} \text{for every } \hat{x} \in S \text{ there exists} \\ x \in T(\hat{x}) \text{ satisfying } F(\hat{x}, x) + d(\hat{x}, x) \leq 0. \end{cases} \tag{7.6}$$

Then, there exists $\bar{x} \in S$ such that $\bar{x} \in T(\bar{x})$.

(d) *(Oettli–Théra theorem):*
Let $D \subset X$ have the property that

$$\begin{cases} \text{for every } \hat{x} \in \hat{S} \setminus D \text{ there exists} \\ x \in X \text{ such that } F(\hat{x}, x) + d(\hat{x}, x) \leq 0, \text{ for all } x \neq \hat{x}, \end{cases} \tag{7.7}$$

Then, there exists $\bar{x} \in \hat{S} \cap D$.

Theorem 7.9 shows that Caristi-type fixed-point theorem would be a fundamental tool for non-linear analysis and optimization. Here, before ending this section, we will show that Ekeland variational principle holds in locally complete convex spaces, as discussed below. But first, we need recall some facts and definition used for locally complete convex spaces.

We know that every (sequentially) complete locally convex space is locally complete and every (sequentially) closed set is locally closed, but none of the converses are true, for further details, we refer to Horváth [27], Pérez Carreras and Bonet [54], and Qiu [55–60] for discussion in details. By using the notion of locally closed sets, we also introduce the following class of functions, which is wider than the class of (sequentially) lower semicontinuous functions.

Definition 7.11. Let X be a locally convex space and $f : X \to (-\infty, +\infty]$ be a proper function (i. e., the set $domf := \{x \in X : f(x) \neq +\infty\} \neq \emptyset$). Then f is said to be locally lower semicontinuous if for each $r \in \mathbb{R}$, the set $\{x \in X : f(x) \leq r\}$ is locally closed in X.

We now recall the following result which is Corollary 3.1 of Qiu [57], and stated it as Lemma 7.1 below.

Lemma 7.1. *Let X be a locally complete convex space, $\{p_\lambda\}_{\lambda \in I}$ be a family of seminorms defining the topology on X and $\{\alpha_\lambda\}_{\lambda \in I}$ be a family of positive real numbers. Let $f : X \to (-\infty, +\infty]$ be a locally lower semicontinuous, bounded from below, proper function and* *let* $\{a_\lambda\}_{\lambda\in I}$ *be a family of positive real numbers. Then we have that for any given* $x_0 \in domf$, *there exists* $z \in X$ *such that for all* $\lambda \in I$,

(i) $f(z) + a_\lambda p_\lambda(z - x_0) \le f(x_0)$, *and*
(ii) *for any* $x \ne z$, *there exists* $\mu \in I$ *such that* $f(z) < f(x) + a_\mu p_\mu(x - z)$.

Proof. It is Theorem 1.1 of Qiu [56] (which is indeed Corollary 3.1 of Qiu [57]). The proof is complete. □

Now, as an application of Lemma 7.1, we first have the following Caristi fixed-point theorem in locally complete convex spaces for a single-valued mapping.

Theorem 7.10. *Let X be a locally complete convex space,* $\{p_\lambda\}_{\lambda\in I}$ *be a family of seminorms defining the topology on X and* $\{a_\lambda\}_{\lambda\in I}$ *be a family of positive real numbers. Let* $f : X \to (-\infty, +\infty]$ *be a locally lower semicontinuous, bounded from below, proper function and let* $\{a_\lambda\}_{\lambda\in I}$ *be a family of positive real numbers. Assume* $T : X \to X$ *is a single-valued mapping with non-empty values, and for each* $x \in X$ *such that for all* $\lambda \in I$, *we have*

$$a_\lambda p_\lambda(x - Tx) + f(Tx) \le f(x),$$

then, there exists $z \in domf \subset X$ *such that* $T(z) = z$, *i. e., T has a fixed point* $z \in X$.

Proof. Take any $x_0 \in domf$, then by Lemma 7.1, we know that there exists $z \in X$ such that for any given $\lambda \in I$, we have

$$f(z) + a_\lambda p_\lambda(z - x_0) \le f(x_0),$$

and for any $x \in X$ with $x \ne z$,

$$f(z) < f(x) + \sup_{\lambda\in I} a_\lambda p_\lambda(x - z).$$

Thus $z \in domf$. We now prove that z is a fixed point of T. If not, $T(z) \ne z$, then by inequality above, we have

$$f(z) < f(T(z)) + \sup_{\lambda\in I} a_{\lambda\in I} p_\lambda(T(z) - z).$$

Thus there exists $\mu \in I$ such that $f(z) < f(T(z)) + a_\mu p_\mu(T(z) - z)$. Now by the assumption for the mapping T, we have

$$f(z) < f(T(z)) + a_\mu p_\mu(T(z) - z) \le f(z),$$

this is a contradiction, thus we must have $T(z) = z$. This completes the proof. □

Now we have the following general Caristi-type fixed-point theorem, which shows that the conclusion of Theorem 7.10 is true for mapping T is a set-valued mapping in locally complete convex spaces, but with weaker conditions on the mapping T by comparing with Theorem 5.A of Qiu [56] (which is Corollary 3.3 of Qiu [57], and see also Theorem 4.1, Theorem 4.2 and Theorem 4.3 of Qiu [55]).

Theorem 7.11. *Let X be a locally complete convex space, $\{p_\lambda\}_{\lambda\in I}$ a family of seminorms defining the topology on X and $\{\alpha_\lambda\}_{\lambda\in I}$ be a family of positive real numbers. Let $f : X \to (-\infty, +\infty]$ be a locally lower semicontinuous, bounded from below, proper function; and let $\{\alpha_\lambda\}_{\lambda\in I}$ be a family of positive real numbers. Assume $T : X \to 2^X$ is a set-valued mapping with non-empty values, and such that for each given $\lambda \in I$, any $x \in X$, there exists $y \in T(x)$, we have*

$$\alpha_\lambda p_\lambda(x - y) + f(y) \leq f(x),$$

then, there exists $z \in domf \subset X$ such that $z \in T(z)$, i. e., T has a fixed point $z \in X$.

Proof. For each $x \in X$, we define a single-valued mapping $T_1 : X \to X$ by $T_1(x) := \{y\}$, where y is an element in $T(x)$ such that for all $\lambda \in I$, the following inequality holds:

$$\alpha_\lambda p_\lambda(x - y) + f(y) \leq f(x).$$

By the assumption for the mapping T, we have $T_1(x) \neq \emptyset$, well-defined, and $T_1(x) \subset T(x)$ for all $x \in X$. In addition, for each $x \in X$, such that for all $\lambda \in I$,

$$\alpha_\lambda p_\lambda(x - T_1 x) + f(T_1 x) \leq f(x).$$

Now applying Theorem 7.10 to the mapping T_1, it implies that T_1 has one fixed point $z \in X$ such that $z = T_1(z) \subset T(z)$, which means z is a fixed point of T. This completes the proof. □

Remark 7.1. We first notes that Theorem 7.11 shows that: 1) Theorem 3.5 and Theorem 3.6 of Ansatri [3] hold in locally complete convex spaces; 2) the general Caristi-type fixed-point theorem holds in locally complete convex spaces; 3) Theorem 7.11 shows the corresponding results for Caristi-type fixed-point theorem hold by weakening the condition on the set-valued mappings established by Qiu [55–58] and references wherein; and 4) in particular, Theorem 7.11 improves Theorem 5.A of Qiu [56] (which is Corollary 3.3 of Qiu [57]).

7.4 Fixed-point theorems of non-self $\mathfrak{F}$ contraction and non-expansive set-valued mappings

The goal of this section is to establish fixed-point theorems for non-self contraction and non-expansive set-valued mappings under the general locally complete convex spaces, using Caristi fixed-point theorem established by Caristi [12] and its generalization given in last section as a tool. In this section, we focus on the study of general fixed-point theorems for non-self contraction and non-expansive set-valued mappings under the framework of locally complete convex spaces, which indeed unify or improve corresponding results in the existing literature.

For the convenience of our discussion, for a given space E, we denote by $K(E)$ the family of all non-empty compact subsets E. For a given Hausdorff locally convex space, awe denote by $\mathfrak{F}$ the family of all continuous seminorms generating the topology of E.

For each $p \in \mathfrak{F}$ and $A, B \in K(E)$, we recall that the metric $\delta(A,B)$ between subsets A and B by

$$\delta(A,B) := \sup\{p(a-b) : a \in A, b \in B\},$$

and the Hausdorff metric $d_{Hp}(A,B)$ by

$$d_{Hp}(A,B) := \max\Big\{\sup_{a\in A}\inf_{b\in B} p(a-b), \sup_{b\in B}\inf_{a\in A} p(a-b)\Big\}.$$

It is clear that when p is only a seminorm in E (thus E is a normed space), d_{Hp} is a traditional Hausdorff metric on $K(E)$ (e. g., see Ko and Tsai [37]). Secondly, by the definition of Hausdorff metric d_{Hp} for each $p \in \mathfrak{F}$, we know that if $A, B \in K(E)$, then each $a \in A$, there exists $b \in B$ such that $p(a-b) \leq d_{Hp}(A,B)$ (e. g., by the definition of Hausdorff metric and the compactness of subsets, as the same idea used by Nadler [44]).

Let E denote a Hausdorff locally convex topological vector space, and $\mathfrak{F}$ to denote the family of continuous seminorms generating the topology of E and K being a non-empty subset of E. We recall that a mapping $T : K \to K(E)$ is said to be a $\mathfrak{F}$ contraction set-valued mapping if for each $p \in \mathfrak{F}$, there exists a constant $k_p \in (0,1)$ such that $d_{Hp}(T(x), T(y)) \leq k_p p(x-y)$.

We note that when $k_p = 1$ for all $p \in \mathfrak{F}$, T is a non-expansive set-valued mapping if for any $x, y \in K$, we have

$$d_{Hp}(T(x), T(y)) \leq p(x-y).$$

By Chen and Singh [14], we recall that for a given locally convex space E, which is said to satisfy the *P-Opial condition* if for each $x \in E$ and every net (x_α) converging weakly to x, then for each $P \in \mathfrak{F}$, we have the following inequality holding for any $y \neq x$,

$$\liminf P(x_\alpha - y) > \liminf P(x_\alpha - x).$$

Here, we like to remark that the notion of P-Opial condition (denoted by the upper case letter P), originally introduced by Chen and Singh [14] in 1992, is given for the locally convex spaces E.

Let K be a non-empty bounded closed convex subset of a normed space $(E, \|\cdot\|)$. We say that K has the fixed point property (FPP) for non-expansive mapping T if for every non-expansive mapping $T : K \to K$ (i. e., $\|Tx - Ty\| \leq \|x - y\|$ for all $x, y \in K$), T has a fixed point $x \in K$ such that $Tx = x$. The space E is said to have fixed point property (FPP) if any non-empty bounded closed convex subset of E has the fixed point property for non-expansive mapping; the space E has the weak fixed point property (WFPP) if any weakly compact convex subset of E has the fixed point property for non-expansive mapping. For a reflexive Banach space, both properties are obviously the same.

The famous question whether a Banach space has the fixed point property (WFPP) had remained open for a long time. It is now well known that not all non-expansive mappings have the FPP or WFPP, and in order to obtain positive results in the problem for the existence of fixed points for non-expansive mappings, it is necessary to impose some restrictions either on mapping T or on the underlying space E.

Now it is time for us to establish fixed-point theorems for non-self contraction, and non-self expansive set-valued mappings by using Caristi fixed-point theorem [12] and its generalization above as tools under the framework of locally complete convex spaces.

Throughout this, section, all locally convex spaces are assume to be locally complete unless specified.

Let M be a subset of a given locally convex space E and assume $T : M \to K(E)$ being a set-valued mapping with non-empty values, here $K(E)$ is the family of all non-empty compact subsets of the space E. Then we say that:

(1) T satisfies the boundary condition (α): if for each $x \in M$, any $y \in T(x)$, we have $(x, y] \cap M \neq \emptyset$, where, $(x, y] := \{(1 - \lambda)x + \lambda y : \lambda \in (0, 1]\}$ (and also, we define $[x, y] := \{(1 - \lambda)x + \lambda y : \lambda \in [0, 1]\}$); and
(2) T is called weakly inward if for each $x \in M$, $T(x) \subset \overline{I_M(x)}$, where, $I_M(x) := \{z \in E : x = x + \lambda(y - x), y \in M,$ and $\lambda \in [1, +\infty)\}$.

Now, we recall the following result, which is the famous Caristi fixed-point theorem in the setting of Hausdorff locally complete convex spaces, which is actually a special case of Theorem 7.10 above in locally complete spaces (see also more from Caristi [12], Kirk and Saliga [35], Cammaroto et al. [11], Fang [23], Latif et al. [39], Qiu [55–60], Qiu and Rolewicz [61], Zhong [81], Zhong et al. [83], and also recent work by Feng and Liu [24], Lau and Yao [40] and references wherein for some new results in other direction not covered by this paper).

Lemma 7.2. *Let E be a locally complete convex space and $T : E \to E$ being any arbitrary (single-valued) mapping with non-empty values. Suppose there exists a lower semicontinuous function $f : E \to [0, +\infty)$, such that for each for each given $p \in \mathfrak{F}$, we have for any $x \in E$, $p(x - T(x)) \leq f(x) - f(T(x))$. Then T has a fixed point.*

Proof. This is a special case of Theorem 7.10 above. The proof is complete. □

Here, we like point out that Lemma 7.2 actually is Theorem 2.1 of Latif et al. [39], which is special form of Theorem 3.2 given by Cammaroto et al. [11] with the original result of Theorem 3.1 established by Fang [23]. In addition, the general form of Lemma 7.2

is also given by Theorem 4.1 or Theorem 4.2 of Qiu [55], for which we state its another special form as Lemma 7.3 below, which includes the above Lemma 7.2 as its special case.

We now have the following result, which is a generalization of Lemma 7.2 above, which is also a variant of Lemma 1.2 given by Zhong et al. [83] (indeed, it is a special case of Theorem 4.1 established by Qiu [55]).

Lemma 7.3. *Let E be a locally complete convex space and $f : E \to [0, +\infty)$ being a bounded below lower semicontinuous function and $h : [0, +\infty)$ be a non-negative continuous non-decreasing function such that $\int_0^{+\infty} \frac{dr}{1+h(r)} = +\infty$. Let $T : E \to E$ be a (single-valued) mapping with non-empty values, such that for any given $x_0 \in E$ and $p \in \mathfrak{F}$, for any $x \in E$, we have $\frac{p(x-f(x))}{1+h(p(x_0-x))} \le f(x) - f(T(x))$. Then T has a fixed point.*

Proof. Actually, this is a special case of Theorem 4.1 of Qiu [55]. Indeed, assume $\psi(s) := \frac{1}{1+h(s)}$, and $\phi(s) := s$ for each $s \in [0, +\infty)$ in Theorem 4.1 of Qiu [55]. Then, for each given $p_\lambda \in \mathfrak{F}$, with properties of h and f defined by Lemma 8.6.2 above, it follows that $\phi_\lambda(p_\lambda(x - T(x))) = p_\lambda(x - T(x))$. Let $\psi_\lambda(f(x)) := \psi_\lambda(p_\lambda(x_0 - x))$ for each $x \in E$. Then we have $\psi_\lambda(f(x)) = \psi_\lambda(p_\lambda(x_0 - x)) = \frac{1}{1+h(p_\lambda(x_0-x))}$ (by definitions of ψ and ϕ) for each $x \in X$. It is clear that all hypotheses of Theorem 4.1 of Qiu [55] are satisfied, thus T has a fixed point by Theorem 4.1 of Qiu [55]. This completes the proof. □

Here we also like to point out that Lemma 7.3 is actually Theorem 2.2 of Latif et al. [39] (which is actually a special case of Theorem 4.1 or Theorem 4.2 by Qiu [55]). Moreover, it is a version of Lemma 1.2 given by Zhong et al. [83] under the framework of locally complete convex spaces, which can be proved by using the same idea used for Theorem 2.1 by Zhong [81]. In addition, the corresponding results on Ekeland principle and Caristi fixed point theorem in locally p-convex spaces for $p \in (0, 1]$ have been also established by Qiu and Rolewicz [61] (but we omit their discussion here for saving spaces).

Now by Lemma 7.2, we first have the following fixed-point theorem for non-self $\mathfrak{F}$ contract set-valued mapping in locally complete convex spaces.

Theorem 7.12. *Let M be a non-empty closed subset of locally complete convex space E and $T : M \to K(E)$ be a $\mathfrak{F}$ contraction set-valued mapping satisfying the boundary condition (α). Then T has a fixed point.*

Proof. Let $p \in \mathfrak{F}$ be any given one seminorm in E. For each $x \in M$, as $T(x)$ is compact, we choose $y \in T(x)$ such that

$$p(x - y) = d_p(x, T(x))(:= \inf\{p(x - z) : z \in T(x)\}).$$

We denote by z_{xp} as a farthest point from x in $[x, y] \cap M$, which means z_{xp} satisfies the following equation:

$$p(x - z_{xp}) = \max\{p(x - w) : w \in [x, y] \cap M \neq \emptyset\}.$$

Then it is easy to verify the following equality holding by using the definition of z_{xp} and the seminorm p's triangle inequality property:

$$p(x-y) = p(x-z_{xp}) + p(z_{xp}-y).$$

Since T is a $\mathfrak{F}$ contraction set-valued mapping with non-empty compact values, it follows that for each $p \in \mathfrak{F}$, there exists $k_p \in (0,1)$, we have

$$\begin{aligned} d_p(z_{xp}, T(z_{xp})) &\le p(z_{xp}-y) + d_p(y, T(z_{xp})) \le p(z_{xp}-y) + d_p(T(x), T(z_{xp})) \\ &\le p(z_{xp}-y) + k_p p(x-z_{xp}) = p(x-y) - p(x-z_{xp}) + k_p p(x-z_{xp}) \\ &= d_p(x, T(x)) - (1-k_p)p(x-z_{xp}). \end{aligned}$$

Thus, we have

$$p(x-z_{xp}) \le (1-k_p)^{-1}\{d_p(x,T(x)) - d_p(z_{xp}, T(z_{xp}))\}.$$

Now we define the mapping $f : M \to M$ by $f_p(x) := z_{xp}$ for each $x \in M$; and also define a non-negative function $\phi_p : M \to [0,+\infty)$ by $\phi(x) := (1-k_p)^{-1} d_p(x, T(x))$ for each $x \in M$. Then by the definition of f and ϕ, it follows that

$$p(x - f_p(x)) \le \phi_p(x) - \phi_p(f_p(x)),$$

for each $x \in M$. By a fact that M is a closed subset of a complete space E, so it is complete and hence by Lemma 7.2, f has a fixed point $u \in M$ such that $u = f_p(u)$. Note that $f_p(u) = u = z_{up}$, and by the definition of z_{up}, it is the farthest point from u in $[u,y] \cap M$, which means $p(u - z_{up}) = \max\{p(u-w) : w \in [u,y] \cap M \neq \emptyset\}$, this will help us to prove $d(u, T(u)) = 0$, which implies u is a fixed point of T, as will be shown below in details.

Now by the boundary condition (α) for the $\mathfrak{F}$ mapping T, we have $(u,y] \cap M \neq \emptyset$. Without loss of the generality, there exists a non-zero positive number $\lambda_0 \in (0,1]$ such that $w_0 = (1-\lambda_0)u + \lambda_0 y$, and $w_0 \in (u,y] \cap M \neq \emptyset$. By the definition of z_{up}, we have $p(u-z_{up}) = \max\{p(u-w) : w \in [u,y] \cap M \neq \emptyset\}$, it follows that $p(u-z_{up}) \ge p(u-w_0) = \lambda_0 p(u-y) = \lambda_0 d(u, T(u))$ (as given in the beginning of the proof above). Since $u = z_{up}$, it follows that

$$\lambda_0 d(u, T(u)) = p(u - z_{up}) = p(z_{up} - z_{up}) = 0,$$

which implies that $d(u, T(u)) = 0$ as $\lambda_0 > 0$. This means u is a fixed point of T. The proof is complete. □

We note that Theorem 7.12 extends Theorem 2 of Massa [41] (see also Massa et al. [42]) for non-self $\mathfrak{F}$ contraction set-valued mappings in Hausdorff locally convex spaces.

Now as an application of Lemma 7.2, we have the following fixed point result for non-self $\mathfrak{F}$ contraction set-valued mappings with different boundary conditions.

Theorem 7.13. *Let M be a non-empty closed subset of locally complete convex space E, and $T : M \to K(E)$ be a $\mathfrak{F}$ contraction set-valued mapping such that for each given $p \in \mathfrak{F}$, for any $x \in M$, we have $\{z \in T(x) : p(x-z) = d_p(x, T(x))\} \cap \overline{I_M(x)} \neq \emptyset$ (also called intersection boundary condition). Then, T has a fixed in M.*

Proof. Suppose that T has no fixed point. Then for any given $p \in \mathfrak{F}$, for any $x \in M$, we have $d_p(x, T(x)) > 0$. As T is a $\mathfrak{F}$ contract set-valued mapping, there exists $k_p \in (0,1)$ such that $d_{Hp}(T(x), T(y)) \leq k_p p(x-y)$. Now, choose a constant $q \in (0,1)$ and denoted by $k := k_p < \frac{1-q}{1+q}$, where we know that $k_p \in (0,1)$. In addition, by the assumption for the boundary condition of the $\mathfrak{F}$ mapping F, the following set is not empty, i. e.,

$$\{z \in T(x) : p(x-z) = d_p(x, T(x))\} \cap \overline{I_M(x)} \neq \emptyset.$$

Since $T(x)$ is also non-empty compact for $x \in M$, there exists $z \in T(x) \cap \overline{I_M(x)}$ such that $d_p(x, T(x)) = p(x-z) > 0$. In addition, by the definition of $I_M(x)$ and $z \in T(x) \cap \overline{I_M(x)}$, it follows that we can choose some $t \in (0,1)$ (see also the proof in details from Theorem 1.2 by Caristi [12]) such that

$$t^{-1} d_p((1-t)x + tz, M) < qp(x-z).$$

Now let $w := (1-t) + tz$, then there exists some $y \in M$ such that

$$p(w-y) < qtp(x-z) = qp((1-t)x + tz - x) = qp(w-x).$$

Since $p(y-x) \leq p(y-w) + p(w-x)$, it follows that

$$p(y-x) - p(w-x) \leq p(w-y) < qp(w-x),$$

which implies

$$p(y-x) < (1+q)p(w-x).$$

Therefore, the following inequality holding:

$$(q-1)p(w-x) < \frac{q-1}{q+1} p(x-y).$$

On the other hand, as $w = (1-t) + tz$, where $t \in (0,1)$, it is easy to verify the following equality is true:

$$p(w-x) + p(w-z) = p(x-z).$$

Since $T(x)$ is a $\mathfrak{F}$ contraction set-valued mapping with non-empty compact values, and by the definition of the Hausdorff metric d_{Hp}, we can choose $u \in T(x)$ and $v \in T(y)$ such that $p(w-u) = d_p(w, T(x))$ and

$$p(u-v) \leq d_{Hp}(T(x), T(y)) \leq k_p p(x-y).$$

Now, by putting all above inequalities and the equality (i. e., $p(w-x)+p(w-z)=p(x-z)$) together, we have the following inequality holding:

$$\begin{aligned}
d_p(y,T(y)) &\le p(y-v)\le p(y-w)+p(w-u)+p(u-v)\\
&\le qp(w-x)+d_p(w,T(x))+kp(x-y)<qp(w-x)+p(w-z)+kp(x-y)\\
&= qp(w-x)+p(x-z)-p(w-x)+kp(x-y)\\
&= (q-1)p(w-x)+p(x-z)+kp(x-y)\\
&< \frac{q-1}{q+1}p(x-y)+p(x-z)+kp(x-y)=d_p(x,T(x))-\left(\frac{1-q}{1+q}-k\right)p(x-y).
\end{aligned}$$

Now let $c=\frac{1-q}{1+q}-k$, we have $d_p(y,T(y))<d_p(x,T(x))-cp(x-y)$. Therefore, we have the following inequality:

$$p(x-y)<\frac{d_p(x,T(x))}{c}-\frac{d_p(y,T(y))}{c}.$$

Now we define two mappings $f:M\to M$, and $\phi:M\to[0,+\infty)$ by $f(x):=y$, and $\phi(x):=\frac{d_p(x,T(x))}{c}$ for each $x\in M$. It is clearly we have

$$p(x-f(x))<\phi(x)-\phi(f(x)).$$

Now by Lemma 7.2, f has a fixed point $x_0\in M$, thus $f(x_0)=x_0$. On the other hand, we have

$$0=p(x_0-f(x_0))<\phi(x_0)-\phi(f(x_0))=\phi(x_0)-\phi(x_0)=0$$

this is impossible. Hence, T must have a fixed point. This completes the proof. □

As a special case of Theorem 7.13, we have the following result.

Corollary 7.1. *Let M be a closed subset of locally complete convex space E and let $T:M\to K(E)$ be a weakly inward $\mathfrak{F}$ contraction mapping, i. e., $T(x)\subset\overline{I_M(x)}$ for each $x\in M$. Then T has a fixed point.*

Remark 7.2. We like to point out that our Theorem 7.13 extends Theorem 3.3 given by Zhang [80]. Secondly, it is worth to mention that the intersection condition of Theorem 7.13 can not be replaced the condition $T(x)\cap\overline{I_M(x)}\neq\emptyset$ for each $x\in M$ even in the setting of Banach spaces, as shown by Zhang [80], Xu [70] and references wherein.

Next, by applying Lemma 7.3 above and combining a contractive condition basically due to Zhong [81], we have the following fixed-point theorem for non-self $\mathfrak{F}$ contract weekly inward set-valued mappings in the setting of locally convex topological vector spaces.

Here, we assume that the mapping $h:[0,+\infty)\to[0,+\infty)$ is a continuous nondecreasing function satisfying: $\int_0^{+\infty}\frac{dr}{1+h(r)}=+\infty$. Now we have the following fixed-point theorem for non-self contract set-valued mappings for spaces satisfying p-Opial condition.

Theorem 7.14. *Let M be a closed subset of a locally complete convex space E, and $T : M \to K(E)$ being a weakly inward mapping, for a given $x_0 \in M$ and a given constant number $\sigma \in (0,1]$. For each given $p \in \mathfrak{F}$, the following condition holding:*

$$d_{Hp}(T(x),T(y)) \le \left(1 - \frac{\sigma}{1+h(p(x_0-x))}\right)p(x-y),$$

for any $x,y \in M$. Then T has a fixed point.

Proof. Suppose that T has no fixed point. Then, $d_p(x,T(x)) > 0$ for all $x \in M$. By the assumption the function h is a continuous nondecreasing function satisfying: $\int_0^{+\infty} \frac{dr}{1+h(r)} = +\infty$, we can choose $c \in (0,\sigma)$, where $\sigma \in (0,1]$, and let $q(x) := \frac{\sigma-c}{2(1+h(p(x_0-x)))}$ such that $q(x) < 1$. Since T is with non-empty values, there exists $z \in T(x)$ such that $d_p(x,T(x)) = p(x-z) > 0$. By the assumption, T is weakly inward, we have $T(x) \subset \overline{I_M(x)}$ for $x \in M$. Thus there exist $y \in M$ and $\lambda \ge 1$ such that $z' := x + \lambda(y-x)$ with the following (approximation) inequality property:

$$P(z-z') = p(z-(x+\lambda(y-x))) < q(x)p(x-z).$$

Let $t := \frac{1}{\lambda}$ and $w := (1-t)x+tz$. Note that $w-y = (1-t)x+tz-y = t(z-(x+\frac{1}{t}(y-x))) = t(z-z')$, it follows that

$$p(w-y) = t(p(z-(x+\lambda(y-x)))) = tp(z-z').$$

Thus, we have: $p(w-y) = tp(z-z') < tq(x)p(x-z)$, $p(w-x) = tp(x-z)$, and $p(w-z) = (1-t)p(x-z)$.

Secondly, as $w = (1-t)x+tz$, it is easy to verify the equality: $p(w-x)+p(w-z) = p(x-z)$.

As $p(y-x) \le p(y-w) + p(w-x)$, it implies that $p(y-x) - p(w-x) \le p(w-y)$, thus we have

$$p(x-y) < p(w-x) + tq(x)p(x-z) = (1+q(x))p(w-x).$$

Therefore, we obtain the following inequality:

$$(q(x)-1)p(w-x) < \frac{q(x)-1}{q(x)+1}p(x-y),$$

where $q(x) \in (0,1)$. Since T is with compact values, and by the definition of Hausdorf metric d_{Hp}, we can choose $u \in T(x)$ and $v \in T(y)$ such that $p(w-u) = d_p(w,T(x))$ and

$$p(u-v) \le d_{Hp}(T(x),T(y)).$$

Now, putting all above inequalities and the equality (i. e., $p(w-x)+p(w-z)=p(x-z)$) together, and also consider the assumption on the mapping T, we have the following inequalities:

$$\begin{aligned} d_p(y,T(y)) &\le p(y-v)\le p(y-w)+p(w-u)+p(u-v)\\ &< q(x)p(w-x)+p(w-z)+d_{Hp}(T(x),T(y))\\ &= q(x)p(w-x)-p(w-x)+p(x-z)+d_{Hp}(T(x),T(y))\\ &= (q(x)-1)p(w-x)+p(x-z)+d_{Hp}(T(x),T(y))\\ &< \left(\frac{q(x)-1}{q(x)+1}+1\right)p(x-y)+p(x-z)-\frac{\sigma}{1+h(p(x_0-x))p(x_0-x))}p(x-y)\\ &< 2\left(\frac{\sigma-c}{2(1+h(p(x_0-x)))}\right)p(x-y)+p(x-z)-\frac{\sigma}{1+h(p(x_0-x))}p(x-y)\\ &< d_p(x,T(x))-\frac{c}{1+h(p(x_0-x))}p(x-y). \end{aligned}$$

Thus it follows

$$\frac{p(x-y)}{1+h(p(x_0-x))}<\frac{d_p(x,T(x))}{c}-\frac{d_p(y,T(y))}{c}.$$

Now, we define two mappings for $f: M\to M$ and $\phi: M\to[0,+\infty)$ by $f(x)=y$ and $\phi(x)=\frac{d_p(x,T(x))}{c}$ for each $x\in M$. Note that ϕ is continuous and non-negative, and

$$\frac{p(x-f(x))}{1+h(p(x_0-x))}<\phi(x)-\phi(f(x)).$$

Now applying Lemma 7.3, f has a fixed point. But, above inequality shows that $0<0$, this is impossible. This contradiction shows that T must have a fixed point. The proof is complete. □

Remark 7.3. We note that Theorem 7.14 extends Theorem 2.5 by Zhong et al. [83] to Hausdorff locally convex spaces. Now, we have the following fixed-point theorem for non-self expansive set-valued mappings with boundary condition (α) in locally complete convex spaces, which satisfy the P-Opial condition introduced by Chen and Singh [14] in 1992.

Theorem 7.15. *Let M be a non-empty weakly compact convex subset of a locally complete convex space E satisfying P-Opial condition. Let $T: M\to K(E)$ be a non-self nonexpansive set-valued mapping with the boundary condition (α) (and thus $T(M)$ is bounded). Then mapping $(I-T)$ is demiclosed, i. e., the set $(I-T)(M)$ is closed in $E_w\times E$, where E_w is E with its weak topology and I is the identify mapping; and T has at least one fixed point.*

Proof. Since E satisfies P-Opial condition, we have that if $x_n\to x$ weekly, and $y\ne x$, then $\limsup\|x_n-x\|<\limsup\|x_n-y\|$. By Theorem 1 of Chen and Singh [14], we have that the set $(I-T)(M)$ is closed in $E_w\times E$, where E_w is E with its weak topology and I is the identify mapping. This fact will help us to prove T has a fixed point. In order to do so, we now show that there is a sequence $\{x_n\} \subset M$ and a sequence $\{y_n\} \subset E$ with $y_n \in T(x_n)$ such that $\lim_{n\to\infty}(x_n - y_n) = 0$, and then $0 \in (I - T)(M)$.

Now let q be a given point in M. For each $n \in \mathbb{N}$, we define a set-valued mapping $T_n : M \to K(E)$ by

$$T_n(x) = \left(1 - \frac{1}{n}\right)Tx + \frac{1}{n}q,$$

for each $x \in M$. Then it is clear that $T_n(x) \subset E$ and T_n satisfies the boundary condition (a), since so does T. As each T_n is a $\mathfrak{F}$ contraction set-valued mapping for $n \in \mathbb{N}$, by Theorem 7.12, there exists $x_n \in M$ such that $x_n \in T_n(x_n)$. This means that there exists $y_n \in T(x_n)$ such that $x_n - y_n = \frac{1}{n}q - \frac{1}{n}y_n$ for $n \in \mathbb{N}$. As $T(M)$ is bounded, therefore $\lim_{n\to\infty}(x_n - y_n) = 0$. Now by Theorem 1 of Chen and Singh [14], we have that the set $(I - T)(M)$ is closed in $E_w \times E$. Then it follows that $0 \in (I - T)K$. This means that there exists $x \in K$ such that $x \in T(x)$, which means T has at least one fixed point in M. This completes the proof. □

Now we have the following fixed point result for non-self non-expansive mappings that are weekly inward in locally convex spaces.

Theorem 7.16. *Let M be a non-empty weakly compact convex subset of a locally complete convex space E, and let $T : M \to K(E)$ being a non-expansive and weakly inward mapping. If $(I - T)$ is demiclosed, then T has a fixed point.*

Proof. The same as above, Leq q be a point in M. For each $n \in \mathbb{N}$, we define a set-valued mapping $T_n : M \to K(X)$ by

$$T_n(x) = \left(1 - \frac{1}{n}\right)Tx + \frac{1}{n}q,$$

for each $x \in M$. Then it is clear that $T_n(x) \subset E$ and T_n is also weekly inward, since so does T. As each T_n is a $\mathfrak{F}$ contraction set-valued mapping for $n \in \mathbb{N}$, by Corollary 7.1, there exists $x_n \in M$ such that $x_n \in T_n(x_n)$. This means that there exists $y_n \in T(x_n)$ such that $x_n = (1 - \frac{1}{n})y_n + \frac{1}{n}q$ for $n \in \mathbb{N}$, which means we have $(x_n - y_n) = \frac{1}{n-1}(q - x_n)$. As M is weakly compact, the sequence $\{q - x_n\}$ is also bounded. In addition, without loss of the generality, assume x_n converges weekly to z in M. Now for any $p \in \mathfrak{F}$, we have

$$\lim_{n\to+\infty} p(x_n - y_n) = \lim_{n\to+\infty} \frac{1}{n-1}p(q - x_n) = 0.$$

By the assumption that $(I - T)$ is demiclosed, it implies that $0 \in (I - T)z$, which means $z \in T(z)$. This completes the proof. □

As an application of Theorem 7.16, we have the following general fixed-point theorem for non-expansive set-valued mappings which are weekly inward in locally complete convex spaces which satisfy P-Opial condition, see more results by Yanagi [71] and related references wherein.

Theorem 7.17. *Let M be a non-empty weakly compact convex subset of a locally complete convex space E, which satisfies P-Opial condition. Let $T : M \to K(X)$ be a non-expansive and weakly inward mapping. Then T has a fixed point.*

Proof. As E satisfies P-Opial condition and T is non-expansive, it implies that the mapping $(I - T)$ is demiclosed by Theorem 7.15 above, or by following the same proof of Lemma 7.3. Then the conclusion follows by Theorem 7.16, and this completes the proof. □

Remark 7.4. We first note that the intersection condition used in Theorem 7.13, and the weekly inward condition used in Theorem 7.16 and Theorem 7.17 could not be replaced by the assumption that $T(x) \cap I_M(x) \neq \emptyset$ for each $x \in M$ as shown by the example provided by Xu [70, p. 705] (see also Zhang [80] and references wherein).

Secondly, we remark that there exist comprehensive references on the study of Leray–Schauder principle for nonexpansive singe- or set-valued mappings, we refer Browder [8–10], Morales [43], Park [46] and related references to interested readers. In particular, see also Park [46–50] for comprehensive references for the new development of fixed point theory and related applications to ordered spaces and associated discussion there.

In addition, we also refer Cobzaş [16], Jachymski et al. [28], Park [49], Park and Rhoades [51] and related references therein to interested readers for the comprehensive review on the discussion of various circumstances in which fixed point results imply completeness for metric spaces involving Ekeland variational principle and of its equivalent, Caristi fixed-point theorem; and other fixed point results having this property in metric spaces, including quasi-metric spaces and in partial metric spaces, for the discussion on topology and order and on fixed points in ordered structures and their completeness properties.

Before ending this paper, we conclude with summary as follows: First, a few of key principles established in nonlinear analysis and optimization are derived as applications of Caristi fixed-point theorem (which are indeed equivalent to each others) in complete metric spaces and locally complete convex spaces, see also discussion given by Almezel et al. [2], Ansatri [3], Ansatri and Sahu [4], Kirk [33], Park [46–50], Penot [53], Yuan [73–78], Zeidler [79] and references wherein for more in details. Second, we establish fixed-point theorems for non-self contraction and non-expansive set-valued mappings under the general locally complete convex spaces by using Caristi fixed-point theorem established by Caristi [12] and its generalization given in this paper as a tool under the framework of locally complete convex spaces, which indeed unify or improve corresponding results in the existing literature. Finally, we know that Caristi fixed-point theorem has been used as a powerful tool by Zhong et al. [82, 83] to study surjective mappings theorems and the existence of fixed points for non-self set-valued contraction mappings, to study optimization problems for set-valued mappings with ordering structures. In particular, Ansari and Sahu [4] give an updated comprehensive discussion on Caristi fixed point and its relation to Ekeland variational principle, and equilibrium problems and extended Ekeland's variational principle, and Yuan [78] for the general study on the new development of fixed point theory in p-vector spaces and its applications, where $p \in (0, 1]$, and a (topological) p-vector space is a topological vector space when $p = 1$.

Bibliography

[1] Agarwal, R. P., Meehan, M. and O'Regan, D., *Fixed Point Theory and Applications*, Cambridge Tracts in Mathematics, vol. 141, Cambridge University Press, Cambridge, 2001.

[2] Almezel, S., Ansari, Q. H. and Khamsi, M. A., *Topics in Fixed Point Theory*, Springer, Cham, 2014.

[3] Ansari, Q. H., Ekeland's variational principle and its extensions with applications, in: Almezel, S., Ansari, Q. H., Khamsi, M. A. (eds.), *Topics in Fixed Point Theory*, pp. 65–100, Springer, Cham, 2014.

[4] Ansari, Q. H. and Sahu, D. R., *Fixed Point Theory and Variational Principles in Metric Spaces*, Cambridge University Press, Cambridge, 2023.

[5] Balachandran, V. K., *Topological Algebras*, North-Holland Mathematics Studies, vol. 185, Elsevier, Amsterdam, 2000.

[6] Bayoumi, A., *Foundations of Complex Analysis in Non Locally Convex Spaces: Function Theory Without Convexity Condition*, North-Holland Mathematics Studies, vol. 193, Elsevier Science B. V., Amsterdam, 2003.

[7] Brézis, H. and Browder, F., *A general principle on ordered sets in nonlinear functional analysis*, Adv. Math. 21 (1976), 355–364.

[8] Browder, F. E., *Non-expansive nonlinear operators in a Banach space*, Proc. Natl. Acad. Sci. USA 54 (1965), 1041–1044.

[9] Browder, F. E., *Convergence of approximants to fixed points of non-expansive nonlinear mappings in Banach spaces*, Arch. Ration. Mech. Anal. 24 (1967), 82–90.

[10] Browder, F. E., *Fixed point theory and nonlinear problems*, Bull. Am. Math. Soc. (N. S.) 9 (1983), 1–39.

[11] Cammaroto, F., Chinni, A. and Sturiale, G., *A remark on Ekeland's principle in locally convex topological vector spaces*, Math. Comput. Model. 30 (1999), 75–79.

[12] Caristi, J., *Fixed-point theorems for mappings satisfying inward conditions*, Trans. Am. Math. Soc. 215 (1976), 241–251.

[13] Caristi, J. and Kirk, W. A., Geometric fixed point theory and inwardness conditions, in: *The Geometry of Metric and Linear Spaces*, Lecture Notes in Math., vol. 490, pp. 74–83, Springer-Verlag, Berlin–New York, 1975.

[14] Chen, Y. K. and Singh, K. L., *Fixed points for non-expansive multivalued mapping and the Opial's condition*, Jñānābha 22 (1992), 107–110.

[15] Cho, Y. J., Survey on metric fixed point theory and applications, in: Ruzhansky, M., Cho, Y. J., Agarwal, P., Area, I. (eds.), *Advances in Real and Complex Analysis with Applications*, Trends in Mathematics, pp. 183–241, Birkhäuser, Singapore, 2017.

[16] Cobzas, S., *Fixed points and completeness in metric and generalized metric spaces*, J. Math. Sci. (N.Y.) 250 (2020), 475–535.

[17] Danes, J., *A geometric theorem useful in nonlinear functional analysis*, Boll. Unione Mat. Ital. 6 (1972), 369–372.

[18] Du, W. S., *A simple proof of Caristi's fixed-point theorem without using Zorn's lemma and transfinite induction*, Thai J. Math. 14 (2016), 259–264.

[19] Du, W. S., *A direct proof of Caristi's fixed-point theorem*, Appl. Math. Sci. 10 (2016), 2289–2294.

[20] Ekeland, I., *Sur les problèmes variationnels*, C. R. Acad. Sci. Paris, Sér. A–B 275 (1972), A1057–A1059.

[21] Ekeland, I., *On the variational principle*, J. Math. Anal. Appl. 47 (1974), 324–353.

[22] Ekeland, I., *On convex minimization problems*, Bull. Am. Math. Soc. 1 (1979), 445–474.

[23] Fang, J., *The variational principle and fixed-point theorems in certain topological spaces*, J. Math. Anal. Appl. 202 (1996), 398–412.

[24] Feng, Y. Q. and Liu, S. Y., *Fixed-point theorems for multi-valued contractive mappings and multi-valued Caristi type mappings*, J. Math. Anal. Appl. 317 (2006), 103–112.

[25] Goebel, K. and Kirk, W. A., *Topics in Metric Fixed Point Theory*, Cambridge Studies in Advanced Mathematics, vol. 28, Cambridge University Press, Cambridge, 1990.

[26] Goebel, K. and Kirk, W. A., *Some problems in metric fixed point theory*, J. Fixed Point Theory Appl. 4 (2008), 13–25.

[27] Horváth, J., *Topological Vector Spaces and Distributions, Vol. 1*, Addison-Wesley, Reading, MA, 1966.

[28] Jachymski, J., Józwik, I. and Terepeta, M., *The Banach fixed-point theorem: Selected topics from its hundred-year history*, Rev. R. Acad. Cienc. Exactas Fís. Nat., Ser. A Mat. 118 (2024), 140.

[29] Jarchow, H., *Locally Convex Spaces*, B. G. Teubner, Stuttgart, 1981.

[30] Kirk, W. A., Caristi's fixed-point theorem and the theory of normal solvability, in: *Fixed Point Theory and Its Applications* (Proc. Sem., Dalhousie Univ., Halifax, N. S., 1975), pp. 109–120, Academic Press, New York–London, 1976.

[31] Kirk, W. A., *Caristi's fixed-point theorem and metric convexity*, Colloq. Math. 36 (1976), 81–86.

[32] Kirk, W. A., *An abstract fixed-point theorem for non-expansive mappings*, Proc. Am. Math. Soc. 82 (1981), 640–642.

[33] Kirk, W. A., *Metric fixed point theory: A brief retrospective*, Fixed Point Theory Appl. 2015 (2015), 215.

[34] Kirk, W. A. and Caristi, J., *Mapping theorems in metric and Banach spaces*, Bull. Acad. Pol. Sci., Sér. Sci. Math. Astron. Phys. 23 (1975), 891–894.

[35] Kirk, W. A. and Saliga, L. M., *The Brézis–Browder order principle and extensions of Caristi's theorem*, Nonlinear Anal. 47 (2001), 2765–2778.

[36] Kirk, W. A. and Shahzad, N., *Fixed Point Theory in Distance Spaces*, Springer, Cham, 2014.

[37] Ko, H. M. and Tsai, Y. H., *Fixed-point theorems for point-to-set mappings in locally convex spaces and a characterization of complete metric spaces*, Bull. Acad. Sin. 6 (1979), 461–470.

[38] Kozlowski, W. M., *A purely metric proof of the Caristi fixed-point theorem*, Bull. Aust. Math. Soc. 95 (2017), 333–337.

[39] Latif, A., Hussain, N. and Kutbi, M. A., *Applications of Caristi's fixed point results*, J. Inequal. Appl. 2012 (2012), 40.

[40] Lau, A. T. M. and Yao, L. J., *Common fixed point properties for a family of set-valued mappings*, J. Math. Anal. Appl. 459 (2018), 203–216.

[41] Massa, S., *Some remarks on Opial spaces*, Boll. Unione Mat. Ital. 6 (1983), 65–70.

[42] Massa, S., Roux, D. and Singh, S. P., *Fixed-point theorems for multifunctions*, J. Pure Appl. Math. 18 (1987), 512–514.

[43] Morales, C. H., *The Leray–Schauder condition for continuous pseudo-contraction mappings*, Proc. Am. Math. Soc. 137 (2009), 1013–1020.

[44] Nadler, S. B., *Multi-valued contraction mappings*, Pac. J. Math. 30 (1969), 475–488.

[45] Oettli, W. and Théra, M., *Equivalents of Ekeland's principle*, Bull. Aust. Math. Soc. 48 (1993), 385–392.

[46] Park, S., *One hundred years of the Brouwer fixed-point theorem*, J. Natl. Acad. Sci., Repub. Korea, Nat. Sci. Ser. 60 (2021), 1–77.

[47] Park, S., *Equivalents of ordered fixed-point theorems of Kirk, Caristi, Nadler, Banach, and others*, Adv. Theory Nonlinear Anal. Appl. 6 (2022), 420–439.
[48] Park, S., *Foundations of ordered fixed point theory*, J. Natl. Acad. Sci., Repub. Korea, Nat. Sci. Ser. 61 (2022), 247–287.
[49] Park, S., *Historical remarks on the Caristi related theorems*, J. Nonlinear Convex Anal. 24 (2023), 2531–2542.
[50] Park, S., *Improving many metric fixed-point theorems*, Lett. Nonlinear Anal. Appl. 2(2) (2024), 35–61, https://lettersinnonlinearanalysis.com/index.php/lnaa
[51] Park, S. and Rhoades, B. E., *Comments on characterizations for metric completeness*, Math. Jpn. 31(1) (1986), 95–97.
[52] Penot, J. P., *fixed-point theorems without convexity*, Bull. Soc. Math. Fr., Mém. 60 (1979), 129–152.
[53] Penot, J. P., *The drop theorem, the petal theorem and Ekeland's variational principle*, Nonlinear Anal. 10(9) (1986), 813–822.
[54] Pérez Carreras, P. and Bonet, J., *Barrelled Locally Convex Spaces*, North-Holland Math. Stud., vol. 131, North-Holland Publishing Co., Amsterdam, 1987.
[55] Qiu, J. H., *Ekeland's variational principle in locally convex spaces and the density of extremal points*, J. Math. Anal. Appl. 360(1) (2009), 317–327.
[56] Qiu, J. H., *The density of extremal points in Ekeland's variational principle*, J. Math. Anal. Appl. 328 (2007), 946–957.
[57] Qiu, J. H., *Local completeness, drop theorem and Ekeland's variational principle*, J. Math. Anal. Appl. 311 (2005), 23–39.
[58] Qiu, J. H., *Ekeland's variational principle in locally complete spaces*, Math. Nachr. 257 (2003), 55–58.
[59] Qiu, J. H., *Local completeness and drop theorem*, J. Math. Anal. Appl. 266 (2002), 288–297.
[60] Qiu, J. H., *Local completeness and dual local quasi-completeness*, Proc. Am. Math. Soc. 129 (2001), 1419–1425.
[61] Qiu, J. H. and Rolewicz, S., *Ekeland's variational principle in locally p-convex spaces and related results*, Stud. Math. 186(3) (2008), 219–235.
[62] Qiu, J. H. and Rolewicz, S., *Local completeness of locally pseudoconvex spaces and Borwein–Preiss variational principle*, Stud. Math. 183 (2007), 99–115.
[63] Rolewicz, S., *On drop property*, Stud. Math. 85(1) (1986), 27–35.
[64] Rolewicz, S., *Metric Linear Spaces*, PWN-Polish Scientific Publishers, Warsaw, Warszawa, 1985.
[65] Ruzhansky, M., Cho, Y. J., Agarwal, P. and Iván, A., *Advances in Real and Complex Analysis with Applications*, Trends Math., Birkhäuser/Springer, Singapore, 2017.
[66] Siegel, J., *New proof of Caristi's fixed-point theorem*, Proc. Am. Math. Soc. 66(1) (1977), 54–56.
[67] Takahashi, W., Existence theorems generalizing fixed-point theorems for multivalued mappings, in: Baillon, J.-B., Théra, M. (eds.), *Fixed Point Theory and Applications*, Pitman Research Notes in Mathematics, vol. 252, pp. 397–406, Longman, Harlow, 1991.
[68] Takahashi, W., *Nonlinear Functional Analysis*, Yokohama Publishers, Yokohama, Japan, 2000.
[69] Wong, C. S., *On a fixed-point theorem of contractive type*, Proc. Am. Math. Soc. 57(2) (1976), 283–285.
[70] Xu, H. K., *Multivalued non-expansive mappings in Banach spaces*, Nonlinear Anal. 43(6) (2001), 693–706.
[71] Yanagi, K., *On some fixed-point theorems for multivalued mappings*, Pac. J. Math. 87(1) (1980), 233–240.
[72] Yuan, G. X., *The study of minimax inequalities and applications to economies and variational inequalities*, Mem. Am. Math. Soc. 132(625) (1998), 1–140.
[73] Yuan, G. X., *KKM Theory and Applications in Nonlinear Analysis*, Monographs and Textbooks in Pure and Applied Mathematics, vol. 218, Marcel Dekker, Inc., New York, 1999.
[74] Yuan, G. X., *Nonlinear analysis in p-vector spaces for single-valued* 1*-set contractive mappings*, Fixed Point Theory Algorithms Sci. Eng. 2022 (2022), 26, https://doi.org/10.1186/s13663-022-00735-6.
[75] Yuan, G. X., *Nonlinear analysis by applying best approximation method in p-vector spaces*, Fixed Point Theory Algorithms Sci. Eng. 2022 (2022), 20, https://doi.org/10.1186/s13663-022-00730-x.

[76] Yuan, G. X., *Fixed-point theorems and applications in p-vector spaces*, Fixed Point Theory Algorithms Sci. Eng. 2023 (2023), 10, https://doi.org/10.1186/s13663-023-00747-w.
[77] Yuan, G. X., *Corrigendum: Fixed-point theorem and related nonlinear analysis by the best approximation method in p-vector spaces*, Numer. Funct. Anal. Optim. 44(10) (2023), 1094–1096.
[78] Yuan, G. X., *Fixed Point Theory in p-Vector Spaces*, World Scientific, Singapore and London, 2025.
[79] Zeidler, E., *Nonlinear Functional Analysis and Its Applications, Vol. I: Fixed-Point Theorems*, Springer Verlag, New York, 1986.
[80] Zhang, S., *Star-shaped sets and fixed points of multivalued mappings*, Math. Jpn. 36(2) (1991), 327–334.
[81] Zhong, C. K., *A generalization of Ekeland's variational principle and application to the study of the relation between the weak P. S. condition and coercivity*, Nonlinear Anal. 29(12) (1997), 1421–1431.
[82] Zhong, C. K. and Zhao, P. H., *Locally Ekeland's variational principle and some surjective mapping theorems*, Chin. Ann. Math., Ser. B 19(3) (1998), 273–280.
[83] Zhong, C. K., Zhu, J. and Zhao, P. H., *An extension of multi-valued contraction mappings and fixed points*, Proc. Am. Math. Soc. 128(8) (1999), 2439–2444.

Santhosh George

8 On the order of convergence of a Traub-type method

Abstract: Traub-like methods with convergence order five and six for solving equations in Banach space are studied. Techniques are developed to prove the convergence order using assumptions up to the second order derivative of the operator involved. Our techniques can be used to obtain the convergence of other similar higher order methods. Numerical examples and basin of attractions are given in this study.

Keywords: Traub-like method, Banach space, convergence order, Taylor expansion, iterative methods

MSC 2020: 49M15, 65J15, 41A25

8.1 Introduction

The problem of determining a solution x^* of equation

$$F(x) = 0, \tag{8.1}$$

where $F : \Omega \subset X \longrightarrow Y$ is an operator acting between Banach spaces X and Y with $\Omega \neq \emptyset$ is an interesting tasks in computational mathematics. Since the closed form solution of (8.1) is possible only in rare cases, iterative methods are considered for solving (8.1). Throughout this paper, $B(x_0, \rho) = \{x \in X : \|x - x_0\| < \rho\}$ and $B[x_0, \rho] = \{x \in X : \|x - x_0\| \leq \rho\}$ for some $\rho > 0$. Convergence order is an important issue, when one deal with iterative methods.

Let sequence $\{x_n\}$ converge to x^* and if there exist a real number p and non-zero constant C such that

$$\|x_{n+1} - x^*\| \leq C\|x_n - x^*\|^p.$$

Then p is called order of convergence of the sequence and C is called an asymptotic error constant or rate of convergence [8].

Acknowledgement: This work is supported by National Board of Higher Mathematics (NBHM), No. 020111/17/2020 NBHM (R. P.)/R & D II/8073.

Santhosh George, Department of Mathematical and Computational Sciences, National Institute of Technology, Karnataka 575 025, India, e-mail: sgeorge@nitk.edu.in

https://doi.org/10.1515/9783112215760-008

In general, convergence order of the iterative methods are obtained using Taylor expansion. So, one require assumptions on the higher-order derivatives of the operator. This reduces the applicability of iterative methods (see [1–5]).

In this paper, we modified the following three-step fifth-order method studied in [11] (see also [12] and [6]), defined for $k = 0, 1, 2, \ldots$, by

$$\begin{aligned} y_k &= x_k - F'(x_k)^{-1}F(x_k), \\ z_k &= y_k - F'(x_k)^{-1}F(y_k), \\ x_{k+1} &= z_k - F'(x_k)^{-1}F(z_k). \end{aligned} \tag{8.2}$$

The convergence order of method (8.2) is obtained using Taylor expansion and assumptions on the derivatives of F up to the order six [12] (see also [6]). In [11], we have proved the convergence of method (8.2) using assumptions only on the first derivative of F. But, in [11] the order of convergence was not proved and used Computational Order of Convergence (COC) defined as

$$\xi = \ln\left(\frac{\|x_{n+1} - x_*\|}{\|x_n - x_*\|}\right) \Big/ \ln\left(\frac{\|x_n - x_*\|}{\|x_{n-1} - x_*\|}\right)$$

or the approximate computational order of convergence (ACOC) defined as

$$\xi_1 = \ln\left(\frac{\|x_{n+1} - x_n\|}{\|x_n - x_{n-1}\|}\right) \Big/ \ln\left(\frac{\|x_n - x_{n-1}\|}{\|x_{n-1} - x_{n-2}\|}\right)$$

to obtain the order of convergence.

It is known [10] that the oscillating behaviour of approximations and slow convergence of iterates in the initial stage may result in inaccurate computation of COC and ACOC.

The main advantage of the approach in [11] was that it extended the applicability of Traub's method and under the same computational effort as in earlier studies using the restricted Lipschitz condition. But as stated above COC and ACOC has some disadvantages in applying to iterative methods having oscillating behaviour of approximations and slow convergence of iterates. So, in this paper we modified the method (8.2) to obtain two methods with convergence order five and six, respectively. The convergence order is obtained by using assumptions up to the second order derivatives of the operator involved.

The new methods of order five and six, respectively are:

$$\begin{aligned} z_k &= x_k + F'(x_k)^{-1}[F(x_k) - F(x_k + F'(x_k)^{-1}F(x_k))], \\ x_{k+1} &= z_k - F'(x_k)^{-1}F(z_k), \quad k = 0, 1, 2, \ldots. \end{aligned} \tag{8.3}$$

and

$$
\begin{aligned}
z_k &= x_k + F'(x_k)^{-1}[F(x_k) - F(x_k + F'(x_k)^{-1}F(x_k))], \\
x_{k+1} &= z_k - F'(z_k)^{-1}F(z_k), \quad k = 0, 1, 2, \ldots .
\end{aligned} \tag{8.4}
$$

Note that the above methods can be written as

$$
\begin{aligned}
y_k &= x_k + F'(x_k)^{-1}F(x_k), \\
z_k &= y_k - F'(x_k)^{-1}F(y_k), \\
x_{k+1} &= z_k - F'(x_k)^{-1}F(z_k), \quad k = 0, 1, 2, \ldots,
\end{aligned} \tag{8.5}
$$

and

$$
\begin{aligned}
y_k &= x_k + F'(x_k)^{-1}F(x_k), \\
z_k &= y_k - F'(x_k)^{-1}F(y_k), \\
x_{k+1} &= z_k - F'(z_k)^{-1}F(z_k), \quad k = 0, 1, 2, \ldots,
\end{aligned} \tag{8.6}
$$

respectively. One can easily see that method (8.5) is obtained from (8.2) by changing the sign in the first step. We prove using assumptions on the derivatives of F of up to second order, that methods (8.5) and (8.6) are, respectively, of fifth and sixth order. Our convergence analysis is not based on Taylor expansion, so we do not need assumptions on the higher than second-order derivatives of the operator involved.

Note that the computational cost of methods (8.5) and (8.6) are almost the same. In this connection, let us recall that (cf. Traub [12]) the informational efficiency of an iterative method I is defined as

$$I = \frac{R}{p},$$

where R is the order of convergence of the method and p is the number of function (and derivative) evaluations. Ostrowski [9] introduced efficiency index or computational efficiency, defined as

$$c_f = R^{\frac{1}{p}}.$$

Thus the informational efficiency and computational efficiency of method (8.5), are $5/4 = 1.25$ and $5^{1/4} = 1.4953$, respectively. But that of method (8.6) are $6/5 = 1.2$ and $6^{1/5} = 1.4310$, respectively. Even though, the order of convergence of method (8.6) is six its informational efficiency and computational efficiency are smaller than that of method (8.5).

The idea used in this paper to obtain convergence order is new and can be used to obtain the convergence order of other similar methods.

The paper contains local convergence analysis in Section 8.2, numerical examples are given in Section 8.3. Finally, the paper ends with basin of attractions in Section 8.4.

8.2 Convergence analysis

The following assumptions are used to prove our results.

There exists $k_0 > 0$ such that for every $x, y \in D(F)$,

$$\|F'(x^*)^{-1}(F'(y) - F'(x^*))\| \le k_0\|y - x\|. \tag{8.7}$$

There exists $c > 0$, $c_1 > 0$ such that for all $x, y \in D(F)$,

$$\|F'(x^*)^{-1}F''(y)\| \le c \tag{8.8}$$

and

$$\|F'(x^*)^{-1}[F''(y) - F''(x^*)]\| \le c_1\|y - x^*\|. \tag{8.9}$$

Remark 8.1. By (8.14), one can see that if $y \in B(x^*, \frac{1}{k_0})$, then

$$\|F'(y)^{-1}F'(x^*)\| \le \frac{1}{1 - k_0\|y - x^*\|} \tag{8.10}$$

and

$$\|F'(x^*)^{-1}F'(y)\| \le 1 + k_0\|y - x^*\|. \tag{8.11}$$

Let $g_1, h_1 : [0, \frac{1}{k_0}) \longrightarrow \mathbb{R}$ be continuous and nondecreasing functions defined by

$$g_1(t) = \frac{c_1}{1 - k_0 t}\left(\frac{1}{2}\left(1 + \frac{2 + k_0 t}{2(1 - k_0 t)}\right)^2 + \frac{1}{2}\left(1 + \frac{2 + k_0 t}{2(1 - k_0 t)}\right) + \frac{1}{6}\left(1 + \frac{2 + k_0 t}{2(1 - k_0 t)}\right)^3\right) + \frac{ck_0}{4(1 - k_0 t)}\left(1 + \frac{2 + k_0 t}{2(1 - k_0 t)}\right)$$

and $h_1(t) = g_1(t)t^2 - 1$. Then, $h_1(0) = -1$ and $h_1(t) \longrightarrow +\infty$ as $t \longrightarrow \frac{1}{k_0}^-$. So, h_1 has a smallest zero r_1 in $(0, \frac{1}{k_0})$.

Let $g_2, h_2 : [0, \frac{1}{k_0}) \longrightarrow \mathbb{R}$ continuous and nondecreasing functions defined by

$$g_2(t) = \left[\frac{c_1}{1 - k_0 t}\left(\frac{1}{2}g_1(t)t^2 + \frac{1}{2} + \frac{1}{6}g_1(t)^2 t^4\right) + \frac{k_0 c}{4(1 - k_0 t)}\left(1 + \left(1 + \frac{2 + k_0 t}{2(1 - k_0 t)}\right)^2\right)\right]g_1(t)$$

and

$$h_2(t) = g_2(t)t^4 - 1.$$

Then, $h_2(0) = -1$ and $h_2(t) \longrightarrow +\infty$ as $t \longrightarrow \frac{1}{k_0}^-$. So, h_2 has a smallest zero r_2 in $(0, \frac{1}{k_0})$.

$$r = \min\{r_1, r_2\}. \tag{8.12}$$

Next, we present convergence analysis for (8.5).

Theorem 8.1. *Let r be as in (8.12). Then the sequence $\{x_n\}$ defined by (8.5) with $x_0 \in B(x^*, r) - \{x^*\}$, converges to x^* and satisfies*

$$\|x_{n+1} - x^*\| \leq g_2(r)\|x_n - x^*\|^5 \tag{8.13}$$

(i. e., order of convergence is five).

Proof. We shall use induction to prove (8.13). First, we obtain estimates for $\|y_0 - x_0\|$ and $\|y_0 - x^*\|$. Note that

$$\begin{aligned}\|y_0 - x_0\| &= \|F'(x_0)^{-1}[F(x_0) - F(x^*)]\|\\ &= \left\|F'(x_0)^{-1}F'(x^*)\int_0^1 F'(x^*)^{-1}F'(x^* + t(x_0 - x^*))dt(x_0 - x^*)\right\|\\ &\leq \frac{2 + k_0\|x_0 - x^*\|}{2(1 - k_0\|x_0 - x^*\|)}\|x_0 - x^*\|,\end{aligned} \tag{8.14}$$

and hence

$$\|y_0 - x^*\| \leq \|y_0 - x_0\| + \|x_0 - x^*\| \leq \left(1 + \frac{2 + k_0\|x_0 - x^*\|}{2(1 - k_0\|x_0 - x^*\|)}\right)\|x_0 - x^*\|. \tag{8.15}$$

We have again by (8.5),

$$\begin{aligned}z_0 - x^* &= y_0 - x^* - F'(x_0)^{-1}[F(y_0) - F(x^*)]\\ &= F'(x_0)^{-1}\int_0^1 [F'(x_0) - F'(x^* + t(y_0 - x^*))](y_0 - x^*)dt\\ &= F'(x_0)^{-1}\int_0^1\Bigg[\int_0^1 F''(x^* + t(y_0 - x^*)\\ &\quad + \theta(x_0 - x^* - t(y_0 - x^*)))d\theta(x_0 - x^* - t(y_0 - x^*))\Bigg](y_0 - x^*)dt,\end{aligned}$$

so

$$\begin{aligned}&\|z_0 - x^*\|\\ &= \|F'(x_0)^{-1}F'(x^*)\int_0^1\int_0^1 F'(x^*)^{-1}[F''(x^* + t(y_0 - x^*)\\ &\quad + \theta(x_0 - x^* - t(y_0 - x^*))) - F''(x^*)]d\theta(x_0 - x^* - t(y_0 - x^*))](y_0 - x^*)dt\end{aligned}$$

$$
\begin{aligned}
&+\left\|F'(x_0)^{-1}F''(x^*)\int_0^1(x_0-x^*-t(y_0-x^*))(y_0-x^*)dt\right\|\\
&\le \frac{c_1}{1-k_0\|x_0-x^*\|}\int_0^1\int_0^1\|(1-\theta)t(y_0-x^*)+\theta(x_0-x^*)d\theta\\
&\quad\times(x_0-x^*-t(y_0-x^*))dt(y_0-x^*)\|+c\left\|\int_0^1(x_0-x^*-t(y_0-x^*))dt(y_0-x^*)\right\|\\
&\le \frac{c_1}{1-k_0\|x_0-x^*\|}\left(\frac{1}{2}\|x_0-x^*\|\|y_0-x^*\|^2+\frac{1}{2}\|x_0-x^*\|^2\|y_0-x^*\|+\frac{1}{6}\|y_0-x^*\|^3\right)\\
&\quad+c\left\|\int_0^1(x_0-x^*-t(y_0-x^*))dt(y_0-x^*)\right\|.
\end{aligned}
\tag{8.16}
$$

Note that

$$
\begin{aligned}
&\int_0^1[x_0-x^*-t(y_0-x^*)]dt(y_0-x^*)\\
&\quad=\int_0^1[x_0-x^*-t(x_0+F'(x_0)^{-1}F(x_0)-x^*)]dt(y_0-x^*)\\
&\quad=\int_0^1[x_0-x^*-t(2(x_0-x^*)+F'(x_0)^{-1}F(x_0)-(x_0-x^*)]dt(y_0-x^*)\\
&\quad=\int_0^1(1-2t)(x_0-x^*)dt-\int_0^1 t(x_0-x^*-F'(x_0)^{-1}F(x_0))dt(y_0-x^*)\\
&\quad=-\int_0^1 tF'(x_0)^{-1}\left(F'(x_0)-\int_0^1 F'(x^*+v(x_0-x^*))\right)dv(x_0-x^*)dt(y_0-x^*)\\
&\quad=-\int_0^1 t\int_0^1 F'(x_0)^{-1}F'(x^*)F'(x^*)^{-1}[F'(x_0)-F'(x^*+v(x_0-x^*))]dv(x_0-x^*)dt(y_0-x^*)
\end{aligned}
$$

and hence

$$
\left\|\int_0^1(x_0-x^*-t(y_0-x^*))dt(y_0-x^*)\right\|\le\frac{k_0}{4(1-k_0\|x_0-x^*\|)}\|x_0-x^*\|^2\|y_0-x^*\|. \tag{8.17}
$$

So, by (8.16) and (8.17), we have

$$
\|z_0-x^*\|\le g_1(\|x_0-x^*\|)\|x_0-x^*\|^3. \tag{8.18}
$$

Since $g_1(r)r^2 < 1$, $z_0 \in B(x^*, r)$. By the third substep of method (8.5), we have

$$\begin{aligned}
x_1 - x^* &= z_0 - x^* - F'(x_0)^{-1} - [F(z_0) - F(x^*)] \\
&= F'(x_0)^{-1}\left[F'(x_0)(z_0 - x^*) - \int_0^1 F'(x^* + t(z_0 - x^*))(z_0 - x^*)dt\right] \\
&= F'(x_0)^{-1}\int_0^1 [F'(x_0) - F'(x^* + t(z_0 - x^*))](z_0 - x^*)dt \\
&= F'(x_0)^{-1}F'(x^*)\int_0^1\int_0^1 F'(x^*)^{-1}[F''(x^* + t(z_0 - x^*) \\
&\quad + \theta(x_0 - x^* - t(z_0 - x^*)) - F''(x^*)]\, d\theta(x_0 - x^* - t(z_0 - x^*))(z_0 - x^*)dt \\
&\quad + F'(x_0)^{-1}F'(x^*)\int_0^1\int_0^1 F'(x^*)^{-1}F''(x^*)d\theta(x_0 - x^* - t(z_0 - x^*))(z_0 - x^*)dt
\end{aligned}$$

and hence, we have

$$\begin{aligned}
\|x_1 - x^*\| &\le \frac{c_1}{1 - k_0\|x_0 - x^*\|} \\
&\quad \times \int_0^1\int_0^1 \|t(z_0 - x^*) + \theta(x_0 - x^* - t(z_0 - x^*))\|d\theta \\
&\quad \times \|(x_0 - x^* - t(z_0 - x^*))(z_0 - x^*)\|dt \\
&\quad + \frac{c}{1 - k_0\|x_0 - x^*\|}\left\|\int_0^1 (x_0 - x^* - t(z_0 - x^*))(z_0 - x^*)dt\right\| \\
&\le \frac{c_1}{1 - k_0\|x_0 - x^*\|}\left(\frac{1}{2}\|z_0 - x^*\|\|x_0 - x^*\| + \frac{1}{2}\|x_0 - x^*\|^2 + \frac{1}{6}\|z_0 - x^*\|^2\right)\|z_0 - x^*\| \\
&\quad + \frac{c}{1 - k_0\|x_0 - x^*\|}\left\|\int_0^1 (x_0 - x^* - t(z_0 - x^*))(z_0 - x^*)dt\right\| \\
&\le \frac{c_1}{1 - k_0\|x_0 - x^*\|}\left(\frac{1}{2}g_1(\|x_0 - x^*\|)\|x_0 - x^*\|^2 + \frac{1}{2} + \frac{1}{6}g_1(\|x_0 - x^*\|)^2\|x_0 - x^*\|^4\right) \\
&\quad \times g_1(\|x_0 - x^*\|)\|x_0 - x^*\|^5 \\
&\quad + \frac{c}{1 - k_0\|x_0 - x^*\|}\left\|\int_0^1 (x_0 - x^* - t(z_0 - x^*)(z_0 - x^*)dt\right\|.
\end{aligned} \tag{8.19}$$

Note that

$$
\begin{aligned}
&\left\|\int_0^1 (x_0 - x^* - t(z_0 - x^*)(z_0 - x^*)dt\right\| \\
&= \left\|\int_0^1 (x_0 - x^* - t(y_0 - F'(x_0)^{-1}F(y_0) - x^*)(z_0 - x^*)dt\right\| \\
&= \left\|\int_0^1 (x_0 - x^* - t(y_0 - x_0 + x_0 - x^* - F'(x_0)^{-1}[F(y_0) - F(x_0) + F(x_0)](z_0 - x^*)dt\right\| \\
&= \Bigg\|\int_0^1 [(1-t)(x_0 - x^*) - tF'(x_0)^{-1}F(x_0)](z_0 - x^*)dt \\
&\quad - \int_0^1 t[y_0 - x_0 - F'(x_0)^{-1}(F(y_0) - F(x_0))](z_0 - x^*)dt\Bigg\| \\
&= \Bigg[\Bigg\|\int_0^1 (1-2t)(x_0 - x^*) + t[x_0 - x^* - F'(x_0)^{-1}F(x_0)](z_0 - x^*)dt \\
&\quad - \int_0^1 tF'(x_0)^{-1}\left[F'(x_0) - \int_0^1 F'(x_0 + \theta(y_0 - x_0))d\theta\right](y_0 - x_0)(z_0 - x^*)dt\Bigg\|\Bigg] \\
&\le \Bigg[\left\|\int_0^1 (1-2t)(x_0 - x^*)(z_0 - x^*)dt\right\| \\
&\quad + \left\|\int_0^1 tF'(x_0)^{-1}\left[F'(x_0) - \int_0^1 F'(x^* + \theta(x_0 - x^*))d\theta\right](x_0 - x^*)(z_0 - x^*)dt\right\| \\
&\quad + \left\|\int_0^1 tF'(x_0)^{-1}\left[F'(x_0) - \int_0^1 F'(x_0 + \theta(y_0 - x_0))d\theta\right](y_0 - x_0)(z_0 - x^*)dt\right\|\Bigg] \\
&\le \Bigg[\left\|\int_0^1 (1-2t)(x_0 - x^*)(z_0 - x^*)dt\right\| + \Bigg\|\int_0^1 t\int_0^1 F'(x_0)^{-1}F'(x^*)F'(x^*)^{-1}[F'(x_0) \\
&\quad - F'(x^* + \theta(x_0 - x^*))d\theta](x_0 - x^*)(z_0 - x^*)dt\Bigg\| \\
&\quad + \left\|\int_0^1 t\int_0^1 F'(x_0)^{-1}F'(x^*)F'(x^*)^{-1}[F'(x_0) - F'(x_0 + \theta(y_0 - x_0))d\theta](y_0 - x_0)(z_0 - x^*)dt\right\|\Bigg] \\
&\le \frac{k_0}{4(1 - k_0\|x_0 - x^*\|)}[\|x_0 - x^*\|^2 + \|x_0 - y_0\|^2]\|z_0 - x^*\| \\
&\le \frac{k_0}{4(1 - k_0\|x_0 - x^*\|)}\left(1 + \left(1 + \frac{2 + k_0\|x_0 - x^*\|}{2(1 - k_0\|x_0 - x^*\|)}\right)^2\right)\|x_0 - x^*\|^2\|z_0 - x^*\|. \qquad (8.20)
\end{aligned}
$$

Therefore, by (8.18) and (8.20), we have

$$\|x_1 - x^*\| \le g_2(\|x_0 - x^*\|)\|x_0 - x^*\|^5. \tag{8.21}$$

Now since $g_2(r)r^4 < 1$, we have $x_1 \in B(x^*, r)$. The induction for (8.13) will be completed, if we simply replace x_0, y_0, z_0, x_1 by x_k, y_k, z_k, x_{k+1} in the preceding arguments. □

Let $g_3, h_2 : [0, \frac{1}{k_0}) \longrightarrow \mathbb{R}$ be continuous and nondecreasing functions defined by

$$g_3(t) = \frac{k_0}{2(1 - k_0 g_1(t)t^2)} g_1(t)^2$$

and

$$h_3(t) = g_3(t)t^5 - 1.$$

Then, h_3 has a smallest zero r_3 in $(0, \frac{1}{k_0})$.

$$\bar{r} = \min\{r_1, r_3\}. \tag{8.22}$$

Next, we present convergence analysis for (8.6).

Theorem 8.2. *Let $\bar{r}$ be as in (8.22). Then the sequence $\{x_n\}$ defined by (8.6) with $x_0 \in B(x^*, \bar{r}) - \{x^*\}$, converges to x^* with order of convergence five, i. e.,*

$$\|x_{n+1} - x^*\| \le g_3(\bar{r})\|x_n - x^*\|^6. \tag{8.23}$$

Proof. Note that, in this case, we have

$$\begin{aligned} x_1 - x^* &= z_0 - x^* - F'(z_0)^{-1}[F(z_0) - F(x^*)] \\ &= F'(z_0)^{-1}F'(x^*)\int_0^1 F'(x^*)^{-1}[F'(z_0) - F'(x^* + t(z_0 - x^*))](z_0 - x^*)dt, \end{aligned}$$

so

$$\begin{aligned} \|x_1 - x^*\| &\le \frac{k_0}{2(1 - k_0\|z_0 - x^*\|)}\|z_0 - x^*\|^2 \\ &\le \frac{c_1}{2(1 - k_0 g_1(\|x_0 - x^*\|)\|x_0 - x^*\|^2)} g_1(\|x_0 - x^*\|)^2\|x_0 - x^*\|^6 \\ &= g_3(\|x_0 - x^*\|)\|x_0 - x^*\|^6 \\ &< \bar{r}. \end{aligned}$$

The rest of the proof follows as in Theorem 8.1. □

Proposition 8.1. *Suppose x^* is a simple solution of equation $F(x) = 0$, and (8.14) holds. Then, the only solution of equation $F(x) = 0$ in the set $S = \Omega \cap B[x^*, R]$ is x^* provided that*

$$k_0 R < 2. \tag{8.24}$$

Proof. Let $p \in S$ be a solution of equation $F(x) = 0$. Set $M = \int_0^1 F'(x^* + t(p - x^*))dt$. Then, by (8.14) and (8.24), we have in turn that

$$\begin{aligned}
\|F'(x^*)^{-1}(M - F'(x^*))\| &\le k_0 \int_0^1 \|x^* + t(p - x^*) - x^*\| dt \\
&\le k_0 \int_0^1 t\|x^* - p\| dt \\
&\le \frac{k_0 R}{2} < 1,
\end{aligned}$$

so $p = x^*$ follows from the invertibility of M and the identity $0 = F(p) - F(x^*) = M(p - x^*)$. □

8.3 Numerical experiments

We compute the radius of convergence for an example in this section.

Example 8.1. Let $X = Y = \mathbb{R}^3$, $D = B[0, 1]$, $x_* = (0, 0, 0)^T$. Define function F on D for $w = (x, y, z)^T$ by

$$F(w) = \left(e^x - 1, \frac{e - 1}{2} y^2 + y, z\right)^T.$$

Then, we get

$$F'(v) = \begin{bmatrix} e^x & 0 & 0 \\ 0 & (e - 1)y + 1 & 0 \\ 0 & 0 & 1 \end{bmatrix},$$

so $k_0 = e - 1$, $c = c_1 = e$. Then, we have

$$r = 0.096174692612445658806502953642559 \quad \text{and}$$

$$\bar{r} = 0.16005956770819789486655153890952.$$

8.4 Basins of attractions

The collection of all initial points from which the iterative method converges to a solution of a given equation are called the basins of attraction or Fatou sets, of an iterative method [7]. And the complement of Fatou set is known as Julia set. Next, we provide the basins of attraction associated to each of the root of the following three systems of equations.

Example 8.2.

$$\begin{cases} x^3 - y = 0 \\ y^3 - x = 0 \end{cases}$$

with solutions $\{(-1, -1), (0, 0), (1, 1)\}$

Example 8.3.

$$\begin{cases} 3x^2y - y^3 = 0 \\ x^3 - 3xy^2 - 1 = 0 \end{cases}$$

with solutions $\{(-\frac{1}{2}, -\frac{\sqrt{3}}{2}), (-\frac{1}{2}, \frac{\sqrt{3}}{2}), (1, 0)\}$.

Example 8.4.

$$\begin{cases} x^2 + y^2 - 4 = 0 \\ 3x^2 + 7y^2 - 16 = 0 \end{cases}$$

with solutions $\{(\sqrt{3}, 1), (-\sqrt{3}, 1), (\sqrt{3}, -1), (-\sqrt{3}, -1)\}$.

We apply method (8.5) and (8.6) to obtain the basins of attraction for each test problem and compare the results. For generating the basin of attraction associated to each root of a given system of nonlinear equations, we consider the rectangular region $\mathcal{R} = \{(x, y) \in \mathbb{R}^2 : -2 \leq x \leq 2, -2 \leq y \leq 2\}$ which contains all the roots of test problems. We consider an equidistant grid of 401×401 points in $\mathcal{R}$ and we choose these points as the initial guess x_0, for the methods (8.5) and (8.6). A fixed tolerance 10^{-8} and a maximum of 50 iterations are used for all the cases. A color is being assigned to each attracting basin corresponding to each root. If we do not obtain the desired tolerance with the fixed iterations, we do not continue and we decide that the iterative method starting at x_0 does not converge to any of the roots and assign black color to those points. In this way, we distinguish the basins of attraction by their respective colors for distinct roots of each method.

Figures 8.1–8.3 demonstrate the basin of attraction corresponding to each root of above examples (Examples 8.2–8.4) for the methods (8.5) and (8.6). The Julia set (black region), which contains all the initial points from which the iterative method does not converge to any of the roots, can easily be observed in the figure.

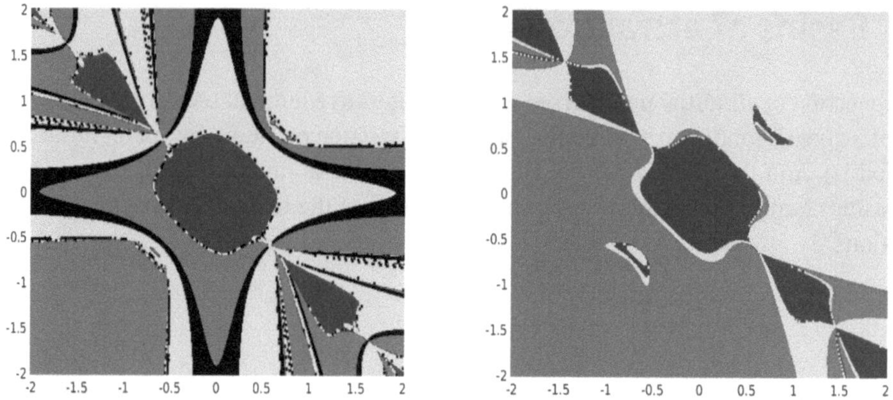

Figure 8.1: Dynamical plane of the methods (8.5) (left) and (8.6) (right) with basins of attraction for the Examples 8.2.

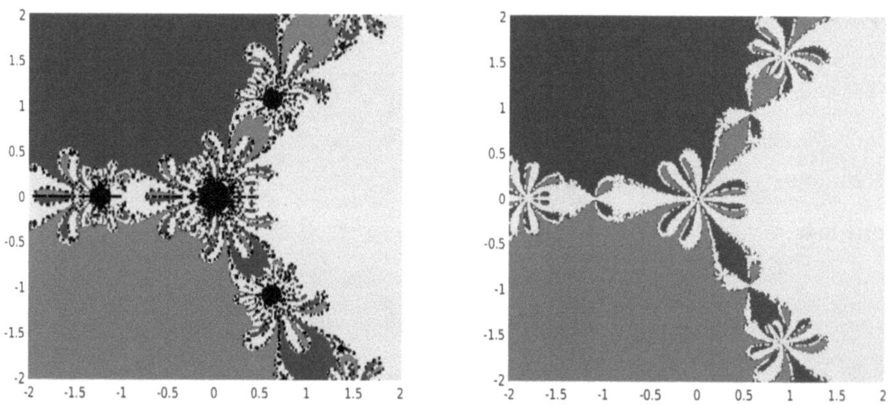

Figure 8.2: Dynamical plane of the methods (8.5) (left) and (8.6) (right) with basins of attraction for the Examples 8.3.

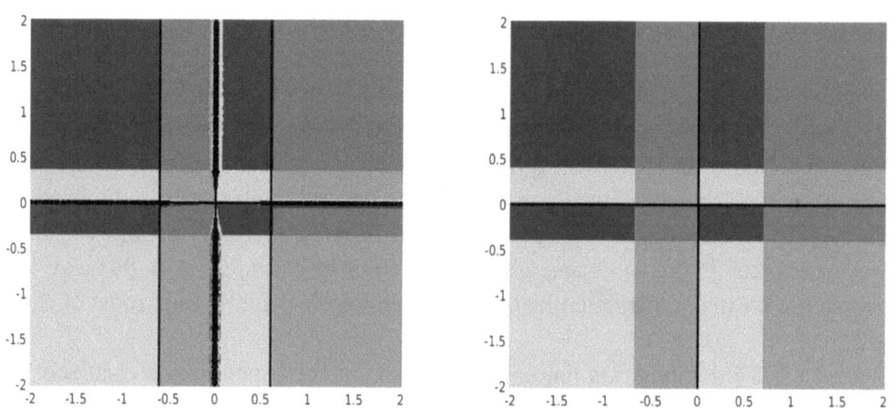

Figure 8.3: Dynamical plane of the methods (8.5) (left) and (8.6) (right) with basins of attraction for the Examples 8.4.

The figures presented in this work is performed in a 4-core 64 bit Windows machine with Intel Core i7-3770 processor using MATLAB programming language.

Bibliography

[1] Argyros, I. K., *On the Newton–Kantorovich hypothesis for solving equations*, J. Comput. Math. 169 (2004), 315–332.

[2] Argyros, I. K., *Computational Theory of Iterative Methods*, C. K. Chui and L. Wuytack (eds.), Studies in Computational Mathematics, vol. 15, Elsevier Publ. Co., New York, U.S.A, 2007.

[3] Argyros, I. K., *Convergence and Applications of Newton-type Iterations*, Springer Verlag, Berlin, Germany, 2008.

[4] Argyros, I. K. and George, S., *On the complexity of extending the convergence region for Traub's method*, J. Complex. 56 (2020), 101423.

[5] Argyros, I. K. and Hilout, S., *On an improved convergence analysis of Newton's method*, Appl. Math. Comput. 225 (2013), 372–386.

[6] Behl, R., Maroju, P., Martinez, E. and Singh, S., *A study of the local convergence of a fifth order iterative method*, Indian J. Pure Appl. Math. 51(2) (2020), 439–455.

[7] Magréñan, A. A. and Gutiérrez, J. M., *Real dynamics for damped Newton's method applied to cubic polynomials*, J. Comput. Appl. Math. 275 (2015), 527–538.

[8] Ortega, J. M. and Rheinboldt, W. C., *Iterative Solution of Nonlinear Equations in Several Variables*, Classics in Applied Mathematics, vol. 30, Society for Industrial and Applied Mathematics (SIAM), Philadelphia, PA, 2000. Reprint of the 1970 original.

[9] Ostrowski, A. M., *Solution of Equations and Systems of equations*, 2nd ed., Pure and Applied Mathematics, vol. 9, Academic Press, New York–London, 1966.

[10] Petković, M. S., Neta, B., Petković, L. D. and Dzunić, J., *Multipoint methods for solving nonlinear equations: A survey*, Appl. Math. Comput. 226 (2014), 635–660.

[11] Regmi, S., Argyros, C. I., Argyros, I. K. and George, S., *On the semi-local convergence of a Traub-type method for solving equations*, Foundations 2 (2022), 114–127, https://doi.org/10.3390/foundations2010006.

[12] Traub, J. F., *Iterative Methods for the Solution of Equations*, Prentice Hall, New Jersey, U.S.A, 1964.

Nora Fetouci and Stojan Radenović

9 Some fixed-point results for expansive mappings in *G*-metric spaces

Abstract: The aim of this paper is to establish some existence results of fixed point for a class of expansive mappings defined on a complete G-metric space. We extend some existing results for mappings satisfying contractive conditions to expansive ones, and we provide a new condition expressed in one variable, under which the existence of fixed point holds.

Keywords: Fixed point, Cauchy sequence, G-metric space, contraction map

MSC 2020: 47H10, 54H25, 54E50

9.1 Introduction and preliminaries

Metric fixed-point theory provides essential tools for solving problems emerging from various areas of mathematical analysis, such as variational and linear inequalities, optimization and approximation theory. It is well known that many differential equations with boundary values can be reformulated for instance as integral equations of Volterra and Fredholm types. In this context, fixed-point theory offers new tools to resolve non linear integral equations that arise in theory of non linear dynamical systems.

In 2006, Z. Mustafa and B. Sims [26] introduced the notion of G-metric spaces as a generalization of ordinary metric spaces, and analysed the topological structure of the G-metric spaces. In this first paper, authors developed the well-known Banach contraction principle, and some fixed-point results for various classes of mappings in the setting of G-metric space. Later on, in this direction, several research articles related to fixed-point theory on G-metric space have appeared, see for instance papers [1, 2, 4, 22, 28–30, 32, 35]. For more relevant literature, we refer to the recent works of Debnath and co-authors [6–17] and several others [3, 5, 18–20, 23, 25, 27, 31, 33, 34].

In this paper, we establish some existence results of fixed point for a class of expansive mappings in G-metric spaces. Our contribution gives new tools to fixed-point theory and their applications.

We recall some definitions and properties for G-metric spaces given by Z. Mustafa and B. Sims.

Nora Fetouci, LMPA Laboratory, Department of Mathematics, Jijel University, Jijel, Algeria, e-mail: n.fetouci@univ-jijel.dz
Stojan Radenović, Faculty of Mechanical Engineering, University of Belgrade, Belgrade, Serbia, e-mail: sradenovic@mas.bg.ac.rs

https://doi.org/10.1515/9783112215760-009

Definition 9.1 ([26]). Let X be a non-empty set and let $G : X \times X \times X \to \mathbb{R}^+$ be a function satisfying the following properties:
(G1) $G(x,y,z) = 0$ if $x = y = z$.
(G2) $0 < G(x,x,y)$ for all $x,y \in X$ with $x \neq y$.
(G3) $G(x,x,y) \leq G(x,y,z)$ for all $x,y,z \in X$ with $y \neq z$.
(G4) $G(x,y,z) = G(x,z,y) = G(y,z,x) = \cdots$ (symmetry in all three variables).
(G5) $G(x,y,z) \leq G(x,a,a) + G(a,y,z)$ for all $x,y,z,a \in X$ (rectangle inequality).

Then the function G is called a generalized metric or, more specifically, a G metric on X, and the pair (X,G) is called a G-metric space.

Remark 9.1. It is clear that from $G(x,y,z) = 0$ it follows that $x = y = z$.

Indeed, suppose the opposite. Let at least two of the three elements of X are different from each other. For example, let $x = y \neq z$. Further, from the axiom (G_3), taking into account that $G(x,y,z) = 0$, it follows that $G(x,x,z) = 0$, which contradicts the axiom (G_2).

Definition 9.2 ([26]). Let (X,G) be a G-metric space. Let (x_n) a sequence of points of X. We say that (x_n) is G-convergent to x if $\lim_{n,m\longrightarrow+\infty} G(x,x_n,x_m) = 0$; that is, for any $\epsilon > 0$, there exists $k \in \mathbb{N}$ such that $G(x,x_n,x_m) < \epsilon$ for all $n,m \geq k$. We call x the limit of the sequence and we write $x_n \longrightarrow x$ or $\lim_{n\longrightarrow+\infty} x_n = x$.

Proposition 9.1 ([26]). *Let (X,G) be a G-metric space. Then the following are equivalent:*
(i) *(x_n) is G-convergent to x.*
(ii) *$G(x_n,x_n,x) \to 0$, as $n \longrightarrow +\infty$.*
(iii) *$G(x_n,x,x) \to 0$ as $n \longrightarrow +\infty$.*
(iv) *$G(x_m,x_n,x) \to 0$, as $m,n \longrightarrow +\infty$.*

Proposition 9.2 ([26]). *Every G-metric space (X,G) will define a metric space (X,d_G) by*

$$d_G(x,y) = G(x,y,y) + G(y,x,x), \quad \text{for all } x,y \in X.$$

Proposition 9.3 ([26]). *A G-metric space (X,G) is G-complete if and only if (X,d_G) is a complete metric space.*

Definition 9.3 ([26]). Let (X,G) be a G-metric space. A sequence (x_n) is called G-Cauchy if for each $\epsilon > 0$, there is $k \in \mathbb{N}$ such that $G(x_n,x_m,x_l) < \epsilon$ for all $n,m,l \geq k$, that is $G(x_n,x_m,x_l) \to 0$ as $n,m,l \to +\infty$.

Proposition 9.4 ([26]). *Let (X,G) be a G-metric space, then the following conditions are equivalent:*
(i) *The sequence (x_n) is G-Cauchy.*
(ii) *For every $\epsilon > 0$, there exists $k \in \mathbb{N}$ such that $G(x_n,x_m,x_m) < \epsilon$, for all $n,m \geq k$.*

Definition 9.4 ([26]). A G-metric space (X,G) is called G-complete if every G-Cauchy sequence is G-convergent in (X,G).

Definition 9.5 ([26]). A G-metric space (X, G) is called symmetric G-metric space if $G(x,y,y) = G(y,x,x)$ for all $x, y \in X$.

We give by the sequel an example of a non-symmetric G-metric space.

Example 9.1 ([29]). Let $X = \{a, b\}$, where $a \neq b$ and let $G : X \times X \times X \to [0, +\infty[$ be the function defined as

$$\begin{aligned} G(a,a,a) &= G(b,b,b) = 0, \\ G(a,a,b) &= G(a,b,a) = G(b,a,a) = 1, \\ G(a,b,b) &= G(b,a,b) = G(b,b,a) = 2. \end{aligned}$$

Then G is a G-metric on X, but it is not symmetric because

$$G(a,a,b) = 1 \neq 2 = G(a,b,b).$$

Proposition 9.5 ([26]). *Let (X, G) be a G-metric space. Then $T : X \to X$ is G-continuous at $x \in X$ if and only if it is G-sequentially continuous at x, that is, whenever (x_n) is G-convergent to x, $(T(x_n))$ is G-convergent to $T(x)$.*

Proposition 9.6 ([26]). *Let (X, G) be a G-metric space. Then the function $G(x,y,z)$ is jointly continuous in all the three of its variables*

Definition 9.6 ([24]). Let (X, G) be a G-metric space and T be a self mapping on X. Then T is called expansive mapping if there exists a constant $\alpha > 1$ such that for all $x, y, z \in X$, we have

$$G(Tx, Ty, Tz) \geq \alpha G(x,y,z).$$

9.2 Main results

We start with the following theorem, which will be used in the example below.

Theorem 9.1. *Let (X, G) be a complete G-metric space. If there exists a constant $c > 1$ and a surjective self mapping T on X, such that for all $x, y, z \in X$*

$$G(Tx, Ty, Tz) \geq cG(x,y,z). \tag{9.1}$$

Then T admits a unique fixed point.

Proof. We have from (9.1), taking $y = z$ it follows

$$G(Tx, Tz, Tz) \geq cG(x,z,z),$$

then taking $y = x$ it follows

$$G(Tx, Tx, Tz) \geq cG(x, x, z).$$

Adding the obtained inequalities, and using the definition of the associated metric d_G, we have

$$d_G(Tx, Tz) \geq c d_G(x, z).$$

According to the result for metric spaces see [36], we deduce that T admits a unique fixed point, which completes the proof. □

The following example shows that T is onto in Theorem 9.1 is a necessary condition.

Example 9.2 ([22]). Let $\mathbb{R}$ be the set of real numbers. Define $G : \mathbb{R} \times \mathbb{R} \times \mathbb{R} \to \mathbb{R}^+$ by

$$G(x, y, z) = |x - y| + |y - z| + |z - x|, \quad \text{for all } x, y, z \in \mathbb{R}.$$

Then, it is obvious that $(\mathbb{R}, G)$ is a G-metric space.

Let $T : \mathbb{R} \to \mathbb{R}$ be a mapping defined as follows:

$$T(x) = \begin{cases} 2x - 1 & \text{for } x \leq 0, \\ 2x + 1 & \text{for } x > 0. \end{cases}$$

We have

$$G(Tx, Ty, Tz) = |Tx - Ty| + |Ty - Tz| + |Tz - Tx|, \tag{9.2}$$

for all $x, y, z \in \mathbb{R}$.

Or

$$\begin{aligned} |Tx - Ty| &= |2x - 1 - 2y - 1| \quad \text{for } x \leq 0 \text{ and } y > 0 \\ &= 2|y + 1 - x| \quad \text{for } x \leq 0 \text{ and } y > 0 \\ &= 2|y - x| \quad \text{for } x \leq 0 \text{ and } y > 0 \\ &= 2|x - y| \quad \text{for } x \leq 0 \text{ and } y > 0. \end{aligned}$$

In addition,

$$|Tx - Ty| = 2|x - y| \quad \text{for } x, y \leq 0,$$

and

$$|Tx - Ty| = 2|x - y| \quad \text{for } x, y > 0,$$

so

$$|Tx - Ty| \geq 2|x - y| \quad \text{for } x, y \in \mathbb{R}.$$

Using the same argument, we obtain

$$|Ty - Tz| \geq 2|y - z| \quad \text{for } x, y \in \mathbb{R},$$

and

$$|Tz - Tx| \geq 2|z - x| \quad \text{for } x, y \in \mathbb{R},$$

then, (9.2) becomes

$$G(Tx, Ty, Tz) \geq 2|x - y| + 2|y - z| + 2|z - x| = 2G(x, y, z),$$

for all $x, y, z \in \mathbb{R}$, which entails that T verifies condition (9.1) in Theorem 9.1.

Also, for all $x, y \in \mathbb{R}$,

$$\begin{aligned} d_G(x, y) &= G(x, y, y) + G(y, x, x) \\ &= 2|y - x| + 2|y - x| \\ &= 4|x - y|. \end{aligned}$$

It is easy to show that (X, d_G) is complete and thus, in view of Proposition 9.3, (X, G) is a complete G-metric space.

However, T does not have a fixed point, since it is not onto.

Our first main result is the following:

Theorem 9.2. *Let (X, G) be a complete G-metric space. Let $T : X \longrightarrow X$, be an onto mapping such that,*

$$G(T(x), T(y), T(z)) \geq sN(x, y, z) \tag{9.3}$$

for all x, y, z, where $s > 2$ and

$$N(x, y, z) = \min \left\{ \begin{array}{c} G(x, Tx, y), G(y, T^2x, Ty), G(Tx, T^2x, Ty), G(y, Tx, Ty), \\ G(x, Tx, z), G(z, T^2x, Tz), G(Tx, T^2x, Tz), G(z, Tx, Ty), \\ G(x, y, z), G(x, Tx, Tx), G(y, Ty, Ty), G(z, Tz, Tz), \\ G(z, Tx, Tx), G(x, Ty, Ty), G(y, Tz, Tz) \end{array} \right\}. \tag{9.4}$$

Then T admits a unique fixed point.

Proof. Let $x_0 \in X$, since T is onto, then there exists an element x_1 satisfying $x_1 \in T^{-1}(x_0)$. Continuing in this way, we get a sequence (x_n), where $x_n \in T^{-1}(x_{n-1})$. If $x_n = x_{n-1}$ for some n, then x_n is a fixed point of T. Assume $x_n \neq x_{n-1}$ for every $n \in \mathbb{N}$, then from (9.4), we have

$$G(T(x_n, x_{n-1}, x_{n-1})) = G(Tx_{n+1}, Tx_n, Tx_n) \geq sN(x_{n+1}, x_n, x_n),$$

where

$$N(x_{n+1},x_n,x_n)=\min\left\{\begin{array}{c}G(x_{n+1},Tx_{n+1},x_n),G(x_n,T^2x_{n+1},Tx_n),G(Tx_{n+1},T^2x_{n+1},Tx_n),\\G(x_n,Tx_{n+1},Tx_n),G(x_{n+1},Tx_{n+1},x_n),G(x_n,T^2x_{n+1},Tx_n),\\G(Tx_{n+1},T^2x_{n+1},Tx_n),G(x_n,Tx_{n+1},Tx_n),G(x_{n+1},x_n,x_n),\\G(x_{n+1},Tx_{n+1},Tx_{n+1}),G(x_n,Tx_n,Tx_n),G(x_n,Tx_n,Tx_n),\\G(x_n,Tx_{n+1},Tx_{n+1}),G(x_{n+1},Tx_n,x_n),G(x_n,Tx_n,Tx_n)\end{array}\right\}, \tag{9.5}$$

so

$$N(x_{n+1},x_n,x_n)=\min\left\{\begin{array}{c}G(x_{n+1},x_n,x_n),G(x_n,x_{n-1},x_{n-1}),G(x_n,x_{n-1},x_{n-1}),\\G(x_n,x_n,x_{n-1}),G(x_{n+1},x_n,x_n),G(x_n,x_{n-1},x_{n-1}),G(x_n,x_{n-1},x_{n-1}),\\G(x_n,x_n,x_{n-1}),G(x_{n+1},x_n,x_n),G(x_{n+1},x_n,x_n),G(x_n,x_{n-1},x_{n-1}),\\G(x_n,x_{n-1},x_{n-1}),G(x_n,x_n,x_n),G(x_{n+1},x_{n-1},x_n),G(x_n,x_{n-1},x_{n-1})\end{array}\right\}, \tag{9.6}$$

then

$$G(x_n,x_{n-1},x_{n-1})\geq s\min\left\{\begin{array}{l}G(x_{n+1},x_n,x_n),G(x_n,x_{n-1},x_{n-1}),\\G(x_n,x_n,x_{n-1}),G(x_{n+1},x_{n-1},x_n),\end{array}\right\}. \tag{9.7}$$

By G-metric property, we have $G(x_{n+1},x_{n-1},x_n)\geq G(x_{n-1},x_{n-1},x_n)$ and $G(x_n,x_{n-1},x_{n-1})\leq 2G(x_{n-1},x_n,x_n)$, so

$$\begin{aligned}G(x_n,x_{n-1},x_{n-1})&\geq s\min\left\{\begin{array}{l}G(x_{n+1},x_n,x_n),G(x_n,x_{n-1},x_{n-1}),\\G(x_n,x_n,x_{n-1}),G(x_{n+1},x_{n-1},x_n),\end{array}\right\}\\&\geq s\min\left\{G(x_{n+1},x_n,x_n),\tfrac{1}{2}G(x_n,x_{n-1},x_{n-1})\right\}.\end{aligned} \tag{9.8}$$

This implies that

$$G(x_n,x_{n-1},x_{n-1})\geq sG(x_{n+1},x_n,x_n),$$

hence

$$G(x_{n+1},x_n,x_n)\leq\frac{1}{s}G(x_n,x_{n-1},x_{n-1}).$$

Set $k=\frac{1}{s}$, then $k<1$. By induction, we obtain

$$\begin{aligned}G(x_{n+1},x_n,x_n)&\leq kG(x_n,x_{n-1},x_{n-1})\\&\leq k^2G(x_{n-1},x_{n-2},x_{n-2})\\&\leq\cdots\\&\leq k^nG(x_1,x_0,x_0).\end{aligned} \tag{9.9}$$

Then, by the rectangle inequality, we have for all $n, m \in \mathbb{N}; n < m$,

$$\begin{aligned} G(x_m, x_n, x_n) &\leq G(x_m, x_{m-1}, x_{m-1}) + G(x_{m-1}, x_{m-2}, x_{m-2}) \\ &\quad G(x_{m-2}, x_{m-3}, x_{m-3}) + \cdots + G(x_{n+1}, x_n, x_n) \\ &\leq (k^{m-1} + k^{m-2} + \cdots + k^n) G(x_1, x_0, x_0) \\ &= \frac{k^n}{1-k} G(x_1, x_0, x_0), \end{aligned} \tag{9.10}$$

so, $G(x_m, x_n, x_n) \longrightarrow 0$ as $n, m \longrightarrow +\infty$ and (x_n) is a G-Cauchy sequence. Since (X, G) is G-complete, then there exists $u \in X$ such that (x_n) is G-convergent to u.

We have to show that $Tu = u$, let $y \in T^{-1}(u)$, for n such that $x_n \neq u$, we have $G(x_n, u, u) = G(Tx_{n+1}, Ty, Ty) \geq sN(x_{n+1}, y, y)$, where

$$N(x_{n+1}, y, y) = \min \left\{ \begin{array}{c} G(x_{n+1}, x_n, y), G(y, x_{n-1}, Ty), G(x_n, x_{n-1}, Ty), \\ G(y, x_n, Ty), G(x_{n+1}, x_n, y), G(y, x_{n-1}, Ty), G(x_n, x_{n-1}, Ty), \\ G(y, x_n, Ty), G(x_{n+1}, y, y), G(x_{n+1}, x_n, x_n), G(y, Ty, Ty), \\ G(y, Ty, Ty), G(y, x_n, x_n), G(x_{n+1}, Ty, Ty), G(y, Ty, Ty) \end{array} \right\}, \tag{9.11}$$

taking the limit as $n \longrightarrow +\infty$, we obtain $G(u, u, y) < 0$, which implies that $Ty = u = y$, hence u is a fixed point of T.

Finally, we show that u is the unique fixed point of T. Assume that contrary to our claim, there exists an other fixed point $v \in X$ with $u \neq v$, then from (9.4), we have $G(u, u, v) = G(Tu, Tu, Tv) \geq sN(u, u, v)$, where

$$N(u, u, v) = \min \left\{ \begin{array}{c} G(u, Tu, u), G(u, T^2u, Tu), G(Tu, T^2u, Tu), \\ G(u, Tu, Tu), G(u, Tu, v), G(v, T^2u, Tv), G(Tu, T^2u, Tv), \\ G(v, Tu, Tu), G(u, u, v), G(u, Tu, Tu), G(u, Tu, Tu), \\ G(v, Tv, Tv), G(v, Tu, Tu), G(u, Tu, Tu), G(u, Tv, Tv) \end{array} \right\}, \tag{9.12}$$

this entails

$$G(u, u, v) \geq \frac{s}{2} G(u, u, v),$$

which is a contradiction since $s > 2$ hence $u = v$. We conclude that T has a unique fixed point. □

Our second main result is given by

Theorem 9.3. *Let (X, G) be a complete symmetric G-metric space and let $T : X \to X$ be an onto G-continuous mapping. Assume that, there exist non negative reals a, b, c, d, e, f, g with $b < 1$, and $a + b + c + d + e + f + g > 1$ such that*

$$\begin{aligned} G^4(Tx, T^2x, T^2x) &\geq aG^4(Tx, Tx, x) + bG^4(Tx, T^2x, Tx) + cG^4(Tx, T^2x, x) \\ &\quad + dG^2(Tx, T^2x, Tx)G^2(Tx, Tx, x) + eG^2(Tx, T^2x, x)G^2(Tx, Tx, x) \end{aligned}$$

$$+fG^2(Tx,T^2x,Tx)G^2(Tx,T^2x,x)$$
$$+gG^2(T^2x,T^2x,Tx)G^2(Tx,Tx,x), \qquad (9.13)$$

for all $x \in X$, then T has a fixed point in X.

Proof. Let $x_0 \in X$, since T is onto there exists $x_1 \in T^{-1}(x_0)$. Continuing in this way, we get a sequence (x_n), where $x_n \in T^{-1}(x_{n-1})$.

If, $x_n = x_{n-1}$ for some n, then we obtain x_n as a fixed point of T. Hence, without loss of generality we may assume that $x_n \neq x_{n-1}$ for every $n \in \mathbb{N}$.

By (9.13), we get

$$\begin{aligned} G^4(x_n,x_{n-1},x_{n-1}) &= G^4(Tx_{n+1},T^2x_{n+1},T^2x_{n+1}) \\ &\geq aG^4(x_n,x_n,x_{n+1}) + bG^4(x_n,x_{n-1},x_n) + cG^4(x_n,x_{n-1},x_{n+1}) \\ &\quad + dG^2(x_n,x_{n-1},x_n)G^2(x_n,x_n,x_{n+1}) + eG^2(x_n,x_{n-1},x_{n+1})G^2(x_n,x_n,x_{n+1}) \\ &\quad + fG^2(x_n,x_{n-1},x_n)G^2(x_n,x_{n-1},x_{n+1}) + gG^2(x_{n-1},x_{n-1},x_n)G^2(x_n,x_n,x_{n+1}), \end{aligned}$$

hence

$$(a+c)G^4(x_n,x_n,x_{n+1})+(d+e+f+g)G(x_n,x_n,x_{n+1})G(x_{n-1},x_{n-1},x_n)-(1-b)G^4(x_n,x_{n-1},x_n) \leq 0, \qquad (9.14)$$

which is equivalent to

$$(a+c)r^4 + (d+e+f+g)r^2 - (1-b) \leq 0, \qquad (9.15)$$

where

$$r = \frac{G(x_n,x_{n+1},x_n)}{G(x_{n-1},x_n,x_{n-1})}.$$

Let $f : [0,+\infty[$ the function given by

$$f(r) = (a+c)r^4 + (d+e+f+g)r^2 - (1-b) = 0,$$

then, using assumptions, we have $f(0) = b-1 < 0$ and $g(1) = a+b+c+d+e+f+g-1 > 0$.

One can deduce that, there exists $k \in \,]0,1[$ for which the inequality (9.15) holds whenever $r \leq k$, and hence

$$\begin{aligned} G(x_n,x_{n+1},x_n) &\leq kG(x_{n-1},x_n,x_{n-1}) \\ &\leq k^2G(x_{n-2},x_{n-1},x_{n-2}) \\ &\leq \cdots \\ &\leq k^nG(x_0,x_1,x_0), \end{aligned}$$

then for all $n,m \in \mathbb{N}$, $m > n$, we have

$$\begin{aligned}G(x_m, x_n, x_n) &\le G(x_m, x_{m-1}, x_{m-1}) + G(x_{m-1}, x_{m-2}, x_{m-2})\\ &\quad G(x_{m-2}, x_{m-3}, x_{m-3}) + \cdots + G(x_{n+1}, x_n, x_n)\\ &\le (k^{m-1} + k^{m-2} + \cdots + k^n)G(x_1, x_0, x_0)\\ &= \frac{k^n}{1-k} G(x_1, x_0, x_0),\end{aligned} \tag{9.16}$$

so $G(x_m, x_n, x_n) \to 0$ as $n, m \longrightarrow +\infty$, and (x_n) is a G-Cauchy sequence, and by the completeness of (X, G), there exists $u \in X$ such that (x_n) G-converges to u.

By G-continuity of T, we have

$$T(x_n) = x_{n-1} \to Tu,$$

thus $u = Tu$. We conclude that u is a fixed point of T. □

We state, by the sequel our third result.

Theorem 9.4. *Let (X, G) be a complete G-metric space and let $T : X \to X$ be an onto G-continuous mapping. Assume that, there exist non-negative reals b_1, b_2, b_3, with $b_1 + b_2 + b_3 > 1$ such that*

$$G(Tx, T^2x, T^2x) \ge b_1 G(x, Tx, Tx) + b_2 G(x, Tx, T^2x) + b_3 G(Tx, T^2x, T^2x), \tag{9.17}$$

for all $x \in X$, then T has a fixed point in X.

Proof. Let $x_0 \in X$, since T is onto there exists $x_1 \in T^{-1}(x_0)$. Continuing in this way, we get a sequence (x_n), where $x_n \in T^{-1}(x_{n-1})$.

If, $x_n = x_{n-1}$ for some n, then we obtain x_n as a fixed point of T. Hence, without loss of generality, we may assume that $x_n \ne x_{n-1}$ for every $n \in \mathbb{N}$.

By (9.17), we get

$$\begin{aligned}G(x_n, x_{n-1}, x_{n-1}) &= G(Tx_{n+1}, T^2x_{n+1}, T^2x_{n+1})\\ &\ge b_1 G(x_{n+1}, Tx_{n+1}, Tx_{n+1}) + b_2 G(x_{n+1}, Tx_{n+1}, T^2x_{n+1})\\ &\quad + b_3 G(Tx_{n+1}, T^2x_{n+1}, T^2x_{n+1})\\ &= b_1 G(x_{n+1}, x_n, x_n) + b_2 G(x_{n+1}, x_n, x_{n-1}) + b_3 G(x_n, x_{n-1}, x_{n-1}),\end{aligned}$$

using the G-metric property, we have $G(x_{n+1}, x_n, x_{n-1}) \ge G(x_{n-1}, x_{n-1}, x_n)$ and thus

$$G(x_n, x_{n-1}, x_{n-1}) \ge b_1 G(x_{n+1}, x_n, x_n) + (b_2 + b_3) G(x_n, x_{n-1}, x_{n-1}), \tag{9.18}$$

this entails

$$(1 - b_2 - b_3) G(x_n, x_{n-1}, x_{n-1}) \ge b_1 G(x_{n+1}, x_n, x_n) \tag{9.19}$$

or

$$G(x_{n+1}, x_n, x_n) \le \frac{(1 - b_2 - b_3)}{b_1} G(x_n, x_{n-1}, x_{n-1}). \tag{9.20}$$

Set $k = \frac{(1-b_2-b_3)}{b_1}$, then $k < 1$ and by repeated use of the previous inequality, we obtain

$$G(x_{n+1}, x_n, x_n) \le k^n G(x_1, x_0, x_0). \tag{9.21}$$

By the rectangle inequality and (9.21), we have for all $n, m \in \mathbb{N}$, $n < m$

$$\begin{aligned} G(x_m, x_n, x_n) &\le G(x_m, x_{m-1}, x_{m-1}) + G(x_{m-1}, x_{m-2}, x_{m-2}) \\ &\quad G(x_{m-2}, x_{m-3}, x_{m-3}) + \cdots + G(x_{n+1}, x_n, x_n) \\ &\le (k^{m-1} + k^{m-2} + \cdots + k^n) G(x_1, x_0, x_0) \\ &= \frac{k^n}{1-k} G(x_1, x_0, x_0), \end{aligned} \tag{9.22}$$

letting $n, m \longrightarrow +\infty$ in (9.22), we get that $G(x_m, x_n, x_n) \to 0$. Hence (x_n) is a G-Cauchy sequence in X, since (X, G) is G-complete, then there exists $u \in X$ such that (x_n) is G-convergent to u. We have to show that $u = Tu$. In fact, By G-continuity of T, we have

$$T(x_n) = x_{n-1} \to Tu,$$

thus $u = Tu$. We conclude that u is a fixed point of T. □

As an application of Theorem (9.17), we have the following result.

Corollary 9.1. *Let (X, G) be a complete G-metric space and let $T : X \to X$ be an onto G-continuous mapping satisfying*

$$\begin{aligned} G(Tx, Ty, Tz) &\ge c_1 G(x, y, z) + c_2 G(x, Tx, Tx) + c_3 G(y, Ty, Ty) \\ &\quad + c_4 G(z, Tz, Tz) + c_5 \left[\frac{G(x, Tx, Ty) + G(y, Ty, Tz) + G(z, Tz, Tx)}{3} \right] \\ &\quad + c_6 \left[\frac{G(Tx, y, Ty) + G(Ty, z, Tz) + G(Tz, x, Tx)}{3} \right] \\ &\quad + c_7 \left[\frac{G(x, y, Tz) + G(y, z, Tx) + G(z, x, Ty)}{3} \right], \end{aligned} \tag{9.23}$$

where $c_1, c_2, c_3, c_4, c_5, c_6, c_7$ are non-negative reals, which verify

$$2(3c_1 + 3c_2 + c_3 + c_4 + 2c_7) + 5(c_5 + c_6) > 6,$$

then, T has a fixed point in X.

Proof. Replacing y and z by Tx in (9.23), we obtain

$$G(Tx, T^2x, T^2x) \ge c_1 G(x, Tx, Tx) + c_2 G(x, Tx, Tx) + c_3 G(Tx, T^2x, T^2x)$$

$$
\begin{aligned}
&+ c_4 G(Tx, T^2x, T^2x) \\
&+ c_5 \left[\frac{G(x, Tx, T^2x) + G(Tx, T^2x, T^2x) + G(Tx, T^2x, Tx)}{3} \right] \\
&+ c_6 \left[\frac{G(Tx, Tx, T^2x) + G(T^2x, Tx, T^2x) + G(T^2x, x, Tx)}{3} \right] \\
&+ c_7 \left[\frac{G(x, Tx, T^2x) + G(Tx, Tx, Tx) + G(Tx, x, T^2x)}{3} \right], \qquad (9.24)
\end{aligned}
$$

without loss of generality, we may assume that $Tx \neq T^2x$, then (9.24) entails

$$
\begin{aligned}
G(Tx, T^2x, T^2x) \geq{}& (c_1 + c_2)G(x, Tx, Tx) + \frac{(c_5 + c_6 + 2c_7)}{3} G(x, Tx, T^2x) \\
&+ \frac{(c_3 + c_4 + c_5 + c_6)}{3} G(Tx, T^2x, T^2x) \\
&+ \frac{(c_5 + c_6)}{3} G(Tx, Tx, T^2x), \qquad (9.25)
\end{aligned}
$$

due to G-metric property, we have $G(Tx, Tx, T^2x) \geq \frac{1}{2} G(Tx, T^2x, T^2x)$, hence

$$
\begin{aligned}
G(Tx, T^2x, T^2x) \geq{}& (c_1 + c_2)G(x, Tx, Tx) + \frac{(c_5 + c_6 + 2c_7)}{3} G(x, Tx, T^2x) \\
&+ \frac{(c_3 + c_4 + c_5 + c_6)}{3} G(Tx, T^2x, T^2x) \\
&+ \frac{(c_5 + c_6)}{6} G(Tx, T^2x, T^2x), \qquad (9.26)
\end{aligned}
$$

which implies that

$$
\begin{aligned}
G(Tx, T^2x, T^2x) \geq{}& (c_1 + c_2)G(x, Tx, Tx) + \frac{(c_5 + c_6 + 2c_7)}{3} G(x, Tx, T^2x) \\
&+ \frac{(2c_3 + 2c_4 + 3c_5 + 3c_6)}{6} G(Tx, T^2x, T^2x), \qquad (9.27)
\end{aligned}
$$

Setting, $b_1 = c_1 + c_2$, $b_2 = \frac{(c_5+c_6+2c_7)}{3}$, and $b_3 = \frac{(2c_3+2c_4+3c_5+3c_6)}{6}$, then by assumption, we obtain $b_1 + b_2 + b_3 > 1$. So, (9.27) becomes condition (9.17), therefore the result follows from Theorem 9.4. □

Now, we give an example illustrating our result in Theorem 9.4.

Example 9.3. Let $\mathbb{R}$ be the set of real numbers. Define $G : \mathbb{R} \times \mathbb{R} \times \mathbb{R} \to \mathbb{R}^+$ by

$$
G(x, y, z) = \max\{|x - y|, |y - z|, |z - x|\}, \quad \text{for all } x, y, z \in \mathbb{R}.
$$

Then, it is obvious that $(\mathbb{R}, G)$ is a G-metric space.

Let $T : \mathbb{R} \to \mathbb{R}$ be a mapping defined by $Tx = 2x$.

We have

$$G(Tx, T^2x, T^2x) = \max\{|Tx - T^2x|, |T^2x - T^2x|, |T^2x - Tx|\} = 2|x|,$$

or

$$G(x, Tx, Tx) = \max\{|x - Tx|, |Tx - x|\} = |x|,$$

and

$$G(x, Tx, T^2x) = \max\{|x - Tx|, |Tx - T^2x|, |T^2x - x|\} = 3|x|.$$

Set $b_1 = \frac{1}{2}$, $b_2 = \frac{1}{4}$, and $b_3 = \frac{3}{8}$, so conditions of Theorem 9.4 are verified, hence T admits a fixed point in $\mathbb{R}$. One can see that the fixed point here is zero.

Remark 9.2. (i) Notice that in Theorem 9.2, we have extended the result of Theorem 3.1 in [21] to the context of expansive mappings.

(ii) In Theorem 9.3 and Theorem 9.4, we have used a different technique to establish existence result. We have shown that the condition in the statement of theorems could be expressed in one variable and some iterates of the mapping.

Bibliography

[1] Abbas, M., Hussain, A., Popović, B. and Radenović, S., *Istratescu–Susuki–Ćirić-type fixed points results in the framework of G-metric spaces*, J. Nonlinear Sci. Appl. 9 (2016), 6077–6095.

[2] Agarwal, R. P., Kadelburg, Z. and Radenović, S., *On coupled fixed-point results in asymmetric G-metric spaces*, J. Inequal. Appl. 2013, 528 (2013).

[3] Agarwal, R. P., Karapinar, E., O'Regan, D. and Roldán-López-de-Hierro, A. F., *Fixed-Point Theory in Metric Type Spaces*, Springer Int. Publ., Switzerland, 2015.

[4] An, T. V., Dung, N. V., Kadelburg, Z. and Radenović, S., *Various generalizations of metric spaces and fixed-point theorems*, Rev. R. Acad. Cienc. Exactas Fís. Nat., Ser. A Mat. 109 (2015), 175–198.

[5] Aydi, H., Shatanawi, W. and Vetro, C., *On generalized weak G-contraction mapping in G-metric spaces*, Comput. Math. Appl. 62 (2011), 4223–4229.

[6] Debnath, P., *Optimization through best proximity points for multivalued F-contractions*, Miskolc Math. Notes 22(1) (2021), 143–151.

[7] Debnath, P., *Set-valued Meir–Keeler, Geraghty and Edelstein type fixed-point results in b-metric spaces*, Rend. Circ. Mat. Palermo (2) 70(3) (2021), 1389–1398.

[8] Debnath, P., *Banach, Kannan, Chatterjea, and Reich-type contractive inequalities for multivalued mappings and their common fixed points*, Math. Methods Appl. Sci. 45(3) (2022), 1587–1596.

[9] Debnath, P., *New common fixed-point theorems for Gornicki-type mappings and enriched contractions*, São Paulo J. Math. Sci. 16(2) (2022), 1401–1408.

[10] Debnath, P., *Common fixed-point and fixed-circle results for a class of discontinuous F-contractive mappings*, Mathematics 10(9) (2022), 1605.

[11] Debnath, P., *A new extension of Kannan's fixed-point theorem via F-contraction with application to integral equations*, Asian-Eur. J. Math. 15(7) (2022), 2250123.

[12] Debnath, P., *Results on discontinuity at fixed point for a new class of contractive mappings*, Sahand Commun. Math. Anal. 20(4) (2023), 21–32.
[13] Debnath, P., *Best proximity points of multivalued Geraghty contractions*, Miskolc Math. Notes 24(1) (2023), 119–127.
[14] Debnath, P., *On some open questions about multivalued enriched interpolative Kannan and Ciric–Reich–Rus type contractions in a Banach space*, Fixed Point Theory 26(2) (2025), 465–476.
[15] Debnath, P., Mitrović, Z. and Radenović, S., *Interpolative Hardy–Rogers and Reich–Rus–Ćirić type contractions in b-metric spaces and rectangular b-metric spaces*, Mat. Vesn. 72 (2020), 368–374.
[16] Debnath, P., Neog, M. and Radenović, S., *Set valued Reich type G-contractions in a complete metric space with graph*, Rend. Circ. Mat. Palermo (2) 69 (2020), 917–924.
[17] Debnath, P. and Srivastava, H. M., *New extensions of Kannan's and Reich's fixed-point theorems for multivalued maps using Wardowski's technique with application to integral equations*, Symmetry 12(7) (2020), 1090.
[18] Jeong, G. S. and Rhoades, B. E., *A comparison between the fixed point sets of T and T^n*, Fixed Point Theory Appl. 6 (to appear).
[19] Jeong, G. S. and Rhoades, B. E., *Maps for which $F(T) = F(T^n)$*, Fixed Point Theory Appl. 6 (2005), 71–105.
[20] Jeong, G. S. and Rhoades, B. E., *More maps for which $F(T) = F(T^n)$*, Demonstr. Math. 40 (2007), 671–680.
[21] Karapinar, E. and Agarwal, R., *Further fixed-point results on G-metric spaces*, Fixed Point Theory Appl. 154 (2013), 1–14.
[22] Mohanta, S. K., *Some fixed-point theorem on G-expansive mappings*, Kyungpook Math. J. 52 (2012), 155–165.
[23] Mustafa, Z., *A new structure for generalized metric spaces with applications to fixed-point theory*, Ph. D. thesis, University of Newcastle, Australia, 2005.
[24] Mustafa, Z., Awawdeh, F. and Shatanawi, W., *Fixed-point theorem for expansive mappings in G-metric spaces*, Int. J. Contemp. Math. Sci. 5 (2010), 2463–2472.
[25] Mustafa, Z., Obiedat, H. and Awawdeh, F., *Some fixed-point theorem for mapping on complete G-metric spaces*, Fixed Point Theory Appl. 2008 (2008), 1–12.
[26] Mustafa, Z. and Sims, B., *A new approach to generalized metric spaces*, J. Nonlinear Convex Anal. 7 (2006), 289–297.
[27] Mustafa, Z. and Sims, B., *Fixed-point theorems for contractive mappings in G-metric spaces*, Fixed Point Theory Appl. 2009 (2009), 1–10.
[28] Nashine, H. K., Kadelburg, Z. and Radenović, S., *Coincidence and fixed-point results under generalized weakly contractive condition in partially ordered G-metric spaces*, Filomat 27 (2013), 1333–1343.
[29] Radenović, S., *Remarks on some recent coupled coincidence point results in symmetric G-metric spaces*, J. Oper. 2013 (2013), 290525.
[30] Rhoades, B. E., *A comparison of various definitions of contractive mappings*, Trans. Am. Math. Soc. 226 (1977), 257–290.
[31] Saadati, R., Vaezpour, S. M., Vetro, P. and Rhoades, B. E., *Fixed-point theorems in generalized partially ordered G-metric spaces*, Math. Comput. Model. 52 (2010), 797–801.
[32] Shatanawi, W., Chauhan, S., Postolache, M., Abbas, M. and Radenović, S., *Common fixed points for contractive mappings of integral type in G-metric spaces*, J. Adv. Math. Stud. 6 (2013), 53–72.
[33] Srinivas Chary, V., Mohan Reddy G, S., Srinivasa Chary, D. and Radenović, S., *Existence of fixed points in G-metric spaces*, Bol. Soc. Parana. Mat. (3s) 41 (2023), 1–18.
[34] Sudhaamsh Mohan Reddy, G., Srinivas Chary, V., Srinivasa Chary, D., Radenović, S. and Mitrović, S., *Coupled fixed-point theorems of JSG-contraction on G-metric spaces*, Bol. Soc. Parana. Mat. (3s) 41 (2023), 1–10.
[35] Vujaković, J., Mitrović, S., Mitrović, Z. and Radenović, S., On Wardowski type results within *G*-metric spaces, in: Debnath, P., Torres, D. F. M. and Cho, Y. J. (eds.), *Advanced Mathematical Analysis and its Applications*, 2023.
[36] Wang, S. Z., Li, B. Y., Gao, Z. M. and Iseki, K., *Some fixed-point theorems on expansion mappings*, Math. Jpn. 29 (1984), 631–636.

Ioannis K. Argyros and Santhosh George

10 Extended Kantorovich's results on Newton's method for solving generalized equations on Hilbert space

Abstract: The applicability of Newton's method for solving generalized equations involving monotone operators is extended so that the convergence domain is extended, the estimates of the error distances are tighter, and the information on the location of the solution is at least as precise as in earlier articles. These advantages are realized by developing more precise majorizing sequences. Moreover, no additional cost is required, since the new majorant functions are specializations of the ones used in earlier articles. The last section contains cases where the earlier results cannot apply, but the ones in this article apply to solve equations.

Keywords: Hilbert space, generalized equation, Majorant function, semi-local convergence

MSC 2020: 54H25, 54E50

10.1 Introduction

In this article, we are concerned with the problem of approximating a solution $x_* \in D$ of the generalized equation

$$f(x) + G(x) \ni 0, \tag{10.1}$$

where $f : D \to H$ is a Fréchet differentiable operator, H is a Hilbert space, $D \subset H$ is an open set and $G : D \rightrightarrows H$ is a set-valued and maximal monotone operator. There is a plethora of applications that can be formulated like a problem (10.1) [8–15, 27, 16–26, 28]. As an example, let $\phi : H \to (-\infty, +\infty]$ be a proper lower semicontinuous convex function and

$$G(x) = \partial\phi(x) = \{v \in H : \phi(y) \geq \phi(x) + \langle v, y - x\rangle\},$$

for each $\phi \in H$. Then, the problem (10.1) reduces to the variational inequality

Ioannis K. Argyros, Department of Mathematical Sciences, Cameron University, Lawton, OK 73505, USA, e-mail: iargyros@cameron.edu

Santhosh George, Department of Mathematical and Computational Sciences, National Institute of Technology Karnataka, Karnataka 575 025, India, e-mail: sgeorge@nitk.edu.in

https://doi.org/10.1515/9783112215760-010

$$f(x) + \partial\phi(x) \ni 0.$$

This class includes linear and nonlinear complementary and other problems [1–3]. Newton's method is the most popular procedure for generating a sequence $\{x_n\}$ for $x_0 \in D$ converging to a solution x_* of (10.1). This method is defined for each $n = 0, 1, 2, \dots$ by

$$f(x_n) + f'(x_n)(x_{n+1} - x_n) + G(x_{n+1}) \ni 0. \tag{10.2}$$

There is a plethora of local as well as semi-local convergence results for Newton's method (10.2) under various controlling conditions on f' and G. These approaches have some limitations in general:
(ℓ_1) The convergence domain is small;
(ℓ_2) The upper bounds on the distances $\|x_n - x_*\|$ are pessimistic; and
(ℓ_3) The information on the uniqueness of the solution can be improved.

These concerns limit the applicability of the method. We address positively these concerns and under more general conditions. In particular, we improve the results in [1–3] and those of the references therein for the semi-local convergence of Newton's method (10.2).

10.2 Mathematical background

To make the article as self-contained as possible, we have reproduced some standard concepts. More detailed information can be found in [1–3]. Let $Q : H \rightrightarrows H$ be a linear operator. Then, Q is said to be a positive operator if it is self-conjugate and $\langle Q(x), x\rangle \geq 0$, for each $x \in H$. The domain and the range of H are defined by $dom(H) = \{u \in H : H(x) \neq \emptyset\}$ and $rge(H) = \{v \in H : \phi \in H(x) \text{ for some } v \in H\}$.

Definition 10.1. Let $Q : H \rightrightarrows H$ be a set-valued operator. Q is said to be monotone if for each $u_1, u_2 \in dom(Q)$ and $v \in Q(u_2)$, $v_1 \in Q(u_1)$ implies

$$\langle v - v_1, u_2 - u_1\rangle \geq 0.$$

A subset of $H \times H$ is monotone if it is the graph of a monotone operator. If $\Psi : H \to (-\infty, +\infty]$ is a proper function, then the subdifferentiable of Ψ is monotone.

Definition 10.2. Let $G : H \rightrightarrows H$ be monotone. Then, G is maximal monotone if for each $u_1, u_2 \in H$, $\langle u_2, v, y - x\rangle \geq 0$, for each $y \in dom(G)$ and $v \in G(y) \implies x \in dom(G)$ and $v \in G(x)$. Let $Q : H \to H$ be a bounded linear operator. We denote $\hat{Q} = \frac{1}{2}(Q + Q^*)$, where Q^* is the conjugate operator of Q. Then, $\hat{Q}$ is a self-conjugate operator. It is convenient to denote $\|\hat{F'}(x_0)^{-1}\|$ by c, where $F'(x_0)$ is a positive definite operator and $\hat{F'}(x_0)^{-1}$ exists.

From now on, we assume that $G : H \rightrightarrows H$ is a set-valued maximal monotone operator and $f : H \to H$ is a Fréchet-differentiable operator.

10.3 Local convergence

Some majorant conditions are developed and compared in this section. Let $R > 0$ and $\bar{\rho} = \sup\{s \in [0, R) : U(x_0, s) \subset D\}$.

Definition 10.3. We say that the operator F' satisfies the center majorant condition with respect to c if there exists a function $g_0 : [0, R) \to (-\infty, +\infty)$, which is twice continuously differentiable so that

(A_1) $c\|F'(x) - F'(X_0)\| \leq g_0'(\|x - x_0\|) - g_0'(0)$, for each $x \in U(x_0, \bar{\rho})$.
Moreover, suppose

(A_2) $\|x_1 - x_0\| \leq g_0(0)$.
Furthermore, suppose

(A_3) There exists a smallest solution $\rho \in (0, \bar{\rho})$ of the equation $g_0(t) - 1 = 0$.

(A_4) $g_0(0) > 0, g_0'(0) = -1$.

(A_5) g_0' is convex and strictly increasing.

(A_6) $g_0(s) = 0$, for some $s \in (0, R)$.

Definition 10.4. We say that the operator F' satisfies the restricted-majorant condition with respect to c if there exists a function $g : [0, \rho) \to (-\infty, +\infty)$, which is twice continuously differentiable so that

(A_7) $c\|F'(y) - F'(x)\| \leq g'(\|x - y\| + \|x - x_0\|) - g'(\|x - x_0\|)$, for each $x, y \in U(x_0, \rho)$.

(A_8) $\|x_1 - x_0\| \leq g(0)$.

(A_9) $g(0) > 0, g'(0) = -1$.

(A_{10}) g' is convex and strictly increasing.

(A_{11}) $g(s) = 0$, for some $s \in (0, \rho)$.

Definition 10.5. We say that the operator F' satisfies the majorant condition with respect to c if there exists a function $g_1 : [0, R) \to (-\infty, +\infty)$ differentiable so that

$(A_7)'$ $c\|F'(y) - F'(x)\| \leq g_1'(\|x - y\| + \|x - x_0\|) - g_1'(\|x - x_0\|)$, for each $x, y \in U(x_0, R)$.

$(A_8)'$ $\|x_1 - x_0\| \leq g_1(0)$.

$(A_9)'$ $g_1(0) > 0, g_1'(0) = -1$.

$(A_{10})'$ g_1' is convex and strictly increasing.

$(A_{11})'$ $g_1(s) = 0$, for some $s \in (0, R)$.

Remark 10.1. It follows by these definitions that

$$g_0'(s) \leq g_1'(s) \tag{10.3}$$

and

$$g'(s) \leq g_1'(s), \tag{10.4}$$

for each $s \in [0, \rho)$.

The Definition 10.5 and the conditions $(A_8)' - (A_{11})'$ are used to show the main Theorem 1 in [3]. In particular, the following estimate is obtained

$$\|\hat{f}'(x)^{-1}\| \leq -\frac{c}{f_1'(s)}, \quad \text{for each } s \in [0, t_*], \tag{10.5}$$

where t_* is the limit of the sequence $\{t_n\}$ defined by

$$t_0 = 0, \quad t_1 = \|x_1 - x_0\|, \quad t_{n+1} = t_n - \frac{g_1(t_n)}{g_1'(t_n)}, \quad n = 0, 1, 2, \ldots. \tag{10.6}$$

But the Definition 10.3 is needed to obtain instead the tighter estimate

$$\|\hat{f}'(x)^{-1}\| \leq -\frac{c}{g_0'}, \quad \text{for each } s \in [0, s_*], \tag{10.7}$$

where s_* is the limit of the sequence $\{s_n\}$ defined by

$$\begin{aligned} s_0 = 0, \quad s_1 = \|x_1 - x_0\|, \quad s_2 = s_1 - \frac{g_0(s_1)}{g_0'(s_0)}, \\ s_{n+1} = s_n - \frac{g(s_n)}{g_0'(s_n)}, \quad n = 2, 3, \ldots. \end{aligned} \tag{10.8}$$

Notice that $g_0 = g(D, \rho)$, $g_1 = g_1(D, R)$ but $g = g(D, \rho)$. The proof in Theorem 1 can be followed with the above changes to arrive at the extended semi-local convergence result for the Newton method (10.2).

Theorem 10.1. *Assume that the conditions $(A_1) - (A_{10})$ hold. Then, the sequence $\{x_n\}$ generated by Newton method (10.2) is well defined in $U(x_0, s_*)$, remains in $U(x_0, s_*)$ and is convergent to some solution of the equation (10.1). Moreover, the following estimates hold*

$$\|x_* - x_n\| \leq s_* - s_n, \quad \|x_* - x_{n+1}\| \leq \frac{s_* - s_n}{(s_* - s_n)^2}\|x_* - x_n\|^2$$

and the sequences $\{s_n\}$ and $\{x_n\}$ convergence Q– linearly so that

$$\|x_* - x_{n+1}\| \leq \frac{1}{2}\|x_* - x_n\|, \quad 0 \leq s_* - s_{n+1} \leq \frac{1}{2}(s_* - s_n),$$

and $\|x_{n+1} - x_n\| \leq s_{n+1} - s_n$. Furthermore, if
(A_{12}) $g'(s_*) < 0$,

then the sequences $\{s_n\}$ and $\{x_n\}$ converge Q-quadratically so that

$$\|x_* - x_2\| \le -\frac{D^{-1}g_0(s_*)}{2g_0'(s_*)}\|x_* - x_1\|^2,$$
$$\|x_* - x_{n+1}\| \le -\frac{D^{-1}g(s_*)}{2g_0'(s_*)}\|x_* - x_n\|^2, \quad n = 2, 3, \ldots$$
$$0 \le s_* - s_2 \le -\frac{D^{-1}g_0(s_*)}{2g_0'(s_*)}\|s_* - s_1\|^2,$$

and

$$0 \le s_* - s_{n+1} \le -\frac{D^{-1}g(s_*)}{2g_0'(s_*)}\|s_* - s_n\|^2, \quad n = 2, 3, \ldots.$$

Remark 10.2. If $g_0 = g = g_1$ the results of the Theorem 10.1 reduce to the ones in Theorem 1 in [3]. Otherwise, in view of (10.3) and (10.4), they constitute an extension. In particular, notice that $0 \le s_n \le t_n$,

$$0 \le s_{n+1} - s_n = -\frac{g(s_n)}{g_0'(s_n)} \le -\frac{g_1(t_n)}{g_1'(t_n)} = t_{n+1} - t_n$$

and $s_* \le t_*$.

Therefore, the new error analysis is tighter. It is also worth noticing that the advantages are obtained under the same computational cost, since in practice the computation of the function g_1 requires that of g_0 and g as special cases. The convergence criteria given in [3] imply the ones here, but not necessarily vice versa. In order to increase the number of classes of equations (10.1) that can be solved using the Newton method (10.2), consider instead of (A_1) and (A_2) the conditions, respectively

(B_1) $c\|f'(x) - f'(x_0)\| \le q_0(\|x - x_0\|)$, for each $x \in U(x_0, R)$; and

(B_2) $c\|f'(y) - f'(x)\| \le q(\|y - x\|)$, for each $x, y \in U(x_0, R)$, where $q_0 : [0, R) \to (-\infty, +\infty)$, $q : [0, R_0) \to (-\infty, +\infty)$ are continuous and non-decreasing function and R_0 is the smallest positive solution of the equation $q_0(t) - 1 = 0$ (if exists);

(B_3) The majorizing sequence $\{r_n\}$ is defined by

$$r_0 = 0, \quad r_1 = \|x_1 - x_0\|, \quad r_2 = r_1 + \frac{\int_0^1 q_0(1-\theta)\,d\theta(r_1 - r_0)}{1 - q_0(r_1)},$$

and

$$r_{n+1} = r_n + \frac{\int_0^1 q(1-\theta)(r_n - r_{n-1})\,d\theta(r_n - r_{n-1})}{1 - q_0(r_n)}, \quad n = 2, 3, \ldots. \tag{10.9}$$

A convergence criterion for this sequence is

(B_4) There exists $r \in [0, R_0)$ so that $q_0(r_n) < 1$ and $r_n \le r$.

It follows by this condition and (10.9) that $0 \le r_n \le r_{n+1} \le r$ and $r_* = \lim_{n\to+\infty} r_n \in [0, r)$. The limit r_* is the unique least upper bound of the sequence $\{r_n\}$. Under these conditions, the proof of Theorem 1 can be repeated to arrive at:

Theorem 10.2. *Assume that the conditions $(B_1) - (B_4)$ hold. Then, the sequence $\{x_n\}$ generated by Newton method (10.2) is well defined in $U(x_*, r_*)$, remains in $U(x_0, r_*)$ and is convergent to a solution $x_* \in U[x_0, s_*]$ of the equation (10.1). Moreover, the following error estimates hold for each $n = 0, 1, 2, \ldots$ $\|x_{n+1} - x_n\| \le r_{n+1} - r_n$ and $\|x_* - x_n\| \le r_* - r_n$.*

Remark 10.3. (1) If the function q_0 and q are chosen to be the right-hand sides of the conditions in (B_1) and (B_2), respectively, then the results of the Theorem 10.2 reduce to the ones in Theorem 10.1. However, if they are tighter, then the results of Theorem 10.2 extend the ones in Theorem 10.1, and consequently the ones in Theorem 1 in [3].

(2) Let us see how the majorizing sequences $\{s_n\}$, $\{t_n\}$ and $\{r_n\}$ become in the Lipschitz case by setting

$$\begin{aligned} g_0(s) &= \frac{M_0}{2}s^2 - s + r, \quad r = \|x_1 - x_0\|, \\ g(s) &= \frac{M}{2}s^2 - s + r, \\ g_1(s) &= \frac{M_1}{2}s^2 - s + r, \\ q_0(s) &= \frac{L_0}{2}s^2 - s + r, \\ q(s) &= \frac{L}{2}s^2 - s + r, \end{aligned}$$

and

$$q_1(s) = \frac{L_1}{2}s^2 - s + r.$$

The corresponding majorizing sequences are

$$\begin{aligned} r_0 = 0, \quad r_1 = \|x_1 - x_0\|, \quad r_2 &= r_1 + \frac{L_0(r_1 - r_0)^2}{2(1 - L_0 r_1)}, \\ r_{n+1} = r_n + \frac{L(r_n - r_{n-1})^2}{2(1 - L_0 r_n)}, &\quad n = 2, 3, \ldots, \\ s_0 = 0, \quad s_1 = \|x_1 - x_0\|, \quad s_2 &= s_1 + \frac{M_0(s_1 - s_0)^2}{2(1 - M_0 s_1)}, \\ s_{n+1} = s_n + \frac{M(s_n - s_{n-1})^2}{2(1 - M_0 s_n)}, &\quad n = 2, 3, \ldots, \end{aligned}$$

and

$$t_0 = 0, \quad t_1 = \|x_1 - x_0\|, \quad t_{n+1} = t_n + \frac{M_1(t_n - t_{n-1})^2}{2(1 - M_1 t_n)}, \quad n = 0, 1, 2, \ldots.$$

The Kantorovich convergence criterion [3] for the sequence $\{t_n\}$ is $P_1 = M_1 r \leq \frac{1}{2}$, the one for the sequence $\{s_n\}$ is $P = \bar{M} r \leq \frac{1}{2}$, where $\bar{M} = \frac{1}{8}(4M_0 + \sqrt{MM_0 + 8M_0^2} + \sqrt{M_0 M})$ and the one for r_n is $P_0 = \bar{L} r \leq \frac{1}{2}$, where $\bar{L} = \frac{1}{8}(4L_0 + \sqrt{L_0 L + 8L_0^2} + \sqrt{L_0 L})$. Notice that $\bar{M} \leq L_1$ and $\bar{L} \leq L_1$. Consequently, $P_1 \leq \frac{1}{2} \implies P \leq \frac{1}{2}$ and $P_1 \leq \frac{1}{2} \implies P_0 \leq \frac{1}{2}$. Notice also that if $\frac{L_0}{M_1} \to 0 \implies \frac{P_0}{P_1} \to 0$ and $\frac{M_0}{M_1} \to 0 \implies \frac{P}{P_1} \to 0$. Therefore, the new convergence criteria are infinitely many times weaker than the ones given in Theorem 1 in [3]. Examples, where the constants L_0, L or M_0, M are smaller than M_1, can be found in [4–7].

Bibliography

[1] Adly, S., Van Ngai, H. and Nguyen, V. V., Newton's method for solving generalized equations: Kantorovich's and Smale's approaches, J. Math. Anal. Appl. 439(1) (2016), 396–418.

[2] Araǵon Artacho, F. J., Belyakov, A., Dontchev, A. L. and López, M., Local convergence of quasi-Newton methods under metric regularity, Comput. Optim. Appl. 58(1) (2014), 225–247.

[3] Aragón Artacho, F. J., Dontchev, A. L., Gaydu, M., Geoffroy, M. H. and Veliov, V. M., Metric regularity of Newton's iteration, SIAM J. Control Optim. 49(2) (2011), 339–362.

[4] Argyros, I. K., On the Newton–Kantorovich hypothesis for solving equations, J. Comput. Math. 169 (2004), 315–332.

[5] Argyros, I. K. and Hilout, S., Weaker conditions for the convergence of Newton's method, J. Complex. 28 (2012), 364–387.

[6] Argyros, I. K. and Hilout, S., On an improved convergence analysis of Newton's method, Appl. Math. Comput. 225 (2013), 372–386.

[7] Argyros, I. K., The Theory and Application of Iteration Methods, 2nd ed., Engineering Series, Boca Raton, Florida, USA, 2022.

[8] Cibulka, R., Dontchev, A. and Geoffroy, M. H., Inexact Newton methods and Dennis–Moŕe theorems for nonsmooth generalized equations, SIAM J. Control Optim. 53(2) (2015), 1003–1019.

[9] Cibulka, R., Dontchev, A., Preininger, J., Veliov, V. and Roubal, T., Kantorovich-type theorems for generalized equations, J. Convex Anal. 25(2) (2018), 459–486.

[10] Dontchev, A. L. and Hager, H., Lipschitzian stability for state constrained nonlinear control, SIAM J. Control Optim. 36(2) (1998), 698–718.

[11] Dontchev, A. L., Local convergence of the Newton method for generalized equations, C. R. Acad. Sci., Sér. 1 Math. 322(4) (1996), 327–331.

[12] Dontchev, A. L. and Rockafellar, R. T., Newton's method for generalized equations: a sequential implicit function theorem, Math. Program. 123(1, Ser. B) (2010), 139–159.

[13] Dontchev, A. L. and Rockafellar, R. T., A view from variational analysis, in: Implicit Functions and Solution Mappings, 2nd ed., Springer Series in Operations Research and Financial Engineering, Springer, New York, 2014.

[14] Ferreira, O. P. and Svaiter, R. B., Kantrovich's majorant principle for Newton's method, Comput. Optim. Appl. 42 (2009), 213–229, https://doi.org/10.1007/s10589-007-9082.

[15] Ferreira, O. P. and Silva, G. N., Local convergence analysis of Newton's method for solving strongly regular generalized equations, SIAM J. Optim. 27(2) (2017), 910–926.
[16] Kantorovich, L. V. and Akilov, G. P., Functional Analysis, 2nd ed., Nauka, Moscow, 1977 (Russian).
[17] Magreńan, A. A. and Argyros, I. K., A Contemporary Study of Iterative Methods: Convergence Dynamics and Applications, Academic Press/Elsevier, 2018.
[18] Nesterov, Y. and Nemirovskii, A., Interior-Point Polynomial Algorithms in Convex Programming, Society for Industrial and Applied Mathematics, 1994.
[19] Ortega, J. M. and Rheinboldt, W. C., Iterative Solution of Nonlinear Equations in Several Variables, Academic Press, New York, 1970.
[20] Proinov, P. D., Semi-local convergence of two iterative methods for simultaneous computation of polynomial zeros, C. R. Acad. Bulgare Sci. 59(7) (2006), 705–712.
[21] Rheinboldt, W. C., An adaptive continuation process for solving systems of nonlinear equations, in: Tikhonov, A. N. et al. (eds.), Mathematical Models and Numerical Methods, pub. 3, 129–142. Banach Center, Warsaw, Poland, 1977.
[22] Shakhno, S. M., Convergence of the two-step combined method and uniqueness of the solution of nonlinear operator equations, J. Comput. Appl. Math. 261(1) (2014), 378–386.
[23] Shakhno, S. M., On an iterative algorithm with superquadratic convergence for solving nonlinear operator equations, J. Comput. Appl. Math. 231 (2009), 222–235.
[24] Shakhno, S. M. and Gnatyshyn, O. P., On an iterative algorithm of order 1.839 ... for solving the nonlinear least squares problems, Appl. Math. Comput. 161(1) (2005), 253–264.
[25] Sharma, J. R. and Arora, A., An efficient derivative free numerical methods for solving systems of nonlinear equations, Appl. Anal. Discrete Math. 7(2) (2013), 390–403.
[26] Sharma, J. R., Guha, R. K. and Sharma, R., An efficient fourth order weighted Newton method for systems of nonlinear equations, Numer. Algorithms 62 (2013), 307–323.
[27] Silva, Gilson N., Kantorovich's theorem on Newton's method for solving generalized equations under the majorant condition, Appl. Math. Comput. 286 (2016), 178–188.
[28] Traub, J., Iterative Methods for Solution of Equations, Prentice-Hall, Englewood Cliffs, New Jersey, USA, 1964.

Vuk Stojiljković and Mehmet Gürdal

11 Integral-type Berezin radius inequalities

Abstract: Berezin inequalities of the vector and Berezin radius types have been found. We achieved several extensions of the well-known Berezin radius type inequalities by using the function h described in the publication by Stojiljković and Dragomir in [32] and convexity characteristics with standard operator theory approaches. In particular, the following integral-type Berezin radius inequality has been obtained, that is

$$\begin{aligned}\operatorname{ber}(\mathfrak{P}^{4k}) \le \operatorname{ber}^{2k}(\mathfrak{P}^2) &\le h(\iota)\left(\int_0^1 \left\| t|\mathfrak{P}|^{4k} + (1-t)|\mathfrak{P}^*|^{4k} \right\|_{\mathrm{ber}}^{\frac{1}{2}} dt\right)^2 \\ &\quad + \frac{h(1-\iota)}{2} w^k(\mathfrak{P}^2) \left\| |\mathfrak{P}|^{2k} + |\mathfrak{P}^*|^{2k} \right\|_{\mathrm{ber}} \\ &\le \|\mathfrak{P}\|_{\mathrm{ber}}^{4k}.\end{aligned}$$

Keywords: Berezin radius, Berezin norm, integral type inequality

MSC 2020: 47A30, 47A63

11.1 Introduction

The discipline of operator theory investigates linear mappings, or operators, between certain function spaces, taking into account features such as their spectral properties. Operator theory may also be used to investigate inequality. In inequality analysis, lower and upper limits for operators are evaluated. In almost every field of research and engineering, mathematical inequalities are the most effective tools for explaining and offering solutions to real-world issues. In operator inequalities, Berezin numbers are employed to give tighter limits. It serves as a handy tool for evaluating operator behavior in Hilbert spaces. They may be used to analyze an operator's spectrum, norm, and other properties. Many researchers find the Berezin symbol of an operator defined on the functional Hilbert space to be important in mathematics and mathematical physics. The Berezin number of an operator has been intensively studied in a variety of disciplines, including differential equations, engineering, quantum computing, and quantum physics. These inequalities are valuable tools for assessing the features of operators and investigating their behavior. Understanding the operator's behavior is dependent on

Vuk Stojiljković, Faculty of Sciences, University of Novi Sad, Novi Sad, Serbia; and Mathematical Grammar School, Kraljice Natalije 37, Belgrade, Serbia, e-mail: vuk.stojiljkovic999@gmail.com
Mehmet Gürdal, Department of Mathematics, Suleyman Demirel University, 32260, Isparta, Turkey, e-mail: gurdalmehmet@sdu.edu.tr

https://doi.org/10.1515/9783112215760-011

our capacity to quantify the magnitude and distribution of eigenvalues, which Berezin numbers provide. To do this, we introduced some essential concepts and properties of Berezin inequalities of the vector and integral type Berezin radius inequality.

With the identity operator $1_{\mathfrak{H}}$ in $\mathfrak{B}(\mathfrak{H})$, let $\mathfrak{B}(\mathfrak{H})$ be the Banach algebra of all bounded linear operators defined on a complex Hilbert space $(\mathfrak{H}, \langle .,. \rangle)$. This research paper mainly revolves around the idea of a reproducing kernel Hilbert space defined on $\mathfrak{O}$ is a Hilbert space $\mathfrak{H}$ that contains complex-valued functions, which is represented by the notation $\mathfrak{H} = \mathfrak{H}_{\mathfrak{O}}$. In this case, the linear evaluation functional $\mathfrak{E}_\gamma : \mathfrak{H} \to \mathbb{C}$ defined by $\mathfrak{E}_\gamma(f) = f(\gamma)$ must be continuous for every $\gamma \in \mathfrak{O}$. To keep things simple, the functional Hilbert space on the set $\mathfrak{O}$ is denoted by $\mathfrak{H}$ throughout this study. However, for any $\gamma \in \mathfrak{O}$, the Riesz representation theorem ensures the presence of a unique element $k_\gamma \in \mathfrak{H}_{\mathfrak{O}}$, such that the equivalence $f(\gamma) = \mathfrak{E}_\gamma(f) = \langle f, k_\gamma \rangle$ holds for all $f \in \mathfrak{H}$. Here, k_γ is referred to as the reproducing kernel, associated with γ, and the collection $\{k_\gamma : \gamma \in \mathfrak{O}\}$ constitutes the reproducing kernel of the functional Hilbert space $\mathfrak{H}$. The normalized reproducing kernel is denoted as $\widehat{k}_\gamma = k_\gamma / \|k_\gamma\|$.

For a bounded linear operator $\mathfrak{P}_1$ on $\mathfrak{H}$, the function $\widetilde{\mathfrak{P}_1}$ is defined on $\mathfrak{O}$ through $\widetilde{\mathfrak{P}_1}(\gamma) = \langle \mathfrak{P}_1 \hat{k}_\gamma, \hat{k}_\gamma \rangle$, which is termed the Berezin symbol of $\mathfrak{P}_1$, introduced by F. Berezin [14]. The set $\mathrm{Ber}(\mathfrak{P}_1) = \{\widetilde{\mathfrak{P}_1}(\gamma) : \gamma \in \mathfrak{O}\}$ is termed as Berezin set of operator $\mathfrak{P}_1$.

Karaev [25] introduced the concept of the Berezin number of operator $\mathfrak{P}_1$, denoted as $\mathrm{ber}(\mathfrak{P}_1)$. This quantity is defined as the supremum of $|\widetilde{\mathfrak{P}_1}(\gamma)|$ over all $\gamma \in \mathfrak{O}$:

$$\mathrm{ber}(\mathfrak{P}_1) = \sup_{\gamma \in \mathfrak{O}} |\widetilde{\mathfrak{P}_1}(\gamma)|.$$

The Berezin set $\mathrm{Ber}(\mathfrak{P}_1)$ is a subset of $W(\mathfrak{P}_1)$ and it is well-known that

$$\mathrm{ber}(\mathfrak{P}_1) \leq w(\mathfrak{P}_1) \leq \|\mathfrak{P}_1\|, \tag{11.1}$$

for any $\mathfrak{P}_1 \in \mathcal{B}(\mathfrak{H})$, where $W(\mathfrak{P}_1)$ and $w(\mathfrak{P}_1)$ denote the numerical range and the numerical radius of $\mathfrak{P}_1$. To explore the basic properties and some recent works on numerical range and the numerical radius, we refer to [3–5, 7, 8, 16–18, 27, 28, 33].

We also have the following power inequality

$$\mathrm{ber}(\mathfrak{P}^n) \leq \mathrm{ber}^n(\mathfrak{P}) \tag{11.2}$$

for any $\mathfrak{P} \in \mathfrak{B}(\mathfrak{H})$ and $n \in \mathbb{N}$. More generally, if $\mathfrak{P}$ is not nilpotent, then

$$\mathfrak{C}_1 \mathrm{ber}^n(\mathfrak{P}) \leq \mathrm{ber}(\mathfrak{P}^n) \leq \mathfrak{C}_2 \mathrm{ber}^n(\mathfrak{P})$$

for some constants $\mathfrak{C}_1, \mathfrak{C}_2 > 0$. In [23], Huban et al. demonstrated that if $\mathfrak{P} \in \mathfrak{B}(\mathfrak{H})$, then

$$\mathrm{ber}(\mathfrak{P}) \leq \frac{1}{2} \| |\mathfrak{P}| + |\mathfrak{P}^*| \|_{\mathrm{ber}} \leq \frac{1}{2} (\|\mathfrak{P}\|_{\mathrm{ber}} + \|\mathfrak{P}^2\|_{\mathrm{ber}}^{\frac{1}{2}}), \tag{11.3}$$

Başaran and Gürdal [12] further obtained the inequality for $\mathrm{ber}^2(.)$, which holds for any $\mathfrak{P} \in \mathfrak{B}(\mathfrak{H})$

$$\mathrm{ber}^2(\mathfrak{P}) \leq \frac{1}{6}\left\| |\mathfrak{P}|^2 + |\mathfrak{P}^*|^2 \right\|_{\mathrm{ber}} + \frac{1}{3}\mathrm{ber}(\mathfrak{P})\left\| |\mathfrak{P}| + |\mathfrak{P}^*| \right\|_{\mathrm{ber}}. \tag{11.4}$$

Furthermore, Huban et al. [23] established a few inequalities that may be expressed as follows:

$$\mathrm{ber}^{\tau}(\mathfrak{Z}) \leq \frac{1}{2}\left\| |\mathfrak{Z}|^{2\tau s} + |\mathfrak{Z}^*|^{2\tau(1-s)} \right\|_{\mathrm{ber}} \tag{11.5}$$

and

$$\mathrm{ber}^{2\tau}(\mathfrak{Z}) \leq \left\| l|\mathfrak{Z}|^{2\tau} + (1-l)|\mathfrak{Z}^*|^{2\tau} \right\|_{\mathrm{ber}}, \tag{11.6}$$

where $Z \in \mathfrak{B}(\mathfrak{H})$, $0 \leq l \leq 1$, and $\tau \geq 1$.

Recently, various interesting inequalities were obtained by Erkan and Gürdal [19], one of them being the following two operator inequality

$$\mathrm{ber}(\mathfrak{T}^*\mathfrak{S}) \leq \frac{\varkappa}{2(\varkappa+1)}\left\| |\mathfrak{T}|^2 + |\mathfrak{S}|^2 \right\|_{\mathrm{ber}} + \frac{1}{4(\varkappa+1)}\|\mathfrak{T} + \mathfrak{S}\|_{\mathrm{ber}}(\|\mathfrak{T}\|_{\mathrm{ber}} + \|\mathfrak{S}\|_{\mathrm{ber}}),$$

for $\mathfrak{T}, \mathfrak{S} \in \mathfrak{B}(\mathfrak{H})$ and $\varkappa \geq 0$.

Motivated by the recent research conducted in the area of Berezin radius inequalities, we obtain further refinements of the well-known inequalities. In particular, the inequality (11.23) given in Section 11.4 and its generalization in (11.25) are given. We also obtain a boundary for the complex Berezin radius, that is (11.43). Generalization of the vector type inequalities has been given by (11.18) and (11.19) of inequalities from Stojiljković and Gürdal [34]. You may read more about Hilbert space inequalities in the publications that follow [9–11, 13, 30–32, 21].

11.2 Preliminaries

Lemma 11.1. *If u_1, u_2, u_3 and are all in $\mathfrak{H}$, then we have*

$$|\langle u_1, u_3\rangle\langle u_3, u_2\rangle| \leq \frac{\|u_3\|^2}{2}(\|u_1\|\|u_2\| + |\langle u_1, u_2\rangle|).$$

The final conclusion is the Cauchy–Schwartz inequality extended by Buzano, see [15].

Lemma 11.2 ([29]). *Suppose that $u_1 \in \mathfrak{H}$ be any unit vector and that $\mathfrak{P} \in \mathfrak{B}(\mathfrak{H})$, $\mathfrak{P} \geq 0$. Afterwards, we get*

$$\langle \mathfrak{P}\mathfrak{u}_1, \mathfrak{u}_1\rangle^{\iota} \le \langle \mathfrak{P}^{\iota}\mathfrak{u}_1, \mathfrak{u}_1\rangle \quad \textit{for } \iota \ge 1, \tag{11.7}$$

$$\langle \mathfrak{P}^{\iota}\mathfrak{u}_1, \mathfrak{u}_1\rangle \le \langle \mathfrak{P}\mathfrak{u}_1, \mathfrak{u}_1\rangle^{\iota} \quad \textit{for } 0 < \iota \le 1. \tag{11.8}$$

The following conclusion is found in [6] and is related to non-negative convex functions.

Lemma 11.3. *Assume that $\mathfrak{P}_1$, $\mathfrak{P}_2$ are positive operators in $\mathfrak{B}(\mathfrak{H})$ and that f is a non-negative convex function on $[0, +\infty)$. It follows that*

$$\left\| f\left(\frac{\mathfrak{P}_1 + \mathfrak{P}_2}{2}\right)\right\| \le \left\|\frac{f(\mathfrak{P}_1) + f(\mathfrak{P}_2)}{2}\right\|. \tag{11.9}$$

Lemma 11.4 ([22]). *In the event when $\iota \ge 1$ and $\mathfrak{P}_1, \mathfrak{P}_2 \in \mathfrak{B}(\mathfrak{H})$,*

$$w^{\iota}(\mathfrak{P}_2^*\mathfrak{P}_1) \le \frac{1}{2}\||\mathfrak{P}_1|^{2\iota} + |\mathfrak{P}_2|^{2\iota}\|. \tag{11.10}$$

Lemma 11.5 ([34]). *Let $\mathfrak{P}_1, \mathfrak{P}_2, \mathfrak{P}_3, \mathfrak{P}_4 \in \mathfrak{B}(\mathfrak{H})$. Then, we have the inequality*

$$|\langle \mathfrak{P}_4\mathfrak{P}_3\mathfrak{P}_2\mathfrak{P}_1\widehat{k}_{\gamma}, \widehat{k}_{\delta}\rangle|^2 \le \langle \mathfrak{P}_1^*|\mathfrak{P}_2|^2\mathfrak{P}_1\widehat{k}_{\gamma}, \widehat{k}_{\gamma}\rangle\langle \mathfrak{P}_4|\mathfrak{P}_3^*|\mathfrak{P}_4^*\widehat{k}_{\delta}, \widehat{k}_{\delta}\rangle. \tag{11.11}$$

Lemma 11.6 ([1]). *Suppose $\mathfrak{u}_1 \in H$ is a unit vector and $\mathfrak{P} \in \mathfrak{B}(\mathfrak{H})$ is a self adjoint operator. If f is a convex function on an interval that contains $\mathfrak{P}_1$'s spectrum, then we have*

$$f(\langle \mathfrak{P}_1\mathfrak{u}_1, \mathfrak{u}_1\rangle) \le \langle f(\mathfrak{P}_1)\mathfrak{u}_1, \mathfrak{u}_1\rangle. \tag{11.12}$$

Lemma 11.7. *Assume that $\min\{\mu_1, \mu_2\} \le \xi_1 \le \xi_2 \le \max\{\mu_1, \mu_2\}$. is satisfied by positive real integers ξ_1, ξ_2 and $\mu_1, \mu_2 > 0$. Then, we get*

$$\frac{\xi_2 + \xi_1}{2\sqrt{\xi_1\xi_2}}\sqrt{\mu_1, \mu_2} \le \frac{\mu_2 + \mu_1}{2}. \tag{11.13}$$

Lemma 11.8. *Given a non-negative increasing convex function ψ on the interval $[0, \infty)$, with $\psi(0) = 0$ and $0 \le \tau \le 1$, let ψ be. After that,*

$$\psi(\tau t) \le \tau\psi(t). \tag{11.14}$$

Lemma 11.9 ([2]). *If $\mathfrak{u}_1$, $\mathfrak{u}_2$, $\mathfrak{u}_3$ and are all in $\mathfrak{H}$, then we have*

$$|\langle \mathfrak{u}_3, \mathfrak{u}_1\rangle|^2 + |\langle \mathfrak{u}_3, \mathfrak{u}_2\rangle|^2 \le \|\mathfrak{u}_3\|^2(|\langle \mathfrak{u}_1, \mathfrak{u}_2\rangle| + \max(\|\mathfrak{u}_1\|^2, \|\mathfrak{u}_2\|^2)). \tag{11.15}$$

Lemma 11.10. *The following relation for* max *holds*

$$\max(\mathfrak{a}_1, \mathfrak{a}_2) = \frac{1}{2}[\mathfrak{a}_1 + \mathfrak{a}_2 + |\mathfrak{a}_2 - \mathfrak{a}_1|] \quad \textit{for any } \mathfrak{a}_1, \mathfrak{a}_2 \in \mathbb{R}. \tag{11.16}$$

Lemma 11.11 ([34]). *If h is a non-negative mapping with the following characteristics: $h:(0,1)\subset J\to\mathbb{R}^{+}$ such that $h(\iota)+h(1-\iota)=1$, and $\mathfrak{P}_1$, $\mathfrak{P}_2$, $\mathfrak{P}_3$, and $\mathfrak{P}_4\in\mathfrak{B}(\mathfrak{H})$, $k\geq 1$, then we have*

$$\begin{aligned}
&|\langle\mathfrak{P}_4\mathfrak{P}_3\mathfrak{P}_2\mathfrak{P}_1\widehat{k}_\gamma,\widehat{k}_\delta\rangle|^{2k}\\
&\quad\leq h(\iota)\langle|\mathfrak{P}_2\mathfrak{P}_1|^2\widehat{k}_\gamma,\widehat{k}_\gamma\rangle^k\langle|(\mathfrak{P}_4\mathfrak{P}_3)^*|^2\widehat{k}_\delta,\widehat{k}_\delta\rangle^k\\
&\qquad+h(1-\iota)|\langle\mathfrak{P}_4\mathfrak{P}_3\mathfrak{P}_2\mathfrak{P}_1\widehat{k}_\gamma,\widehat{k}_\delta\rangle|^k\langle|\mathfrak{P}_2\mathfrak{P}_1|^2\widehat{k}_\gamma,\widehat{k}_\gamma\rangle^{\frac{k}{2}}\langle|(\mathfrak{P}_4\mathfrak{P}_3)^*|^2\widehat{k}_\delta,\widehat{k}_\delta\rangle^{\frac{k}{2}}\\
&\quad\leq\langle|\mathfrak{P}_2\mathfrak{P}_1|^2\widehat{k}_\gamma,\widehat{k}_\gamma\rangle^k\langle|(\mathfrak{P}_4\mathfrak{P}_3)^*|^2\widehat{k}_\delta,\widehat{k}_\delta\rangle^k
\end{aligned}\tag{11.17}$$

11.3 Main results

Lemma 11.12. *Let $\mathfrak{P}_1,\mathfrak{P}_2,\mathfrak{P}_3,\mathfrak{P}_4\in\mathbb{B}(\mathfrak{H})$ and the conditions from (11.17) hold, also if f is an increasing convex function such that $f:[0,+\infty)\to\mathbb{R}$, then the following inequality holds*

$$\begin{aligned}
&f(|\langle\mathfrak{P}_4\mathfrak{P}_3\mathfrak{P}_2\mathfrak{P}_1\widehat{k}_\gamma,\widehat{k}_\delta\rangle|^{2k})\\
&\quad\leq h(\iota)f(\langle\mathfrak{P}_1^*|\mathfrak{P}_2|^2\mathfrak{P}_1\widehat{k}_\gamma,\widehat{k}_\gamma\rangle^k\langle\mathfrak{P}_4|\mathfrak{P}_3^*|^2\mathfrak{P}_4^*\widehat{k}_\delta,\widehat{k}_\delta\rangle^k)\\
&\qquad+h(1-\iota)f(|\langle\mathfrak{P}_4\mathfrak{P}_3\mathfrak{P}_2\mathfrak{P}_1\widehat{k}_\gamma,\widehat{k}_\delta\rangle|^k\langle\mathfrak{P}_1^*|\mathfrak{P}_2|^2\mathfrak{P}_1\widehat{k}_\gamma,\widehat{k}_\gamma\rangle^{\frac{k}{2}}\langle\mathfrak{P}_4|\mathfrak{P}_3^*|^2\mathfrak{P}_4^*\widehat{k}_\delta,\widehat{k}_\delta\rangle^{\frac{k}{2}})\\
&\quad\leq f(\langle\mathfrak{P}_1^*|\mathfrak{P}_2|^2\mathfrak{P}_1\widehat{k}_\gamma,\widehat{k}_\gamma\rangle^k\langle\mathfrak{P}_4|\mathfrak{P}_3^*|^2\mathfrak{P}_4^*\widehat{k}_\delta,\widehat{k}_\delta\rangle^k)\\
&\quad\leq\langle h(p)f(|\mathfrak{P}_2\mathfrak{P}_1|^{\frac{2k}{h(p)}})\widehat{k}_\gamma,\widehat{k}_\gamma\rangle+\langle h(1-p)f(|(\mathfrak{P}_4\mathfrak{P}_3)^*|^{\frac{2k}{h(1-p)}})\widehat{k}_\delta,\widehat{k}_\delta\rangle.
\end{aligned}\tag{11.18}$$

Proof. Let $\gamma,\delta\in\mathfrak{D}$ be arbitrary. Think about the following,

$$\begin{aligned}
&|\langle\mathfrak{P}_4\mathfrak{P}_3\mathfrak{P}_2\mathfrak{P}_1\widehat{k}_\gamma,\widehat{k}_\delta\rangle|^{2k}\\
&\quad\leq h(\iota)\langle|\mathfrak{P}_2\mathfrak{P}_1|^2\widehat{k}_\gamma,\widehat{k}_\gamma\rangle^k\langle|(\mathfrak{P}_4\mathfrak{P}_3)^*|^2\widehat{k}_\delta,\widehat{k}_\delta\rangle^k\quad\text{(by (11.17))}\\
&\qquad+h(1-\iota)|\langle\mathfrak{P}_4\mathfrak{P}_3\mathfrak{P}_2\mathfrak{P}_1\widehat{k}_\gamma,\widehat{k}_\delta\rangle|^k\langle|\mathfrak{P}_2\mathfrak{P}_1|^2\widehat{k}_\gamma,\widehat{k}_\gamma\rangle^{\frac{k}{2}}\langle|(\mathfrak{P}_4\mathfrak{P}_3)^*|^2\widehat{k}_\delta,\widehat{k}_\delta\rangle^{\frac{k}{2}}.
\end{aligned}$$

Applying f on both sides and using the fact that f is convex, we get

$$\begin{aligned}
&f(|\langle\mathfrak{P}_4\mathfrak{P}_3\mathfrak{P}_2\mathfrak{P}_1\widehat{k}_\gamma,\widehat{k}_\delta\rangle|^{2k})\\
&\quad\leq h(\iota)f(\langle|\mathfrak{P}_2\mathfrak{P}_1|^2\widehat{k}_\gamma,\widehat{k}_\gamma\rangle^k\langle|(\mathfrak{P}_4\mathfrak{P}_3)^*|^2\widehat{k}_\delta,\widehat{k}_\delta\rangle^k)\\
&\qquad+h(1-\iota)f(|\langle\mathfrak{P}_4\mathfrak{P}_3\mathfrak{P}_2\mathfrak{P}_1\widehat{k}_\gamma,\widehat{k}_\delta\rangle|^k\langle|\mathfrak{P}_2\mathfrak{P}_1|^2\widehat{k}_\gamma,\widehat{k}_\gamma\rangle^{\frac{k}{2}}\langle|(\mathfrak{P}_4\mathfrak{P}_3)^*|^2\widehat{k}_\delta,\widehat{k}_\delta\rangle^{\frac{k}{2}}).
\end{aligned}$$

Using the four operator inequality on $\mathfrak{P}_4\mathfrak{P}_3\mathfrak{P}_2\mathfrak{P}_1$ term and using the reality that f is increasing, that is for $a<b$, we have $f(a)<f(b)$, from Lemma 11.5, we get

$$\le f(\langle|\mathfrak{P}_2\mathfrak{P}_1|^2\widehat{k}_\gamma,\widehat{k}_\gamma\rangle^k\langle|(\mathfrak{P}_4\mathfrak{P}_3)^*|^2\widehat{k}_\delta,\widehat{k}_\delta\rangle^k).$$

By using Lemmas 11.2 and 11.6 and weighted AM-GM inequality, we now turn to obtaining the right-hand side chain of inequalities

$$\begin{aligned}
&\langle|\mathfrak{P}_2\mathfrak{P}_1|^2\widehat{k}_\gamma,\widehat{k}_\gamma\rangle^k|\langle|(\mathfrak{P}_4\mathfrak{P}_3)^*|^2\widehat{k}_\delta,\widehat{k}_\delta\rangle^k\\
&\quad\le\langle|\mathfrak{P}_2\mathfrak{P}_1|^{2k}\widehat{k}_\gamma,\widehat{k}_\gamma\rangle\langle|(\mathfrak{P}_4\mathfrak{P}_3)^*|^{2k}\widehat{k}_\delta,\widehat{k}_\delta\rangle\\
&\quad\le h(p)\langle|\mathfrak{P}_2\mathfrak{P}_1|^{\frac{2k}{h(p)}}\widehat{k}_\gamma,\widehat{k}_\gamma\rangle+h(1-p)\langle|(\mathfrak{P}_4\mathfrak{P}_3)^*|^{\frac{2k}{h(1-p)}}\widehat{k}_\delta,\widehat{k}_\delta\rangle\\
&\quad\le h(p)\langle f(|\mathfrak{P}_2\mathfrak{P}_1|^{\frac{2k}{h(p)}})\widehat{k}_\gamma,\widehat{k}_\gamma\rangle+h(1-p)\langle f(|(\mathfrak{P}_4\mathfrak{P}_3)^*|^{\frac{2k}{h(1-p)}})\widehat{k}_\delta,\widehat{k}_\delta\rangle.
\end{aligned}$$ □

The following inequalities are consequences of (11.18).

Remark 11.1. Setting $f=I$ we get an inequality given by Stojiljković and Gürdal in [34], namely

$$\begin{aligned}
&|\langle\mathfrak{P}_4\mathfrak{P}_3\mathfrak{P}_2\mathfrak{P}_1\widehat{k}_\gamma,\widehat{k}_\delta\rangle|^{2k}\\
&\quad\le h(\iota)\langle\mathfrak{P}_1^*|\mathfrak{P}_2|^2\mathfrak{P}_1\widehat{k}_\gamma,\widehat{k}_\gamma\rangle^k\langle\mathfrak{P}_4|\mathfrak{P}_3^*|^2\mathfrak{P}_4^*\widehat{k}_\delta,\widehat{k}_\delta\rangle^k\\
&\qquad+h(1-\iota)|\langle\mathfrak{P}_4\mathfrak{P}_3\mathfrak{P}_2\mathfrak{P}_1\widehat{k}_\gamma,\widehat{k}_\delta\rangle|^k\langle\mathfrak{P}_1^*|\mathfrak{P}_2|^2\mathfrak{P}_1\widehat{k}_\gamma,\widehat{k}_\gamma\rangle^{\frac{k}{2}}\langle\mathfrak{P}_4|\mathfrak{P}_3^*|^2\mathfrak{P}_4^*\widehat{k}_\delta,\widehat{k}_\delta\rangle^{\frac{k}{2}}\\
&\quad\le\langle\mathfrak{P}_1^*|\mathfrak{P}_2|^2\mathfrak{P}_1\widehat{k}_\gamma,\widehat{k}_\gamma\rangle^k\langle\mathfrak{P}_4|\mathfrak{P}_3^*|^2\mathfrak{P}_4^*\widehat{k}_\delta,\widehat{k}_\delta\rangle^k.
\end{aligned}$$

We also have the following inequality as a consequence of (11.18).

The inequality

$$\begin{aligned}
&f(|\langle\mathfrak{T}|\mathfrak{T}|^{\omega-1}\mathfrak{T}|\mathfrak{T}|^{\tau-1}\widehat{k}_\gamma,\widehat{k}_\delta\rangle|^{2k})\\
&\quad\le h(\iota)f(\langle|\mathfrak{T}|^{2\tau}\widehat{k}_\gamma,\widehat{k}_\gamma\rangle^k\langle|\mathfrak{T}|^{2\omega}\widehat{k}_\delta,\widehat{k}_\delta\rangle^k)\\
&\qquad+h(1-\iota)f(|\langle\mathfrak{T}|\mathfrak{T}|^{\omega-1}\mathfrak{T}|\mathfrak{T}|^{\tau-1}\widehat{k}_\gamma,\widehat{k}_\delta\rangle|^k\langle|\mathfrak{T}|^{2\tau}\widehat{k}_\gamma,\widehat{k}_\gamma\rangle^{\frac{k}{2}}\langle|\mathfrak{T}|^{2\omega}\widehat{k}_\delta,\widehat{k}_\delta\rangle^{\frac{k}{2}})\\
&\quad\le f(\langle|\mathfrak{T}|^{2\tau}\widehat{k}_\gamma,\widehat{k}_\gamma\rangle^k\langle|\mathfrak{T}|^{2\omega}\widehat{k}_\delta,\widehat{k}_\delta\rangle^k)\\
&\quad\le\langle h(p)f(|\mathfrak{T}|^{\frac{2k\tau}{h(p)}})\widehat{k}_\gamma,\widehat{k}_\gamma\rangle+\langle h(1-p)f(|\mathfrak{T}|^{\frac{2k\omega}{h(1-p)}})\widehat{k}_\delta,\widehat{k}_\delta\rangle
\end{aligned}\tag{11.19}$$

holds for every operator $\mathfrak{T}\in\mathfrak{B}(\mathfrak{H})$ and any $\tau,\omega\ge 1$.

Proof. Following the proof of (eq 24 in [34]) the result follows, therefore, we omit the details. □

Setting $f=I$ in (11.19) and considering the first two inequalities from the left, we get the inequality given by Stojiljković and Gürdal in (eq. 24 in [34]), that is

$$\begin{aligned}
&|\langle\mathfrak{T}|\mathfrak{T}|^{\omega-1}\mathfrak{T}|\mathfrak{T}|^{\tau-1}\widehat{k}_\gamma,\widehat{k}_\delta\rangle|^{2k}\\
&\quad\le h(\iota)\langle|\mathfrak{T}|^{2\tau}\widehat{k}_\gamma,\widehat{k}_\gamma\rangle^k\langle|\mathfrak{T}|^{2\omega}\widehat{k}_\delta,\widehat{k}_\delta\rangle^k
\end{aligned}$$

$$+ h(1-\iota)\left|\langle \mathfrak{T}|\mathfrak{T}|^{\omega-1}\mathfrak{T}|\mathfrak{T}|^{\tau-1}\widehat{k}_\gamma, \widehat{k}_\delta\rangle\right|^k \langle |\mathfrak{T}|^{2\tau}\widehat{k}_\gamma, \widehat{k}_\gamma\rangle^{\frac{k}{2}} \langle |\mathfrak{T}|^{2\omega}\widehat{k}_\delta, \widehat{k}_\delta\rangle^{\frac{k}{2}}$$
$$\le \langle |\mathfrak{T}|^{2\tau}\widehat{k}_\gamma, \widehat{k}_\gamma\rangle^k \langle |\mathfrak{T}|^{2\omega}\widehat{k}_\delta, \widehat{k}_\delta\rangle^k.$$

Lemma 11.13. *Let f be a non-negative increasing convex function on $[0,+\infty)$ and $\mathfrak{P}_1$, $\mathfrak{P}_2$, $\mathfrak{P}_3$ and $\mathfrak{P}_4 \in \mathfrak{B}(\mathfrak{H})$, also let the conditions from (11.17) hold. If*

$$0 < |\mathfrak{P}_2\mathfrak{P}_1|^2 \le \xi \le \eta \le |(\mathfrak{P}_4\mathfrak{P}_3)^*|^2$$

or

$$0 < |(\mathfrak{P}_4\mathfrak{P}_3)^*|^2 \le \xi \le \eta \le |\mathfrak{P}_2\mathfrak{P}_1|^2,$$

then

$$\begin{aligned}
& f(|\langle \mathfrak{P}_4\mathfrak{P}_3\mathfrak{P}_2\mathfrak{P}_1\widehat{k}_\gamma, \widehat{k}_\delta\rangle|^{2k}) \\
&\quad \le h(\iota) f(\langle |\mathfrak{P}_2\mathfrak{P}_1|^2\widehat{k}_\gamma, \widehat{k}_\gamma\rangle^k \langle |(\mathfrak{P}_4\mathfrak{P}_3)^*|^2\widehat{k}_\delta, \widehat{k}_\delta\rangle^k) \\
&\qquad + h(1-\iota) f(|\langle \mathfrak{P}_4\mathfrak{P}_3\mathfrak{P}_2\mathfrak{P}_1\widehat{k}_\gamma, \widehat{k}_\delta\rangle|^k \langle |\mathfrak{P}_2\mathfrak{P}_1|^2\widehat{k}_\gamma, \widehat{k}_\gamma\rangle^{\frac{k}{2}} \langle |(\mathfrak{P}_4\mathfrak{P}_3)^*|^2\widehat{k}_\delta, \widehat{k}_\delta\rangle^{\frac{k}{2}}) \\
&\quad \le f(\langle |\mathfrak{P}_2\mathfrak{P}_1|^2\widehat{k}_\gamma, \widehat{k}_\gamma\rangle^k \langle |(\mathfrak{P}_4\mathfrak{P}_3)^*|^2\widehat{k}_\delta, \widehat{k}_\delta\rangle^k) \\
&\quad \le \frac{2\sqrt{\xi\eta}}{\xi+\eta}\left(\frac{\langle f(|\mathfrak{P}_2\mathfrak{P}_1|^{4k})\widehat{k}_\gamma, \widehat{k}_\gamma\rangle + \langle f(|(\mathfrak{P}_4\mathfrak{P}_3)^*|^{4k})\widehat{k}_\delta, \widehat{k}_\delta\rangle}{2}\right).
\end{aligned} \tag{11.20}$$

Proof. Let $\gamma, \delta \in \mathfrak{O}$ be arbitrary. Using the similar reasoning as in (11.18) and Lemmas 11.6–11.8, we continue with the following

$$\begin{aligned}
& \langle |\mathfrak{P}_2\mathfrak{P}_1|^{2k}\widehat{k}_\gamma, \widehat{k}_\gamma\rangle \langle |(\mathfrak{P}_4\mathfrak{P}_3)^*|^{2k}\widehat{k}_\delta, \widehat{k}_\delta\rangle \\
&\quad \le \frac{2\sqrt{\xi\eta}}{\xi+\eta}\left(\frac{\langle |\mathfrak{P}_2\mathfrak{P}_1|^{4k}\widehat{k}_\gamma, \widehat{k}_\gamma\rangle + \langle |(\mathfrak{P}_4\mathfrak{P}_3)^*|^{4k}\widehat{k}_\delta, \widehat{k}_\delta\rangle}{2}\right).
\end{aligned}$$

By using the convexity of f, we arrive

$$\begin{aligned}
& f(\langle |\mathfrak{P}_2\mathfrak{P}_1|^{2k}\widehat{k}_\gamma, \widehat{k}_\gamma\rangle \langle |(\mathfrak{P}_4\mathfrak{P}_3)^*|^{2k}\widehat{k}_\delta, \widehat{k}_\delta\rangle) \\
&\quad \le \frac{2\sqrt{\xi\eta}}{\xi+\eta} f\left(\frac{\langle |\mathfrak{P}_2\mathfrak{P}_1|^{4k}\widehat{k}_\gamma, \widehat{k}_\gamma\rangle + \langle |(\mathfrak{P}_4\mathfrak{P}_3)^*|^{4k}\widehat{k}_\delta, \widehat{k}_\delta\rangle}{2}\right).
\end{aligned}$$

So,

$$\begin{aligned}
&\le \frac{2\sqrt{\eta\xi}}{\xi+\eta}\left(\frac{f(\langle |\mathfrak{P}_2\mathfrak{P}_1|^{4k}\widehat{k}_\gamma, \widehat{k}_\gamma\rangle) + f(\langle |(\mathfrak{P}_4\mathfrak{P}_3)^*|^{4k}\widehat{k}_\delta, \widehat{k}_\delta\rangle)}{2}\right) \\
&\le \frac{2\sqrt{\xi\eta}}{\xi+\eta}\left(\frac{\langle f(|\mathfrak{P}_2\mathfrak{P}_1|^{4k})\widehat{k}_\gamma, \widehat{k}_\gamma\rangle + \langle f(|(\mathfrak{P}_4\mathfrak{P}_3)^*|^{4k})\widehat{k}_\delta, \widehat{k}_\delta\rangle}{2}\right)
\end{aligned}$$

is obtained by once more utilizing the fact that f is convex. □

Remark 11.2. Setting $f(\mathfrak{x}) = \mathfrak{x}$, we obtain

$$\begin{aligned}
&|\langle \mathfrak{P}_4\mathfrak{P}_3\mathfrak{P}_2\mathfrak{P}_1\widehat{k}_\gamma, \widehat{k}_\delta\rangle|^{2k} \\
&\leq h(\iota)\langle |\mathfrak{P}_2\mathfrak{P}_1|^2\widehat{k}_\gamma, \widehat{k}_\gamma\rangle^k \langle |(\mathfrak{P}_4\mathfrak{P}_3)^*|^2\widehat{k}_\delta, \widehat{k}_\delta\rangle^k \\
&\quad + h(1-\iota)|\langle \mathfrak{P}_4\mathfrak{P}_3\mathfrak{P}_2\mathfrak{P}_1\widehat{k}_\gamma, \widehat{k}_\delta\rangle|^k \langle |\mathfrak{P}_2\mathfrak{P}_1|^2\widehat{k}_\gamma, \widehat{k}_\gamma\rangle^{\frac{k}{2}} \langle |(\mathfrak{P}_4\mathfrak{P}_3)^*|^2\widehat{k}_\delta, \widehat{k}_\delta\rangle^{\frac{k}{2}} \\
&\leq \langle |\mathfrak{P}_2\mathfrak{P}_1|^2\widehat{k}_\gamma, \widehat{k}_\gamma\rangle^k \langle |(\mathfrak{P}_4\mathfrak{P}_3)^*|^2\widehat{k}_\delta, \widehat{k}_\delta\rangle^k \\
&\leq \frac{2\sqrt{\xi\eta}}{\xi+\eta}\left(\frac{\langle |\mathfrak{P}_2\mathfrak{P}_1|^{4k}\widehat{k}_\gamma, \widehat{k}_\gamma\rangle + \langle |(\mathfrak{P}_4\mathfrak{P}_3)^*|^{4k}\widehat{k}_\delta, \widehat{k}_\delta\rangle}{2}\right).
\end{aligned} \tag{11.21}$$

Setting $\mathfrak{P}_4 = \mathfrak{S}^*, \mathfrak{P}_1 = \mathfrak{T}, \mathfrak{P}_2 = \mathfrak{P}_3 = I$, we obtain

$$\begin{aligned}
|\langle \mathfrak{S}^*\mathfrak{T}\widehat{k}_\gamma, \widehat{k}_\delta\rangle|^{2k} &\leq h(\iota)\langle |\mathfrak{T}|^2\widehat{k}_\gamma, \widehat{k}_\gamma\rangle^k \langle |\mathfrak{S}|^2\widehat{k}_\delta, \widehat{k}_\delta\rangle^k \\
&\quad + h(1-\iota)|\langle \mathfrak{S}^*\mathfrak{T}\widehat{k}_\gamma, \widehat{k}_\delta\rangle|^k \langle |\mathfrak{T}|^2\widehat{k}_\gamma, \widehat{k}_\gamma\rangle^{\frac{k}{2}} \langle |\mathfrak{S}|^2\widehat{k}_\delta, \widehat{k}_\delta\rangle^{\frac{k}{2}} \\
&\leq \langle |\mathfrak{T}|^2\widehat{k}_\gamma, \widehat{k}_\gamma\rangle^k \langle |\mathfrak{S}|^2\widehat{k}_\delta, \widehat{k}_\delta\rangle^k \\
&\leq \frac{2\sqrt{\xi_1\xi_2}}{\xi_1+\xi_2}\left(\frac{\langle |\mathfrak{T}|^{4k}\widehat{k}_\gamma, \widehat{k}_\gamma\rangle + \langle |\mathfrak{S}|^{4k}\widehat{k}_\delta, \widehat{k}_\delta\rangle}{2}\right),
\end{aligned} \tag{11.22}$$

when

$$0 < |\mathfrak{T}|^2 \leq \xi_1 \leq \xi_2 \leq |\mathfrak{S}|^2$$

or

$$0 < |\mathfrak{S}|^2 \leq \xi_1 \leq \xi_2 \leq |\mathfrak{T}|^2.$$

11.4 Berezin radius type inequalities

Lemma 11.14. *Suppose $\mathfrak{P}_4$, $\mathfrak{P}_3$, $\mathfrak{P}_2$, and $\mathfrak{P}_1 \in \mathfrak{B}(\mathfrak{H})$. Let the conditions from (11.18) hold. Then, the inequality shown below is true:*

$$f(\mathrm{ber}^2(\mathfrak{P}_4\mathfrak{P}_3\mathfrak{P}_2\mathfrak{P}_1)) \leq \|\varsigma f(|\mathfrak{P}_2\mathfrak{P}_1|^{\frac{2}{\varsigma}}) + (1-\varsigma)f(|(\mathfrak{P}_4\mathfrak{P}_3)^*|^{\frac{2}{1-\varsigma}})\|_{\mathrm{ber}} \tag{11.23}$$

for any $0 < \varsigma < 1$. In particular,

$$\mathrm{ber}^{2k}(\mathfrak{P}_4\mathfrak{P}_3\mathfrak{P}_2\mathfrak{P}_1) \leq \|\varsigma|\mathfrak{P}_2\mathfrak{P}_1|^{\frac{2k}{\varsigma}} + (1-\varsigma)|(\mathfrak{P}_4\mathfrak{P}_3)^*|^{\frac{2k}{1-\varsigma}}\|_{\mathrm{ber}} \tag{11.24}$$

for every $k \geq 1$.

Proof. Let $\gamma \in \mathfrak{O}$ be arbitrary. By using the inequality in [34, Lemma 3.1], Lemma 11.2 for concavity of $f(\tau) = \tau^{\varsigma}$ and weighted AM–GM inequality, then we get

$$\begin{aligned}|\langle \mathfrak{P}_4\mathfrak{P}_3\mathfrak{P}_2\mathfrak{P}_1\widehat{k}_\gamma, \widehat{k}_\gamma\rangle|^2 &\le \langle |\mathfrak{P}_2\mathfrak{P}_1|^2\widehat{k}_\gamma, \widehat{k}_\gamma\rangle\langle |(\mathfrak{P}_4\mathfrak{P}_3)^*|^2\widehat{k}_\gamma, \widehat{k}_\gamma\rangle \\ &\le \langle |\mathfrak{P}_2\mathfrak{P}_1|^{\frac{2}{\varsigma}}\widehat{k}_\gamma, \widehat{k}_\gamma\rangle^{\varsigma}\langle |(\mathfrak{P}_4\mathfrak{P}_3)^*|^{\frac{2}{1-\varsigma}}\widehat{k}_\gamma, \widehat{k}_\gamma\rangle^{1-\varsigma} \\ &\le \varsigma\langle |\mathfrak{P}_2\mathfrak{P}_1|^{\frac{2}{\varsigma}}\widehat{k}_\gamma, \widehat{k}_\gamma\rangle + (1-\varsigma)\langle |(\mathfrak{P}_4\mathfrak{P}_3)^*|^{\frac{2}{1-\varsigma}}\widehat{k}_\gamma, \widehat{k}_\gamma\rangle\end{aligned}$$

and

$$\sup_{\gamma\in\mathfrak{O}}|\langle \mathfrak{P}_4\mathfrak{P}_3\mathfrak{P}_2\mathfrak{P}_1\widehat{k}_\gamma, \widehat{k}_\gamma\rangle|^2 \le \sup_{\gamma\in\mathfrak{O}} \varsigma\langle |\mathfrak{P}_2\mathfrak{P}_1|^{\frac{2}{\varsigma}}\widehat{k}_\gamma, \widehat{k}_\gamma\rangle + \sup_{\gamma\in\mathfrak{O}}(1-\varsigma)\langle |(\mathfrak{P}_4\mathfrak{P}_3)^*|^{\frac{2}{1-\varsigma}}\widehat{k}_\gamma, \widehat{k}_\gamma\rangle.$$

We deduce that

$$\mathrm{ber}^2(\mathfrak{P}_4\mathfrak{P}_3\mathfrak{P}_2\mathfrak{P}_1) \le \|\varsigma|\mathfrak{P}_2\mathfrak{P}_1|^{\frac{2}{\varsigma}} + (1-\varsigma)|(\mathfrak{P}_4\mathfrak{P}_3)^*|^{\frac{2}{1-\varsigma}}\|_{\mathrm{ber}}.$$

On account of assumptions on f and from the inequality in [34, Lemma 3.1], we can write

$$\begin{aligned}f(\mathrm{ber}^2(\mathfrak{P}_4\mathfrak{P}_3\mathfrak{P}_2\mathfrak{P}_1)) &\le f(\|\varsigma|\mathfrak{P}_2\mathfrak{P}_1|^{\frac{2}{\varsigma}} + (1-\varsigma)|(\mathfrak{P}_4\mathfrak{P}_3)^*|^{\frac{2}{1-\varsigma}}\|_{\mathrm{ber}}) \\ &\le \|\varsigma f(|\mathfrak{P}_2\mathfrak{P}_1|^{\frac{2}{\varsigma}}) + (1-\varsigma)f(|(\mathfrak{P}_4\mathfrak{P}_3)^*|^{\frac{2}{1-\varsigma}})\|_{\mathrm{ber}}.\end{aligned}$$ □

Theorem 11.1. *Suppose $k \ge 1$ and $\mathfrak{P}_4$, $\mathfrak{P}_3$, $\mathfrak{P}_2$, and $\mathfrak{P}_1 \in \mathfrak{B}(\mathfrak{H})$. Let the conditions from (11.18) hold. Then, the inequality shown below is true:*

$$f(\mathrm{ber}^{2k}(\mathfrak{P}_4\mathfrak{P}_3\mathfrak{P}_2\mathfrak{P}_1)) \le \|h(p)f(|\mathfrak{P}_2\mathfrak{P}_1|^{\frac{2k}{h(p)}}) + h(1-p)f(|(\mathfrak{P}_4\mathfrak{P}_3)^*|^{\frac{2k}{h(1-p)}})\|_{\mathrm{ber}}. \quad (11.25)$$

Proof. We get the necessary inequality by setting $\gamma = \delta \in \mathfrak{O}$ and obtaining the supremum over all $\gamma \in \mathfrak{O}$ in (11.18). □

Remark 11.3. The inequality that was previously discovered extends the inequality (11.23). When we set $k = 1$, $h = I$, we get

$$f(\mathrm{ber}^2(\mathfrak{P}_4\mathfrak{P}_3\mathfrak{P}_2\mathfrak{P}_1)) \le \|pf(|\mathfrak{P}_2\mathfrak{P}_1|^{\frac{2}{p}}) + (1-p)f(|(\mathfrak{P}_4\mathfrak{P}_3)^*|^{\frac{2}{1-p}})\|_{\mathrm{ber}},$$

which is the above inequality (11.23). In particular, by setting $f = I$, $h = I$ we obtain the inequality (11.24). Setting $f = I$, $h = I$, $\mathfrak{P}_2 = I$, $\mathfrak{P}_4 = I$, $\mathfrak{P}_3 = \mathfrak{Q}^*$, $p = \frac{1}{2}$ we obtain the inequality in [12, Theorem 3.1] for the case when the powers are even, that is

$$\mathrm{ber}^{2k}(\mathfrak{Q}^*\mathfrak{P}_1) \le \frac{1}{2}\||\mathfrak{P}_1|^{4k} + |\mathfrak{Q}|^{4k}\|_{\mathrm{ber}}. \quad (11.26)$$

Theorem 11.2. *Let $\mathfrak{P}_1, \mathfrak{P}_2, \mathfrak{P}_3, \mathfrak{P}_4 \in \mathfrak{B}(\mathfrak{H})$ and let f be a non-negative increasing convex function on $[0, +\infty)$. If*

$$0 < |\mathfrak{P}_2\mathfrak{P}_1|^2 \le \xi \le \eta \le |(\mathfrak{P}_4\mathfrak{P}_3)^*|^2$$

or

$$0 < |(\mathfrak{P}_4\mathfrak{P}_3)^*|^2 \le \xi \le \eta \le |\mathfrak{P}_2\mathfrak{P}_1|^2,$$

then

$$f(\mathrm{ber}^{2k}(\mathfrak{P}_4\mathfrak{P}_3\mathfrak{P}_2\mathfrak{P}_1)) \le \frac{2\sqrt{\xi\eta}}{\xi+\eta}\left(\left\|\frac{f(|\mathfrak{P}_2\mathfrak{P}_1|^{4k}) + f(|(\mathfrak{P}_4\mathfrak{P}_3)^*|^{4k})}{2}\right\|_{\mathrm{ber}}\right). \tag{11.27}$$

Proof. We obtain the requisite inequality by taking the supremum over all $\gamma \in \mathfrak{O}$ in (11.20), □

Remark 11.4. Setting $f(\tau) = \tau$ in (11.27), we obtain

$$\mathrm{ber}^{2k}(\mathfrak{P}_4\mathfrak{P}_3\mathfrak{P}_2\mathfrak{P}_1) \le \frac{2\sqrt{\xi\eta}}{\xi+\eta}\left(\left\|\frac{|\mathfrak{P}_2\mathfrak{P}_1|^{4k} + |(\mathfrak{P}_4\mathfrak{P}_3)^*|^{4k}}{2}\right\|_{\mathrm{ber}}\right). \tag{11.28}$$

Setting $\mathfrak{P}_4 = \mathfrak{S}^*$, $\mathfrak{P}_1 = \mathfrak{T}$, $\mathfrak{P}_2 = \mathfrak{P}_3 = I$, we obtain

$$\mathrm{ber}^{2k}(\mathfrak{S}^*\mathfrak{T}) \le \frac{2\sqrt{\xi\eta}}{\xi+\eta}\left(\left\|\frac{|\mathfrak{T}|^{4k} + |\mathfrak{S}|^{4k}}{2}\right\|_{\mathrm{ber}}\right), \tag{11.29}$$

when

$$0 < |\mathfrak{T}|^2 \le \xi \le \eta \le |\mathfrak{S}|^2$$

or

$$0 < |\mathfrak{S}|^2 \le \xi \le \eta \le |\mathfrak{T}|^2.$$

Setting $\mathfrak{P}_3 = \mathfrak{P}_4 = \mathfrak{P}_2 = I$ and $\mathfrak{P}_1 = \mathfrak{T}$, $f = I$ in (11.27), we obtain

$$\mathrm{ber}^{2k}(\mathfrak{T}) \le \frac{2\sqrt{\xi\eta}}{\xi+\eta}\left(\left\|\frac{|\mathfrak{T}|^{4k} + I}{2}\right\|_{\mathrm{ber}}\right), \tag{11.30}$$

when

$$0 < |\mathfrak{T}|^2 \le \xi \le \eta \le I$$

or

$$0 < I \le \xi \le \eta \le |\mathfrak{T}|^2.$$

Further setting $\xi = \eta = 1$ in (11.29), we obtain (11.26) for the even powers, namely

$$\mathrm{ber}^{2k}(\mathfrak{S}^*\mathfrak{T}) \le \frac{1}{2}\||\mathfrak{T}|^{4k} + |\mathfrak{S}|^{4k}\|_{\mathrm{ber}} \tag{11.31}$$

when

$$0 < |\mathfrak{T}|^2 \le 1 \le 1 \le |\mathfrak{S}|^2$$

or

$$0 < |\mathfrak{S}|^2 \le 1 \le 1 \le |\mathfrak{T}|^2.$$

Using the same configuration but setting $\xi = 3$, $\eta = 5$ in (11.29), we obtain a refinement of (11.26) in the case of even powers, namely

$$\mathrm{ber}^{2k}(\mathfrak{S}^*\mathfrak{T}) \le \frac{\sqrt{15}}{8}\||\mathfrak{T}|^{4k} + |\mathfrak{S}|^{4k}\|_{\mathrm{ber}}, \tag{11.32}$$

which obviously is a refinement of (11.26) since

$$\begin{aligned} \mathrm{ber}^{2k}(\mathfrak{S}^*\mathfrak{T}) &\le \frac{\sqrt{15}}{8}\||\mathfrak{T}|^{4k} + |\mathfrak{S}|^{4k}\|_{\mathrm{ber}} \\ &< \frac{\sqrt{16}}{8}\||\mathfrak{T}|^{4k} + |\mathfrak{S}|^{4k}\|_{\mathrm{ber}} = \frac{1}{2}\||\mathfrak{T}|^{4k} + |\mathfrak{S}|^{4k}\|_{\mathrm{ber}}, \end{aligned} \tag{11.33}$$

when

$$0 < |\mathfrak{T}|^2 \le 3 \le 5 \le |\mathfrak{S}|^2$$

or

$$0 < |\mathfrak{S}|^2 \le 3 \le 5 \le |\mathfrak{T}|^2.$$

Theorem 11.3. *Let $\tau, \omega \ge 1$ and $\mathfrak{T} \in \mathfrak{B}(\mathfrak{H})$, then we get the following inequality*

$$f(\mathrm{ber}^{2k}(\mathfrak{T}|\mathfrak{T}|^{\omega-1}\mathfrak{T}|\mathfrak{T}|^{\tau-1})) \le \|h(p)f(|\mathfrak{T}|^{\frac{2k\tau}{h(p)}}) + h(1-p)f(|\mathfrak{T}^*|^{\frac{2k\omega}{h(1-p)}})\|_{\mathrm{ber}}. \tag{11.34}$$

Proof. We obtain the requisite inequality by taking the supremum over all $y \in \mathfrak{O}$ in (11.19). □

Corollary 11.1. *We get*

$$\mathrm{ber}^{2k}(\mathfrak{T}|\mathfrak{T}|\mathfrak{T}|\mathfrak{T}|) \le \|p|\mathfrak{T}|^{\frac{4k}{p}} + (1-p)|\mathfrak{T}^*|^{\frac{4k}{(1-p)}}\|_{\mathrm{ber}}$$

by letting $f = I$, $h = I$, $\tau = \omega = 2$ in (11.34). Further, setting $k = 2$, $p = \frac{1}{2}$, we obtain

$$\mathrm{ber}^4(\mathfrak{T}|\mathfrak{T}|\mathfrak{T}|\mathfrak{T}|) \le \frac{1}{2}\||\mathfrak{T}|^{16} + |\mathfrak{T}^*|^{16}\|_{\mathrm{ber}}.$$

Theorem 11.4. *In the event where $\mathfrak{P}_1 \in \mathfrak{B}(\mathfrak{H})$ and h satisfy the criteria in (11.18), and ψ is a non-negative increasing convex sub-multiplicative function on $[0, +\infty)$, the following inequality holds for $k \geq 1$:*

$$\psi(\mathrm{ber}^{2k}(\mathfrak{P}_1^2)) \leq h(\iota)\psi\left(\left(\int_0^1 \|t|\mathfrak{P}_1|^{4k} + (1-t)|\mathfrak{P}_1^*|^{4k}\|^{\frac{1}{2}} dt\right)^2\right)$$
$$+ h(1-\iota)\psi(\mathrm{ber}^k(\mathfrak{P}_1^2))\left\|\frac{\psi(|\mathfrak{P}_1|^{2k}) + \psi(|\mathfrak{P}_1^*|^{2k})}{2}\right\|_{\mathrm{ber}}. \tag{11.35}$$

Proof. Let $y \in \mathfrak{O}$ be arbitrary. Then, take into account

$$\begin{aligned}
&\psi(|\langle \mathfrak{P}_1\widehat{k}_y, \mathfrak{P}_1^*\widehat{k}_y\rangle|^{2k}) \\
&\leq h(\iota)\psi(\|\mathfrak{P}_1\widehat{k}_y\|^{2k}\|\mathfrak{P}_1^*\widehat{k}_y\|^{2k}) \\
&\quad + h(1-\iota)\psi(\|\mathfrak{P}_1\widehat{k}_y\|^{k}\|\mathfrak{P}_1^*\widehat{k}_y\|^{k}|\langle \mathfrak{P}_1\widehat{k}_y, \mathfrak{P}_1^*\widehat{k}_y\rangle|^{k}) \quad \text{(Lemma 11.17)} \\
&\leq h(\iota)\psi\left(\left(\int_0^1 \|\mathfrak{P}_1\widehat{k}_y\|^{2tk}\|\mathfrak{P}_1^*\widehat{k}_y\|^{2k(1-t)} dt\right)^2\right) \quad \text{(Logarithmic mean)} \\
&\quad + h(1-\iota)\psi(|\langle \mathfrak{P}_1^2\widehat{k}_y, \widehat{k}_y\rangle|^{k})\psi(\|\mathfrak{P}_1\widehat{k}_y\|^{k}\|\mathfrak{P}_1^*\widehat{k}_y\|) \quad \text{(sub-multiplicativity of } \psi) \\
&\leq h(\iota)\psi\left(\left(\int_0^1 \|\mathfrak{P}_1\widehat{k}_y\|^{2tk}\|\mathfrak{P}_1^*\widehat{k}_y\|^{2k(1-t)} dt\right)^2\right) \\
&\quad + h(1-\iota)\psi(|\langle \mathfrak{P}_1^2\widehat{k}_y, \widehat{k}_y\rangle|^{k})\psi\left(\frac{\langle|\mathfrak{P}_1|^{2k}\widehat{k}_y, \widehat{k}_y\rangle + \langle|\mathfrak{P}_1^*|^{2k}\widehat{k}_y, \widehat{k}_y\rangle}{2}\right) \quad \text{(AG)} \\
&\leq h(\iota)\psi\left(\left(\int_0^1 (\langle|\mathfrak{P}_1|^{4k}\widehat{k}_y, \widehat{k}_y\rangle^{t}\langle|\mathfrak{P}_1^*|^{4k}\widehat{k}_y, \widehat{k}_y\rangle^{1-t})^{\frac{1}{2}} dt\right)^2\right) \\
&\quad + h(1-\iota)\psi(|\langle \mathfrak{P}_1^2\widehat{k}_y, \widehat{k}_y\rangle|^{k})\frac{\psi(\langle|\mathfrak{P}_1|^{2k}\widehat{k}_y, \widehat{k}_y\rangle) + \psi(\langle|\mathfrak{P}_1^*|^{2k}\widehat{k}_y, \widehat{k}_y\rangle)}{2} \\
&\leq h(\iota)\psi\left(\left(\int_0^1 (t\langle|\mathfrak{P}_1|^{4k}\widehat{k}_y, \widehat{k}_y\rangle + (1-t)\langle|\mathfrak{P}_1^*|^{4k}\widehat{k}_y, \widehat{k}_y\rangle)^{\frac{1}{2}} dt\right)^2\right) \quad \text{(weighted AG inequality)} \\
&\quad + h(1-\iota)\psi(|\langle \mathfrak{P}_1^2\widehat{k}_y, \widehat{k}_y\rangle|^{k})\frac{\langle(\psi(|\mathfrak{P}_1|^{2k}) + \psi(|\mathfrak{P}_1^*|^{2k}))\widehat{k}_y, \widehat{k}_y\rangle}{2} \quad \text{(Lemma 11.6)} \\
&\leq h(\iota)\psi\left(\left(\int_0^1 \|t|\mathfrak{P}_1|^{4k} + (1-t)|\mathfrak{P}_1^*|^{4k}\|^{\frac{1}{2}} dt\right)^2\right) \\
&\quad + h(1-\iota)\psi(|\langle \mathfrak{P}_1^2\widehat{k}_y, \widehat{k}_y\rangle|^{k})\frac{\langle(\psi(|\mathfrak{P}_1|^{2k}) + \psi(|\mathfrak{P}_1^*|^{2k}))\widehat{k}_y, \widehat{k}_y\rangle}{2}.
\end{aligned}$$

We get the required inequality when we take sup over all $y \in \mathfrak{O}$. □

Corollary 11.2. *The following chain of inequalities for $\mathfrak{P}_1 \in \mathfrak{P}_2(H)$ holds:*

$$\begin{aligned}\operatorname{ber}^{2k}(\mathfrak{P}_1^2) &\le h(\iota)\left(\int_0^1 \left\| t|\mathfrak{P}_1|^{4k} + (1-t)|\mathfrak{P}_1^*|^{4k}\right\|_{\mathrm{ber}}^{\frac{1}{2}} dt\right)^2 \\ &\quad + \frac{h(1-\iota)}{2}\operatorname{ber}^k(\mathfrak{P}_1^2)\left\| |\mathfrak{P}_1|^{2k} + |\mathfrak{P}_1^*|^{2k}\right\|_{\mathrm{ber}} \\ &\le \|\mathfrak{P}_1\|_{\mathrm{ber}}^{4k}. \end{aligned} \tag{11.36}$$

Proof. We will bound two quantities, first the term inside the integral, that is

$$\left\| t|\mathfrak{P}_1|^{4k} + (1-t)|\mathfrak{P}_1^*|^{4k}\right\|_{\mathrm{ber}} \le \|\mathfrak{P}_1\|_{\mathrm{ber}}^{4k},$$

we will also use the following bound $\operatorname{ber}^k(\mathfrak{P}_1^2) \le \|\mathfrak{P}_1^2\|_{\mathrm{ber}}^k \le \|\mathfrak{P}_1\|_{\mathrm{ber}}^{2k}$, then use the elementary inequality $\| |\mathfrak{P}_1|^{2k} + |\mathfrak{P}_1^*|^{2k}\|_{\mathrm{ber}} \le 2\|\mathfrak{P}_1\|_{\mathrm{ber}}^{2k}$. Combining everything, we get

$$h(\iota)\|\mathfrak{P}_1\|_{\mathrm{ber}}^{4k} + h(1-\iota)\|\mathfrak{P}_1\|_{\mathrm{ber}}^{4k} = \|\mathfrak{P}_1\|_{\mathrm{ber}}^{4k}.$$ □

The inequality (11.36) is sharp. To see this, let us suppose that $\mathfrak{P}_1$ is a normal operator, that is $\mathfrak{P}_1^*\mathfrak{P}_1 = \mathfrak{P}_1\mathfrak{P}_1^*$, we also have $\operatorname{ber}(\mathfrak{P}_1^2) = \operatorname{ber}^2(\mathfrak{P}_1) = \|\mathfrak{P}_1\|_{\mathrm{ber}}^2 = \|\mathfrak{P}_1^2\|_{\mathrm{ber}}$, from that, we get

$$\begin{aligned}\|\mathfrak{P}_1\|_{\mathrm{ber}}^{4k} &\le h(\iota)\left(\int_0^1 \left\| t|\mathfrak{P}_1|^{4k} + (1-t)|\mathfrak{P}_1^*|^{4k}\right\|^{\frac{1}{2}} dt\right)^2 \\ &\quad + \frac{h(1-\iota)}{2}\operatorname{ber}^k(\mathfrak{P}_1^2)\left\| |\mathfrak{P}_1|^{2k} + |\mathfrak{P}_1^*|^{2k}\right\|_{\mathrm{ber}} \\ &= h(\iota)\|\mathfrak{P}_1\|_{\mathrm{ber}}^{4k} + \frac{h(1-\iota)\|\mathfrak{P}_1\|_{\mathrm{ber}}^{2k}}{2}\cdot\left(2\| |\mathfrak{P}_1|^2\|_{\mathrm{ber}}^k\right) \\ &= h(\iota)\|\mathfrak{P}_1\|_{\mathrm{ber}}^{4k} + h(1-\iota)\|\mathfrak{P}_1\|_{\mathrm{ber}}^{2k}\cdot\|\mathfrak{P}_1\|_{\mathrm{ber}}^{2k} \\ &= \|\mathfrak{P}_1\|_{\mathrm{ber}}^{4k}. \end{aligned}$$

Further, since by (11.2) we have $\operatorname{ber}(\mathfrak{P}_1^2) \le \operatorname{ber}^2(\mathfrak{P}_1)$, it may be shown that

$$\begin{aligned}\operatorname{ber}(\mathfrak{P}_1^{4k}) \le \operatorname{ber}^{2k}(\mathfrak{P}_1^2) &\le h(\iota)\left(\int_0^1 \left\| t|\mathfrak{P}_1|^{4k} + (1-t)|\mathfrak{P}_1^*|^{4k}\right\|_{\mathrm{ber}}^{\frac{1}{2}} dt\right)^2 \\ &\quad + \frac{h(1-\iota)}{2}\operatorname{ber}^k(\mathfrak{P}_1^2)\left\| |\mathfrak{P}_1|^{2k} + |\mathfrak{P}_1^*|^{2k}\right\|_{\mathrm{ber}} \\ &\le \|\mathfrak{P}_1\|_{\mathrm{ber}}^{4k}. \end{aligned} \tag{11.37}$$

Setting $h(\iota) = \frac{1}{2}(\frac{1}{2} + \iota)$, $\iota \in [0,1]$, $k = 1$, we obtain

$$\begin{aligned}\operatorname{ber}(\mathfrak{P}_1^4) \le \operatorname{ber}^2(\mathfrak{P}_1^2) &\le \frac{1}{2}\left(\frac{1}{2}+\iota\right)\left(\int_0^1 \| t|\mathfrak{P}_1|^4 + (1-t)|\mathfrak{P}_1^*|^4\|_{\operatorname{ber}}^{\frac{1}{2}} dt\right)^2 \\ &\quad + \frac{\frac{1}{2}(\frac{3}{2}-\iota)}{2}\operatorname{ber}(\mathfrak{P}_1^2)\| |\mathfrak{P}_1|^2 + |\mathfrak{P}_1^*|^2\|_{\operatorname{ber}} \\ &\le \|\mathfrak{P}_1\|_{\operatorname{ber}}^{4k}. \end{aligned} \tag{11.38}$$

Theorem 11.5. *If h is such that $\mathfrak{P}_1 \in \mathfrak{B}(\mathfrak{H})$ and the criteria from (11.20) are met, then the following inequality holds for $k \ge 2$:*

$$\begin{aligned}\operatorname{ber}(\mathfrak{P}_1^{2k}) &\le h(\iota)\int_0^1 \| t|\mathfrak{P}_1|^{4\tau k} + (1-t)|\mathfrak{P}_1^*|^{4k(1-\tau)}\|_{\operatorname{ber}} dt \\ &\quad + h(1-\iota)\frac{\| |\mathfrak{P}_1|^{2k} + |\mathfrak{P}_1^*|^{2k}\|_{\operatorname{ber}}}{4} + \frac{h(1-\iota)}{2}\operatorname{ber}^k(\mathfrak{P}_1^2). \end{aligned} \tag{11.39}$$

Proof. Let $y \in \mathfrak{O}$ be arbitrary. Using the Lemma 11.2 and weighted AM–GM inequality, we get

$$\begin{aligned} |\langle \mathfrak{P}_1\hat{k}_y, \hat{k}_y\rangle|^{2k} &= h(\iota)|\langle \mathfrak{P}_1\hat{k}_y, \hat{k}_y\rangle|^{2k} + h(1-\iota)|\langle \mathfrak{P}_1\hat{k}_y, \hat{k}_y\rangle\langle \mathfrak{P}_1\hat{k}_y, \hat{k}_y\rangle|^k \\ &\le h(\iota)\langle |\mathfrak{P}_1|^{2\tau k}\hat{k}_y, \hat{k}_y\rangle\langle |\mathfrak{P}_1^*|^{2k(1-\tau)}\hat{k}_y, \hat{k}_y\rangle \\ &\quad + \frac{h(1-\iota)}{2}(\|\mathfrak{P}_1\hat{k}_y\|^k\|\mathfrak{P}_1^*\hat{k}_y\|^k + |\langle \mathfrak{P}_1\hat{k}_y, \mathfrak{P}_1^*\hat{k}_y\rangle|^k) \\ &\le h(\iota)\int_0^1 \langle |\mathfrak{P}_1|^{2\tau k}\hat{k}_y, \hat{k}_y\rangle^{2t}\langle |\mathfrak{P}_1^*|^{2k(1-\tau)}\hat{k}_y, \hat{k}_y\rangle^{2(1-t)} dt \\ &\quad + \frac{h(1-\iota)}{2}|\langle \mathfrak{P}_1^2\hat{k}_y, \hat{k}_y\rangle|^k + \frac{h(1-\iota)}{4}\langle (|\mathfrak{P}_1|^{2k} + |\mathfrak{P}_1^*|^{2k})\hat{k}_y, \hat{k}_y\rangle \\ &\le h(\iota)\int_0^1 (t\langle |\mathfrak{P}_1|^{4\tau k}\hat{k}_y, \hat{k}_y\rangle + (1-t)\langle |\mathfrak{P}_1^*|^{4k(1-\tau)}\hat{k}_y, \hat{k}_y\rangle) dt \\ &\quad + \frac{h(1-\iota)}{2}|\langle \mathfrak{P}_1^2\hat{k}_y, \hat{k}_y\rangle|^k + \frac{h(1-\iota)}{4}\langle (|\mathfrak{P}_1|^{2k} + |\mathfrak{P}_1^*|^{2k})\hat{k}_y, \hat{k}_y\rangle \end{aligned}$$

and

$$\begin{aligned} \sup_{y\in\mathfrak{O}}|\langle \mathfrak{P}_1\hat{k}_y, \hat{k}_y\rangle|^{2k} &\le \sup_{y\in\mathfrak{O}} h(\iota)\int_0^1 (t\langle |\mathfrak{P}_1|^{4\tau k}\hat{k}_y, \hat{k}_y\rangle + (1-t)\langle |\mathfrak{P}_1^*|^{4k(1-\tau)}\hat{k}_y, \hat{k}_y\rangle) dt \\ &\quad + \sup_{y\in\mathfrak{O}}\frac{h(1-\iota)}{2}|\langle \mathfrak{P}_1^2\hat{k}_y, \hat{k}_y\rangle|^k + \sup_{y\in\mathfrak{O}}\frac{h(1-\iota)}{4}\langle (|\mathfrak{P}_1|^{2k} + |\mathfrak{P}_1^*|^{2k})\hat{k}_y, \hat{k}_y\rangle. \end{aligned}$$

Hence, we obtain the desired inequality

$$\operatorname{ber}(\mathfrak{P}_1^{2k}) \le h(\iota) \int_0^1 \| t|\mathfrak{P}_1|^{4\tau k} + (1-t)|\mathfrak{P}_1^*|^{4k(1-\tau)} \|_{\mathrm{ber}} dt$$
$$+ h(1-\iota) \frac{\| |\mathfrak{P}_1|^{2k} + |\mathfrak{P}_1^*|^{2k} \|_{\mathrm{ber}}}{4} + \frac{h(1-\iota)}{2} \operatorname{ber}^k(\mathfrak{P}_1^2). \qquad \square$$

It can be shown that (11.39) is sharp for $\tau = \frac{1}{2}$ when assuming that $\mathfrak{P}_1$ is normal, it can also be shown that for $\tau = \frac{1}{2}$

$$\operatorname{ber}^{2k}(\mathfrak{P}_1) \le \int_0^1 \| t|\mathfrak{P}_1|^{2k} + (1-t)|\mathfrak{P}_1^*|^{2k} \|_{\mathrm{ber}} dt$$
$$+ h(1-\iota) \frac{\| |\mathfrak{P}_1|^{2k} + |\mathfrak{P}_1^*|^{2k} \|_{\mathrm{ber}}}{4} + \frac{h(1-\iota)}{2} \operatorname{ber}^k(\mathfrak{P}_1^2)$$
$$\le \|\mathfrak{P}_1\|_{\mathrm{ber}}^{2k}.$$

The following theorem gives us a boundary for $\operatorname{ber}^2(\mathfrak{P}_1)$ and as a corollary we obtain an inequality for a complex numerical radius.

Theorem 11.6. *Suppose that $\mathfrak{P}_1 \in \mathfrak{B}(\mathfrak{H})$. Let ψ be a non-negative increasing convex function on $[0, +\infty)$ and let the mapping h satisfy the conditions of (11.20). Then for any $\tau \in [0, 1]$, the following inequality holds*

$$\psi(\operatorname{ber}^2(\mathfrak{P}_1)) \le \frac{1}{2}\psi(\operatorname{ber}(|\mathfrak{P}_1^*|^{2(1-\tau)}|\mathfrak{P}_1|^{2\tau}))$$
$$+ \frac{1}{4}(\psi(\| |\mathfrak{P}_1|^{4\tau} + |\mathfrak{P}_1^*|^{4(1-\tau)} \|_{\mathrm{ber}}) + \psi(\operatorname{ber}(|\mathfrak{P}_1|^{4\tau} - |\mathfrak{P}_1^*|^{4(1-\tau)}))). \qquad (11.40)$$

Proof. Let $y \in \mathfrak{O}$ be arbitrary. Setting $u_1 = |\mathfrak{P}_1|^{2\tau}$, $u_2 = |\mathfrak{P}_1^*|^{2(1-\tau)}$ and $u_3 = \widehat{k}_y$ in Lemma 11.9, while taking u_3 to be unit vector, we obtain

$$|\langle |\mathfrak{P}_1|^{2\tau}\widehat{k}_y, \widehat{k}_y\rangle|^2 + |\langle |\mathfrak{P}_1^*|^{2(1-\tau)}\widehat{k}_y, \widehat{k}_y\rangle|^2$$
$$\le |\langle |\mathfrak{P}_1^*|^{2(1-\tau)}|\mathfrak{P}_1|^{2\tau}\widehat{k}_y, \widehat{k}_y\rangle| + \max(\| |\mathfrak{P}_1|^{2\tau}\widehat{k}_y\|^2, \| |\mathfrak{P}_1^*|^{2(1-\tau)}\widehat{k}_y\|^2).$$

Now, realizing the following

$$2|\langle \mathfrak{P}_1\widehat{k}_y, \widehat{k}_y\rangle|^2 \le 2\langle |\mathfrak{P}_1|^{2\tau}\widehat{k}_y, \widehat{k}_y\rangle\langle |\mathfrak{P}_1^*|^{2(1-\tau)}\widehat{k}_y, \widehat{k}_y\rangle \le \langle |\mathfrak{P}_1|^{2\tau}\widehat{k}_y, \widehat{k}_y\rangle^2 + \langle |\mathfrak{P}_1^*|^{2(1-\tau)}\widehat{k}_y, \widehat{k}_y\rangle^2,$$

and using Lemma 11.10, we obtain

$$2\langle \mathfrak{P}_1\widehat{k}_y, \widehat{k}_y\rangle|^2 \le |\langle |\mathfrak{P}_1^*|^{2(1-\tau)}|\mathfrak{P}_1|^{2\tau}\widehat{k}_y, \widehat{k}_y\rangle|$$
$$+ \frac{1}{2}[\langle (|\mathfrak{P}_1|^{4\tau} + |\mathfrak{P}_1^*|^{4(1-\tau)})\widehat{k}_y, \widehat{k}_y\rangle + |\langle (|\mathfrak{P}_1|^{4\tau} - |\mathfrak{P}_1^*|^{4(1-\tau)})\widehat{k}_y, \widehat{k}_y\rangle|].$$

It gives rise to the inequality

$$|\langle \mathfrak{P}_1 \widehat{k}_y, \widehat{k}_y\rangle|^2 \leq \frac{1}{2}|\langle |\mathfrak{P}_1^*|^{2(1-\tau)}|\mathfrak{P}_1|^{2\tau}\widehat{k}_y, \widehat{k}_y\rangle| \\ + \frac{1}{4}[\langle (|\mathfrak{P}_1|^{4\tau} + |\mathfrak{P}_1^*|^{4(1-\tau)})\widehat{k}_y, \widehat{k}_y\rangle + |\langle (|\mathfrak{P}_1|^{4\tau} - |\mathfrak{P}_1^*|^{4(1-\tau)})\widehat{k}_y, \widehat{k}_y\rangle|]$$

as follows. Taking ψ on both sides and using the convexity properties, we obtain the following

$$\psi(|\langle \mathfrak{P}_1 \widehat{k}_y, \widehat{k}_y\rangle|^2) \\ \leq \psi\left(\frac{(|\langle |\mathfrak{P}_1^*|^{2(1-\tau)}|\mathfrak{P}_1|^{2\tau}\widehat{k}_y, \widehat{k}_y\rangle|)}{2} + \frac{\frac{\langle (|\mathfrak{P}_1|^{4\tau}+|\mathfrak{P}_1^*|^{4(1-\tau)})\widehat{k}_y,\widehat{k}_y\rangle + |\langle (|\mathfrak{P}_1|^{4\tau}-|\mathfrak{P}_1^*|^{4(1-\tau)})\widehat{k}_y,\widehat{k}_y\rangle|}{2}}{2}\right) \\ \leq \frac{1}{2}\psi(|\langle |\mathfrak{P}_1^*|^{2(1-\tau)}|\mathfrak{P}_1|^{2\tau}\widehat{k}_y, \widehat{k}_y\rangle|) \\ + \frac{1}{2}\psi\left(\frac{\langle (|\mathfrak{P}_1|^{4\tau} + |\mathfrak{P}_1^*|^{4(1-\tau)})\widehat{k}_y, \widehat{k}_y\rangle + |\langle (|\mathfrak{P}_1|^{4\tau} - |\mathfrak{P}_1^*|^{4(1-\tau)})\widehat{k}_y, \widehat{k}_y\rangle|}{2}\right) \\ \leq \frac{1}{2}\psi(|\langle |\mathfrak{P}_1^*|^{2(1-\tau)}|\mathfrak{P}_1|^{2\tau}\widehat{k}_y, \widehat{k}_y\rangle|) \\ + \frac{\psi(\langle (|\mathfrak{P}_1|^{4\tau} + |\mathfrak{P}_1^*|^{4(1-\tau)})\widehat{k}_y, \widehat{k}_y\rangle) + \psi(|\langle (|\mathfrak{P}_1|^{4\tau} - |\mathfrak{P}_1^*|^{4(1-\tau)})\widehat{k}_y, \widehat{k}_y\rangle|)}{4}$$

and

$$\sup_{y\in\mathfrak{O}} \psi(|\langle \mathfrak{P}_1 \widehat{k}_y, \widehat{k}_y\rangle|^2) \\ \leq \sup_{y\in\mathfrak{O}} \frac{1}{2}\psi(|\langle |\mathfrak{P}_1^*|^{2(1-\tau)}|\mathfrak{P}_1|^{2\tau}\widehat{k}_y, \widehat{k}_y\rangle|) \\ + \sup_{y\in\mathfrak{O}} \frac{\psi(\langle (|\mathfrak{P}_1|^{4\tau} + |\mathfrak{P}_1^*|^{4(1-\tau)})\widehat{k}_y, \widehat{k}_y\rangle) + \psi(|\langle (|\mathfrak{P}_1|^{4\tau} - |\mathfrak{P}_1^*|^{4(1-\tau)})\widehat{k}_y, \widehat{k}_y\rangle|)}{4}.$$

Then, we deduce that

$$\psi(\mathrm{ber}^2(\mathfrak{P}_1)) \leq \frac{1}{2}\psi(\mathrm{ber}(|\mathfrak{P}_1^*|^{2(1-\tau)}|\mathfrak{P}_1|^{2\tau})) \\ + \frac{1}{4}(\psi(\||\mathfrak{P}_1|^{4\tau} + |\mathfrak{P}_1^*|^{4(1-\tau)}\|_{\mathrm{ber}}) + \psi(\mathrm{ber}(|\mathfrak{P}_1|^{4\tau} - |\mathfrak{P}_1^*|^{4(1-\tau)}))). \qquad \square$$

Remark 11.5. Using (11.26) and the fact that $\mathrm{ber}(|\mathfrak{P}_1|^{4\tau} - |\mathfrak{P}_1^*|^{4(1-\tau)}) \leq \mathrm{ber}(|\mathfrak{P}_1|^{4\tau} + |\mathfrak{P}_1^*|^{4(1-\tau)})$, we can obtain the following bound while setting $\psi = I$

$$\mathrm{ber}^2(\mathfrak{P}_1) \leq \frac{1}{2}\mathrm{ber}(|\mathfrak{P}_1^*|^{2(1-\tau)}|\mathfrak{P}_1|^{2\tau}) + \frac{1}{4}(\||\mathfrak{P}_1|^{4\tau} + |\mathfrak{P}_1^*|^{4(1-\tau)}\|_{\mathrm{ber}} + \mathrm{ber}(|\mathfrak{P}_1|^{4\tau} - |\mathfrak{P}_1^*|^{4(1-\tau)})) \tag{11.41}$$

$$\leq \frac{3}{4}\||\mathfrak{P}_1|^{4\tau} + |\mathfrak{P}_1^*|^{4(1-\tau)}\|_{\mathrm{ber}}. \tag{11.42}$$

Following the proof of (11.40) and using the fact $\langle |\mathfrak{T}|^{2\tau}\widehat{k}_y, \widehat{k}_y\rangle^2 + \langle |\mathfrak{T}^*|^{2(1-\tau)}\widehat{k}_y, \widehat{k}_y\rangle^2 = |\langle |\mathfrak{T}|^{2\tau}\widehat{k}_y, \widehat{k}_y\rangle + i\langle |\mathfrak{T}^*|^{2(1-\tau)}\widehat{k}_y, \widehat{k}_y\rangle|^2 = |\langle (|\mathfrak{T}|^{2\tau} + i|\mathfrak{T}^*|^{2(1-\tau)})\widehat{k}_y, \widehat{k}_y\rangle|^2$ and taking sup over all $y \in \mathfrak{O}$, we get

$$\begin{aligned}&\mathrm{ber}^2(|\mathfrak{P}_1|^{2\tau} + i|\mathfrak{P}_1^*|^{2(1-\tau)})\\ &\quad\le \mathrm{ber}(|\mathfrak{P}_1^*|^{2(1-\tau)}|\mathfrak{P}_1|^{2\tau}) + \frac{1}{2}(\||\mathfrak{P}_1|^{4\tau} + |\mathfrak{P}_1^*|^{4(1-\tau)}\|_{\mathrm{ber}} + \mathrm{ber}(|\mathfrak{P}_1|^{4\tau} - |\mathfrak{P}_1^*|^{4(1-\tau)})).\end{aligned} \tag{11.43}$$

Corollary 11.3. *Using a similar reasoning to the one in* (11.42)*, we obtain the following chain of inequalities*

$$\begin{aligned}&\mathrm{ber}^2(|\mathfrak{P}_1|^{2\tau} + i|\mathfrak{P}_1^*|^{2(1-\tau)})\\ &\quad\le \mathrm{ber}(|\mathfrak{P}_1^*|^{2(1-\tau)}|\mathfrak{P}_1|^{2\tau}) + \frac{1}{2}(\||\mathfrak{P}_1|^{4\tau} + |\mathfrak{P}_1^*|^{4(1-\tau)}\|_{\mathrm{ber}} + \mathrm{ber}(|\mathfrak{P}_1|^{4\tau} - |\mathfrak{P}_1^*|^{4(1-\tau)}))\\ &\quad\le \frac{3}{2}\||\mathfrak{P}_1|^{4\tau} + |\mathfrak{P}_1^*|^{4(1-\tau)}\|_{\mathrm{ber}}.\end{aligned} \tag{11.44}$$

11.5 Conclusion

Various vector and numerical radius type inequalities have been obtained. Utilizing the function h defined in the paper written by Stojiljković and Dragomir in [32] and convexity properties with standard operator theory techniques, we obtained various refinements of the well-known Berezin radius type inequalities. Namely, we have sharpened the inequality given by Huban et al. (11.10) by (11.33). Further, inequality for the complex Berezin radius of the absolute value of a bounded operator has been given by (11.43). Variation of (11.1) has been given by (11.37) and (11.39). It has also been shown that (11.36) and (11.37) are sharp inequalities for arbitrary function h as mentioned before under the condition that the operator is normal.

Bibliography

[1] Mond, B. and Pečarić, J., *On Jensen's inequality for operator convex functions*, Houston J. Math. 21(4) (1995), 739–753.
[2] Bellman, R., *Almost orthogonal series*, Bull. Amer. Math. Soc. 50(8) (1944), 517–519.
[3] Pouladi Najafabadia, F. and Moradia, H. R., *Advanced refinements of numerical radius inequalities*, Int. J. Math. Model. Comput. 11(04) (2021), 1–10.
[4] Nayak, R. K., *Advancement of numerical radius inequalities of operators and product of operators*, Iran J. Sci. 48(4) (2024), 649–657, https://doi.org/10.1007/s40995-024-01603-1.
[5] Abu-Omar, A. and Kittaneh, F., *Upper and lower bounds for the numerical radius with an application to involution operators*, Rocky Mt. J. Math. 45(4) (2015), 1055–1064, https://doi.org/10.1216/RMJ-2015-45-4-1055.

[6] Aujla, J. and Silva, F., *Weak majorization inequalities and convex functions*, Linear Algebra Appl. 369 (2003), 217–233, https://doi.org/10.1016/S0024-3795(02)00720-6.

[7] Al-Dolat, M. and Jaradat, I., *A refinement of the Cauchy–Schwarz inequality accompanied by new numerical radius upper bounds*, Filomat 37(3) (2023), 971–977, https://doi.org/10.2298/FIL2303971A.

[8] Alomari, M. W., *On Cauchy–Schwarz type inequalities and applications to numerical radius inequalities*, Ric. Mat. 73 (2022), 1493–1510, https://doi.org/10.1007/s11587-022-00689-2.

[9] Bakherad, M. and Garayev, M. T., *Berezin number inequalities for operators*, Concr. Oper. 6 (2019), 33–43.

[10] Başaran, H., Gürdal, M., *Berezin number inequalities via Young inequality*, Honam Math. J. 43(3) (2021), 523–537.

[11] Başaran, H., Gürdal, M. and Güncan, A. N., *Some operator inequalities associated with Kantorovich and Hölder–McCarthy inequalities and their applications*, Turk. J. Math. 43(1) (2019), 523–532.

[12] Başaran, H. and Gürdal, V., *On Berezin radius inequalities via Cauchy–Schwarz type inequalities*, Malaya J. Mat. 11(02) (2023), 127–141, https://doi.org/https://doi.org/doi:10.26637/mjm1102/002.

[13] Başaran, H., Huban, M. B. and Gürdal, M., *Inequalities related to Berezin norm and Berezin number of operators*, Bull. Math. Anal. Appl. 14(2) (2022), 1–11.

[14] Berezin, F. A., *Covariant and contravariant symbols for operators*, Math. USSR-Izv. 6(5) (1972), 1117–1151.

[15] Buzano, M. L., *Generalizzazione della diseguaglianza di Cauchy–Schwarz (Italian)*, Rend. Sem. Mat. Politec. Torino 31 (1974), 405–409.

[16] Mohammad, H. M. R. and Bani-Ahmad, F., *An estimate for the numerical radius of the Hilbert space operators and a numerical radius inequality*, AIMS Math. 8(11) (2023), 26384–26405, https://doi.org/10.3934/math.20231347.

[17] Dragomir, S. S., *Power inequalities for the numerical radius of a product of two operators in Hilbert spaces*, Sarajevo J. Math. 5(18) (2009), 269–278.

[18] El-Haddad, M. and Kittaneh, F., *Numerical radius inequalities for Hilbert space operators II*, Studia Math. 182(2) (2007), 133–140, https://doi.org/10.4064/sm182-2-3.

[19] Erkan, G. G. and Gürdal, M., Operatörlerin Berezin yarıçap eşitsizliklerinin ilerletilmesi, in: *3rd Int. Conf. Eng. Nat. Soc. Sci.*, pp. 463–470, Konya, Turkey, May 16–17, 2024.

[20] Furuta, T., *An extension of the Heinz–Kato theorem*, Proc. Amer. Math. Soc. 120(3) (1994), 785–787, https://doi.org/10.2307/2160471.

[21] Gürdal, M. and Stojiljković, V., *Berezin radius inequalities for finite sums of functional Hilbert space operators*, Gulf J. Math. 17(1) (2024), 101–109, https://doi.org/10.56947/gjom.v17i1.1885.

[22] Huban, M. B., Başaran, H. and Gürdal, M., *New upper bounds related to the Berezin number inequalities*, J. Inequal. Spec. Funct. 12 (2021), 1–12.

[23] Huban, M. B., Başaran, H. and Gürdal, M., *Some new inequalities via Berezin numbers*, Turk. J. Math. Comput. Sci. 14 (2022), 129–137.

[24] Huban, M. B., Başaran, H. and Gürdal, M., *Berezin number inequalities via convex functions*, Filomat 36(7) (2022), 2333–2344.

[25] Karaev, M. T., Berezin set and Berezin number of operators and their applications, in: *8th Workshop on Numerical Ranges and Numerical Radii (WONRA-06)*, Univ. Bremen 7, pp. 983–1018, 2013.

[26] Karaev, M. T., *Reproducing kernels and Berezin symbols techniques in various questions of operator theory*, Complex Anal. Oper. Theory 7 (2013), 983–1018.

[27] Kittaneh, F., *A numerical radius inequality and an estimate for the numerical radius of the Frobenius companion matrix*, Studia Math. 158(1) (2003), 11–17.

[28] Kittaneh, F., *Norm inequalities for sums and differences of positive operators*, Linear Algebra Appl. 383 (2004), 85–91, https://doi.org/10.1016/j.laa.2003.11.023.

[29] McCarthy, C. A., C_p, Isr. J. Math. 5 (1967), 249–271, https://doi.org/10.1007/BF02771613.

[30] Stojiljković, V., *Twice differentiable Ostrowski type tensorial norm inequality for continuous functions of selfadjoint operators in Hilbert spaces*, Eur. J. Pure Appl. Math. 16(3) (2023), 1421–1433, https://doi.org/10.29020/nybg.ejpam.v16i3.4843.

[31] Stojiljković, V., *Twice differentiable Ostrowski type tensorial norm inequality for continuous functions of selfadjoint operators in Hilbert spaces*, Electron. J. Math. Anal. Appl. 11(2) (2023), 1–15, https://doi.org/10.21608/ejmaa.2023.199881.1014.

[32] Stojiljković, V. and Dragomir, S. S., *Refinement of the Cauchy–Schwartz inequality with refinements and generalizations of the numerical radius type inequalities for operators*, Ann. Math. Comput. Sci. 21(2) (2024), 33–43, https://doi.org/10.56947/amcs.v21.246.

[33] Stojiljković, V. and Gürdal, M., *Generalization of the Buzano's inequality and numerical radius inequalities*, J. Appl. Pure Math. 6(3–4) (2024), 191–200, https://doi.org/10.23091/japm.2024.191.

[34] Stojiljković, V. and Gürdal, M., *Berezin radius type inequalities for functional Hilbert space operators*, Electron. J. Math. 7 (2024), 35–44, https://doi.org/10.47443/ejm.2024.017.

Ioannis K. Argyros and Santhosh George

12 Local convergence analysis of a hybrid Gauss–Newton method under a majorant condition

Abstract: The implementation of the Gauss–Newton method for solving least squares problems requires the inversion of a certain linear operator at each step of the iteration, which is an expensive or impossible task in general. We propose a hybrid Gauss–Newton method where the inverse is approximated by a finite sum of linear operators. the local convergence analysis is provided for the hybrid method based on generalized continuity assumptions that control the derivative. The analysis includes the radius of convergence, error distances and uniqueness results for the stationary point.

Keywords: Convergence analysis, Gauss–Newton method, majorant condition

MSC 2020: 54H25, 54E50

12.1 Introduction

Let H_1 and H_2 denote real or complex Hilbert spaces, $D \subset$ be an open set and $F : D \longrightarrow H_2$ be a continuously Fréchet differentiable operator. We are concerned with the problem of finding a solution $s \in D$ of the nonlinear least squares problem

$$\min_{x \in D} \|F(x)\|^2. \tag{12.1}$$

A plethora of applications in this form have appeared in mathematical programming and optimization, e. g., penalization methods, nonlinear inclusions, minimax, and goal programming. If $H_1 = \mathbb{R}^i$ and $H_2 = \mathbb{R}^j$, where j is the number of observations and i is the number of parameters, then we deal with a data fitting problem [1, 6–31].

If the operator $F'(x)$ is injective, finding the solution s, i. e., a solution of the nonlinear equation

$$F'(x)^* F(x) = 0, \tag{12.2}$$

is equivalent to finding a least squares solution for the overdetermined equation

$$F(x) = 0, \tag{12.3}$$

Ioannis K. Argyros, Department of Mathematical Sciences, Cameron University, Lawton, OK 73505, USA, e-mail: iargyros@cameron.edu
Santhosh George, Department of Mathematical and Computational Sciences, National Institute of Technology Karnataka, Karnataka 575 025, India, e-mail: sgeorge@nitk.edu.in

https://doi.org/10.1515/9783112215760-012

where T^* denotes the adjoint of the operator T. There is an extensive literature on the solution of (12.2) or (12.3) [2–5, 12, 24, 25]. The solution s can be found in analytical form only in special cases. That is why researchers and practitioners rely mostly on iterative method approximating s. Whenever the operator $F'(x)$ is injective with closed graph for each $x \in D$, the popular Gauss–Newton method determines a stationary point of (12.2) [8, 18, 20].

The Gauss–Newton method is defined for each $n = -0, 1, 2, \ldots$ by

$$x_0 \in D, \quad x_{n+1} = x_n - \left[F'(x_n)^* F'(x_n)\right]^{-1} F'(x_n)^* F(x_n). \tag{12.4}$$

If $F'(x)$ has a closed image in D, then the Moore–Penrose inverse $F'(x)^+ : H_2 \longrightarrow H_1$ is defined by

$$F'(x)^+ = \left(F'(x)^* F'(x)\right)^{-1} F'(x)^*. \tag{12.5}$$

Consequently, method (12.4) can also be written in a condensed form as

$$x_0 \in D, \quad x_{n+1} = x_n - F'(x_n)^+ F(x_n). \tag{12.6}$$

The implementation of (12.4), required the inversion of the operator $F'(x_n)^* F'(x_n)$ at each step of the iteration. However, this is a computationally expensive or impossible task in general. That is why we propose a hybrid Gauss–Newton method which avoids this inversion. The development of this method is as follows: Let M be an invertible operator such that the linear operator $I_M^{-1}(M - F'(x_n)^* F'(x_n))$ is also invertible. Such a condition can be

$$\alpha = \|M^{-1}(M - F'(x)^* F'(x))\| < 1 \tag{12.7}$$

for each $x \in D$.

Then, we have

$$\left(F'(x^*)F'(x)\right)^{-1} = \left[I - M^{-1}(M - F'(x)^* F'(x))\right]^{-1} M^{-1}. \tag{12.8}$$

Set $A = M^{-1}(M - F'(x)^* F'(x))$,

$$B_\infty = I + A + A^2 + \cdots + A^k + \cdots$$

and

$$B_k = I + A + A^2 + \cdots + A^k,$$

where k is a natural number. Then, the bracket in (12.8) can be replaced by B_∞ provided that (12.7) holds and (12.8) can be written as

$$\left(F'(x^*)F'(x)\right)^{-1} = B_\infty^{-1} M^{-1}.$$

However, the inversion of the operator M and the computing of B_∞ are still needed for the implementation of the method (12.4). But what if we replace B_∞ by $B := B_k$ in (12.4). This way, we obtain the hybrid Gauss–Newton method

$$x_0 \in D, \quad x_{n+1} = x_n - B_k M^{-1} F'(x_n)^* F(x_n). \tag{12.9}$$

Notice that $\lim_{k \longrightarrow +\infty} B_k = B_\infty$. Thus, the hybrid method (12.9) is a useful alternative for the method (12.4). A choice for $M + I$, but it may not be the best possible. More choices for M can be $M = F'(x)$ or $M = F'(s)$ or $M = F'(x)^*$. Other selections are possible provided that the conditions (H3) and (H4) hold (see Theorem 12.1). We develop the local convergence analysis for method (12.9) based on generalized continuity assumptions used to control $F'(x)$. The analysis includes a radius of convergence for the method (12.9), a priori error estimates on the distances $\|x_n - s\|$ and uniqueness of the solution results. The technique is very general. So, it can be used on other methods using inverses [2–7, 9–31].

The rest of the study is organized as follows: Section 12.2 contains the necessary mathematical back ground listed to make this study as self as possible. The local convergence analysis of the method (12.9) can be found in Section 12.3.

12.2 Preliminaries

More details on the standard definitions and results can be found in [2, 18–20]. The set $U(x, \rho)$ denotes an open ball in H_1 with center $x \in H_1$ and of radius $\rho > 0$. The set $U[x, \rho]$ denotes the closure of $U(x, \rho)$.

Lemma 12.1 (Banach Perturbation Lemma [2, 19]). *Let $Q : H_1 \longrightarrow H_1$ be a continuous linear operator and $I : H_1 \longrightarrow H_1$ be the identity operator. If $\|Q - I\| < 1$, then Q is invertible and $\|Q^{-1}\| \le \frac{1}{1 - \|Q - I\|}$.*

Lemma 12.2. *Let $P, Q : H_1 \longrightarrow H_2$ be continuous linear operators with closed images. If P is injective, $G := Q - P$ and $\|GP^+\| < 1$, then the operator Q is injective.*

Proof. The condition $\|GP^+\| < 1$, Lemma 12.1 and the estimate $Q = P + G = (I + GP^+)P$ imply that the operator $I + GP^+$ is invertible. Hence, the operator Q is injective. □

Lemma 12.3 ([18, 20]). *Let $P, Q : H_1 \longrightarrow H_2$ be continuous and injective linear operators with closed images. In addition, suppose $G = Q - P$ and $\|P^+\|\|G\| < 1$. Then, the following assertions hold*

$$\|Q^+\| \le \frac{\|P^+\|}{1 - \|P^+\|\|G\|}$$

and

$$\|Q^+ - P^+\| \le \frac{\sqrt{2}\|P^+\|^2\|G\|}{1 - P^+\|\|G\|}.$$

12.3 Local convergence analysis for the method (12.9)

Let $R > 0$. Define the parameter λ by

$$\lambda = \sup\{t \in [0,R) : U(s,t) \subset D\}.$$

Suppose

(H1) There exists $s \in D$ such that $F'(s)^* F(s) = 0$ and the operator $F'(s)$ is injective. Define $\mu = \|F'(s)^+\|$ and $\gamma = \|F(s)\|$.

(H2) There exists a continuous and nondecreasing function $\varphi_0 : [0,\lambda) \longrightarrow [0,+\infty)$ such that for each $x \in U(s,\lambda)$

$$\|F'(x) - F'(s)\| \le \varphi_0(\|x-s\|)\|x-s\|.$$

(H3) The equation $\mu\varphi_0(t)t - 1 = 0$ has a smallest solution in the interval $(0,\lambda)$. Denote such solution by λ_0.

(H4) There exist an invertible operator M and continuous and nondecreasing functions $\varphi : [0,\lambda) \longrightarrow [0,+\infty)$, $\varphi_1 : [0,\lambda) \longrightarrow [0,+\infty)$, $\varphi_2 : [0,\lambda) \longrightarrow [0,+\infty)$, and $\alpha : [0,\lambda) \longrightarrow [0,1)$ such that for all $x \in D_0$ and $\theta \in [0,1)$

$$\|F'(s+\theta(x-s)) - F'(x)\| \le \varphi((1-\theta)\|x-s\|),$$
$$\|M^{-1}(F'(x) - M)\| \le \varphi_1(\|x-s\|),$$
$$\|F(x)\| \le \varphi_2(\|x-s\|)\|x-s\|$$

and

$$\|M^{-1}(M - F'(x)^* F'(x))\| \le \alpha(\|x-s\|) \le \alpha.$$

Define $\delta = \frac{\alpha^{k+1}}{1-\alpha}$.
Define the function $g : [0,\lambda_0) \longrightarrow [0,+\infty)$ by

$$g(t) = \frac{\mu\int_0^1 \varphi((1-\theta)t)d\theta}{1-\mu\varphi_0(t)t} + \frac{\sqrt{2}\gamma\mu^2\varphi_0(t)}{1-\mu\varphi_0(t)t} + \delta(1+\varphi_1(t))\varphi_2(t).$$

(H5) The equation $g(t)-1 = 0$ has a smallest solution in the interval $(0,\lambda_0)$. Denote such solution by r.

Define the set $T = [0,r)$. It follows by this definition that for each $t \in T$

$$0 \le \varphi_0(t)t < 1, \tag{12.10}$$
$$0 \le \alpha(t) < 1 \tag{12.11}$$

and

$$0 \le g(t) < 1. \tag{12.12}$$

The parameter r is shown to be a radius of convergence for the method (12.9) in the Theorem 12.1. But first, we need a perturbation result for the derivative $F'(x)$.

Lemma 12.4. *Let $x \in D$. Suppose that $x \in U(s,r)$ and (H1)–(H3) hold. Then, the operator $F'(x)^{\perp} * F'(x)$ is invertible. Moreover, the following assertions hold.*

$$\|F'(x)^{+}\| \le \frac{\mu}{1 - \mu\varphi_0(\|x-s\|)\|x-s\|} \tag{12.13}$$

and

$$\|F'(x)^{+} - F'(s)\| \le \frac{\sqrt{2}\mu^2\varphi_0(\|x-s\|)\|x-s\|}{1 - \mu\varphi_0(\|x-s\|)\|x-s\|}. \tag{12.14}$$

Proof. Let $x \in U(s,r)$. By the definition of μ and the conditions (H1)–(H3), we have in turn that

$$\begin{aligned}\|F'(s)^{+}\|\|F'(x) - F'(s)\| &\le \mu\varphi_0(\|x-s\|)\|x-s\| \\ &\le \mu\varphi_0(r)r < 1.\end{aligned} \tag{12.15}$$

It is convenient to define the operators:

$$P = F'(s), \quad Q = F'(x) \quad \text{and} \quad G = Q - P. \tag{12.16}$$

Then, from (12.15) and (12.16), we get

$$\|GP^{+}\| \le G\|\|P^{+}\| < 1,$$

which, together with the hypothesis that $F'(s)$ is injective, implies by Lemma 12.2 that the operator $F'(x)$ is injective. Thus, the operator $F'(x)^{*}F'(x)$ is invertible for each $x \in U(s,r)$. Finally, the conclusions hold by (12.16) and Lemma 12.3. □

Next, we are in a position to present the local convergence analysis of the method (12.2).

Theorem 12.1. *Suppose that the conditions (H1)–(H5) hold. Then, the sequence $\{x_n\}$ generated for $x_0 \in U(s,r) - \{s\}$ by the method (12.9) is well defined in $U(s,r)$, remains in $U(s,r)$ for each $n = 0$ and converges to s. Moreover, the following error assertion holds for each $n = 0, 1, 2, \ldots$.*

$$\|x_{n+1} - s\| \le g(\|x_n - s\|)\|x_n - s\| \le \|x_n - s\| < r. \tag{12.17}$$

Proof. Let $x \in U(s,r)$. Then, the estimates (12.13) and (12.14) hold. It follows that the iterate x_1 is well defined by the method (12.9). We need some estimates. Using the last condition of (H4) and the definition of B_∞, B_k, α and δ, one has

$$\begin{aligned}\|B_\infty - B_k\| &\leq \|A\|^{k+1} + \|A\|^{k+2} + \cdots \\ &\leq \alpha^{k+1}(1+\alpha+\cdots) \\ &\leq \alpha^{k+1}\frac{1}{1-\alpha} = \delta. \end{aligned} \tag{12.18}$$

In view of (12.4), (12.7)–(12.9), (H4), (12.11)–(12.14), (12.18) and some algebraic manipulation, we obtain in turn

$$\begin{aligned} x_{n+1} - s = {} & F'(x_n)^+[F'(x_n)(x_n - s) - F(x_n) + F(s)] \\ & + (F'(s)^+F(s) - F'(x)^+F(s)) \\ & + (B_\infty - B_k)M^{-1}(F'(x_n)^* - M + M)F(x_n), \end{aligned}$$

so

$$\begin{aligned} \|x_{n+1} - s\| \leq {} & \left[\frac{\mu \int_0^1 \varphi((1-\theta)\|x_n - s\|)d\theta}{1 - \mu\varphi_0(\|x_n - s\|)\|x_n - s\|} \frac{\sqrt{2}\gamma\mu^2\varphi_0(\|x_n - s\|)}{1 - \mu\varphi_0(\|x_n - s\|)\|x_n - s\|}\right. \\ & \left. + \delta(1 + \varphi_1(\|x_n - s\|))\left(1 + \int_0^1 \varphi_0(\theta\|x_n - s\|)d\theta\right)\right]\|x_n - s\| \\ = {} & g(\|x_n - s\|)\|x_n - s\| \\ \leq {} & \|x_n - s\| < r, \end{aligned} \tag{12.19}$$

where we also used that estimates

$$\begin{aligned} \|M^{-1}(F'(x_n)^* - M + M)\| &\leq \|M^{-1}M\| + \|M^{-1}(F'(x_n)^* - M)\| \\ &\leq 1 + \varphi_1(\|x_n - s\|) \end{aligned}$$

and

$$\|F(x_n)\| \leq \varphi_2(\|x_n - s\|)\|x_n - s\|.$$

Thus, the iterate $x_{n+1} \in U(s,r)$ and the induction for the assertion (12.17) is completed. Then, from the estimation

$$\|x_{n+1} - s\| \leq c\|x_n - s\| \leq c^{n=1}\|x_0 - s < r, \tag{12.20}$$

where $c = g(\|x_0 - s\|) \in [0,1)$, we conclude that $\lim_{n\longrightarrow\infty} x_n = s$. □

Next, a set is determined inside, which s is the unique stationary point of $F'(x)^*F(x)$.

Proposition 12.1. *Suppose that the condition of the Theorem* 12.1 *hold. Then, s is the unique stationary point of $F'(x)^*F(x)$ in the ball $U(s,r)$.*

Proof. Suppose that $s_1 \in U(s,r)$ is a stationary point in $U(s,r)$ such that $s \neq s_1$. Thus, we can write

$$s_1 - s = s_1 - s - F'(s_1)^+F(s_1).$$

Then, as in (12.19) for x_{n+1} replaced by s_1, we get

$$\|s_1 - s\| \leq g(\|s_1 - s\|)\|s_1 - s < \|s_1 - s\|,$$

showing that $s_1 = s$. □

The choice $M = F'(x_n)^*$ significantly simplifies the method (12.9) and the conditions (H3) and (H4).

12.4 Numerical example

In this section. we present a concrete example, where the convergence conditions are validated.

Example 12.1. *Let $H_1 = H_2 = \mathbb{R}^3$, $k = 1$, $D = U(0,1)$. Define the operator $F : D \longrightarrow \mathbb{R}^3$ for $x = (q_1, q_2, q_3)^{tr}$, $(q_i \in \mathbb{R}, i = 1,2,3)$ by*

$$F(x) = \begin{bmatrix} e^{q_1} + q_2 + q_3 - 1 \\ \frac{1}{2}(e-1)q_2^2 + q_2 + q_1 + q_3 \\ q_1 + q_2 + q_3 \end{bmatrix}.$$

It follows by this definition that for $s = (0,0,0)^{tr}$, $F(s) = 0$. Moreover, the Fréchet derivative $F'(x)$ is

$$F'(x) = \begin{bmatrix} e^{q_1} & 0 & 0 \\ 0 & (e-1)q_2 + 1 & 0 \\ 0 & 0 & 1 \end{bmatrix}.$$

In particular, $F'(x^*) = I = F'(x)^+$, $\mu = 1$, $\lambda = \frac{1}{e-1}$, and $\gamma = 0$. The conditions (H1)–(H4) are satisfied for $M = I$ provided that $\varphi_0(t) = e-1$, $\varphi(t) = e^{\frac{1}{e-1}}t$, $\varphi_1(t) = (e-1)t$, $\varphi_2(t) = 1+(e-1)t$, and $a(t) = (e-1)t$. Then, using the definition of the function g in (H4), we see that the radius of convergence

$$r = 0.62898952721015917510933235674744.$$

12.5 Concluding remarks

The important issue with the implementation of methods using inverses or Moore–Penrose inverses of linear operators is addressed by replacing them with finite sums of linear operators. In particular, the local convergence analysis of the hybrid Gauss–Newton method (12.9) is provided to solve the problem (12.1).

A numerical example is used to validate the convergence conditions. The new approach is very general. Therefore, it can be used to simplify the implementation of other methods along the same lines [1, 6–31].

Bibliography

[1] Aragon Artacho, F. J., Dontchev, A. L., Gaydu, M., Geoffroy, M. H. and Veliov, V. M., *Metric regularity of Newton's iteration*, SIAM J. Control Opti. 49 (2011), 339–362.

[2] Argyros, I. K., *The Theory and Applications of Iteration Methods with Applications*, 2nd ed., Engineering Series, CRC Press, Taylor and Francis Publ., Boca Raton, Florida, USA, 2022.

[3] Argyros, I. K. and George, S., *On a unified convergence analysis for Newton-type methods solving generalized equations with the Aubin property*, Journal of Complexity 81 (2024), 101817.

[4] Argyros, I. K., George, S., Regmi, S. and Argyros, C. I., *Hybrid Newton-like inverse free algorithms for solving nonlinear equations*, Algorithms 17 (2024), 154. https://doi.org/10.3390/a17040154

[5] Argyros, I. K., George, S., Shakhno, S., Regmi, S., Havdiak, M. and Argyros, M. I., *Asymptotically Newton-type methods without inverses for solving equations*, Mathematics 12 (2024), 1069. https://doi.org/10.3390/math12071069

[6] Argyros, I. K. and Hilout, S., *Weaker conditions for the convergence of Newton's method*, J. Complexity 28 (2012), 364–387.

[7] Blum, L., Cucker, F., Shub, M. and Smale, S., *Complexity and Real Computation*, Springer-Verlag, New York, 1998. With a foreword by Richard M. Karp.

[8] Burke, J. V. and Ferris, M. C., *A Gauss–Newton method for convex composite optimization*, Math. Program. 71 (1995), 179–194.

[9] Chen, J., *The convergence analysis of inexact Gauss–Newton methods for nonlinear problems*, Comput. Optim. Appl. 40(1) (2008), 97–118.

[10] Chen, J. and Li, W., *Convergence of Gauss–Newton's method and uniqueness of the solution*, Appl. Math. Comput. 170(1) (2005), 686–705.

[11] Dedieu, J. P. and Kim, M. H., *Newton's method for analytic systems of equations with constant rank derivatives*, J. Complexity 18(1) (2002), 187–209.

[12] Dennis, J. E. Jr. and Schnabel, R. B., *Numerical Methods for Unconstrained Optimization and Nonlinear Equations*, Classics in Applied Mathematics vol. 16, SIAM, Philadelphia, PA, 1996. Corrected reprint of the 1983 original.

[13] Deuflhard P., N., *Methods for Nonlinear Problems: Affine Invariance and Adaptive Algorithms*, Springer, Berlin, Heidelberg, 2004.

[14] Deuflhard, P. and Heindl, G., *Affine invariant convergence theorems for Newton's method and extensions to related methods*, SIAM J. Numer. Anal. 16 (1979), 1–10.

[15] Dontchev, A. L. and Rockafellar, R. T., *Implicit Functions and Solution Mappings: A View from Variational Analysis*, Springer Monogr. Math., Springer, Dordrecht, 2009.

[16] Ferreira, O. P. and Goncalves, M. L. N., *Local convergence analysis of inexact Newton-like methods under majorant condition*, Comput. Optim. Appl. 48 (2011), 1–21.

[17] Ferreira, O. P., Goncalves, M. L. N. and Oliveira, P. R., *Local convergence analysis of the Gauss–Newton method under a majorant condition*, J. Complexity 27 (2011), 111–125.
[18] Ferreira, O. P., Goncalves, M. L. N. and Oliveira, P. R., *Convergence of the Gauss–Newton method for convex composite optimization under a majorant condition*, SIAM J. Optim. 23(3) (2013), 1757–1783.
[19] Kantorovich, L. V. and Akilov, G. P., *Functional Analysis*, 2nd ed., Pergamon Press, Oxford, 1982 (translated from Russian by Howard L. Silcock).
[20] Li, C. and Ng, K. F., *Majorizing functions and convergence of the Gauss–Newton method for convex composite optimization*, SIAM J. Optim. 18 (2007), 613–642.
[21] Li, C. and Ng, K. F., *Convergence analysis of the Gauss–Newton method for convex inclusion and convex-composite optimization problems*, J. Math. Anal. Appl. 389 (2012), 469–485.
[22] Li, C. and Wang, Y. H., *On convergence of the Gauss–Newton method for convex composite optimization*, Math. Program. 91 (2002), 349–356.
[23] Nocedal, J. and Wright, S. J., *Numerical Optimization*, Springer Series in Operations Research, Springer-Verlag, New York, 1999.
[24] Proinov, P. D., *General local convergence theory for a class of iterative processes and its applications to Newton's method*, J. Complexity 25(1) (2009), 38–62.
[25] Robinson, S. M., *Extension of Newton's method to nonlinear function with values in a cone*, Numer. Math. 19 (1972), 341–347.
[26] Robinson, S. M., *Normed convex process*, Trans. Amer. Math. Soc. 174 (1972), 127–140.
[27] Rockafellar, R. T., *Monotone Processes of Convex and Concave Type*, Mem. Amer. Math. Soc., vol. 77, AMS, 1967.
[28] Rockafellar, R. T., *Convex Analysis*, Princeton University Press, Princeton, NJ, 1970.
[29] Smale, S., Newton's method estimates from data at one point, in: *The Merging of Disciplines: New Directions in Pure, Applied, and Computational Mathematics*, Laramine, Wyo., 1985, pp. 185–196, Springer, New York, 1986.
[30] Stewart, G. W., *On the continuity of the generalized inverse*, SIAM J. Appl. Math. 17 (1969), 33–45.
[31] Wedin, P. A., *Perturbation theory for pseudo-inverses*, Nord. Tidskr. Informationsbehandling (BIT) 13 (1973), 217–232.

Ioannis K. Argyros, Stepan Shakhno, Samundra Regmi, and Halyna Yarmola

13 Extended convergence region for a family of King–Traub methods for solving nonlinear equations

Abstract: King–Traub-like methods (KTLM) are very important tools utilized to compute solutions of nonlinear equations. The convergence order at least three is determined using Taylor series. The proofs require the existence of derivatives of high order not on KTLM. Thus, the results are restricted to solving nonlinear equations containing functions that are more than two times differentiable. But the methods may converge even if the high order derivatives do not exist. This observation indicates that the convergence conditions can be weakened. In this article, a methodology is introduced that shows the convergence of KTLM. But the convergence conditions rely exclusively on the functions and their derivatives on KTLM. Other drawbacks of the Taylor series technique which are addressed in this article are the lack of a priori error estimates for the error distances and isolation of the solution results. Hence, the applicability of the methods is extended in cases not possible before. The convergence conditions are verified using numerical experiments.

Keywords: King–Traub's method, root, semi-local convergence

MSC 2020: 65G99, 47H99, 49M15, 65H10

13.1 Introduction

The letter S is used to denote the real line or the complex plane. Moreover, D stands for a convex set, which may be open or closed. Mathematical modeling is mostly employed to convert problems from different disciplines [2, 7, 26, 44] in an equation like

$$G(x) = 0, \tag{13.1}$$

where for $D \subset S$, and the function $G : D \to S$ is continuous or even differentiable on D. The analytic form of a root $x^* \in D$ is obtained only in special cases. That explains

Ioannis K. Argyros, Department of Computing and Mathematical Sciences, Cameron University, Lawton, OK 73505, USA, e-mail: iargyros@cameron.edu
Stepan Shakhno, Department of Theory of Optimal Processes, Ivan Franko National University of Lviv, Lviv, Ukraine, e-mail: stepan.shakhno@lnu.edu.ua
Samundra Regmi, Department of Mathematics, University of Houston, Houston, TX 77204, USA, e-mail: sregmi5@uh.edu
Halyna Yarmola, Department of Computational Mathematics, Ivan Franko National University of Lviv, Lviv, Ukraine, e-mail: halyna.yarmola@lnu.edu.ua

https://doi.org/10.1515/9783112215760-013

why researches rely mostly on iterative methods generating a sequence $\{x_m\}$ such that $\lim_{m\to\infty} x_m = x^*$. The Picard and Newton's method have been used extensively [2, 44]. But these methods are only of order one and two, respectively [2, 44]. Researchers have used Halley-type, Chebyshev-type and Jarratt-type, and other multistep methods to increase the convergence order to three or even higher [6–19, 21, 20, 22–27, 29–36, 38, 40–43].

A popular way to determine the convergence order seems to be using Taylor series. However, these are some disadvantages with this approach:

(1) The existence of derivatives high order is assumed which are not on the iterative method. Thus, if these derivatives do not exist, the results in the aforementioned references cannot assume the convergence of the iterative method. However, it is possible that the method converges even if these derivatives do not exist (see also the example that follows).
(2) A priori error computable estimates on $|x^* - x_m|$ or $|x_{m+1} - x_m|$ are not available. Hence, there is no advance knowledge on how many iterates should be computed to reach a pre-decided error tolerance.
(3) No information is provided on the uniqueness of x^* is a neighborhood containing it.
(4) The results are mainly local.

Therefore, the concerns (1)–(4) constitute our motivation for writing this article. These concerns are addressed for a popular iterative method.

In this article, we study generalized King–Traub-type methods (KTLM) [26, 28, 44]:

$$\begin{aligned} y_m &= x_m - G'(x_m)^{-1}G(x_m), \\ x_{m+1} &= y_m - V_m^{-1}Q_m G'(x_m)^{-1}G(y_m), \\ V_m &= aG(x_m) + bG(y_m), \\ Q_m &= cG(x_m) + dG(y_m), \end{aligned} \tag{13.2}$$

where $a, b, c, d \in S$. Although the new methodology is demonstrated for KTLM, it is also applicable on other iterative methods in a similar way. Here is how we handle these concerns:

(1)' The convergence conditions involve only G and G', i. e., only the functions on KTLM. In order to emphasize the difference between (1) and (1)', let us look at an example. Let $D = [0, \frac{3}{2}]$. Define $G : D \to \mathbb{R}$ as

$$G(x) = \begin{cases} \alpha x^3 \log x + \beta x^5 + \gamma x^4, & x \neq 0, \\ 0, & x = 0, \end{cases}$$

where $\alpha \neq 0$ and $\beta + \gamma = 0$ and α, β, γ are real parameters.

It follows by this definition that $x^* = 1 \in D$ solves the equation $G(x) = 0$. However, the function G''' is not continuous at $x = 0 \in D$. Consequently, the results in [26] cannot assure the convergence of (13.2) to x^*. However, the sequence $\{x_m\}$ converges to x^* provided, e. g., if $x_0 = 0.95$ for $a = b = c = 1$ and $d = -1$. Thus, the convergence conditions requiring the existence of G''' can be weakened.

(2)' In this article, computable a priori estimates on $|x^* - x_m|$ and $|x_{m+1} - x_m|$ become available. So, the number of iterates to be computed is known in advance.

(3)' A domain is specified that contains only x^* as a solution of the equation $G(x) = 0$.

(4)' The results in this article are semi-local which are more interesting and challenging than the local ones.

The convergence analysis relies on Lipschitz conditions employed to control the derivative G' as well as majorizing sequences. We also use the method of recurrent polynomials initiated by us in [2] to weaken the convergence conditions of Newton's methods [3].

The rest of the article is organized as: The majorizing sequence is introduced and its convergence is studied in Section 13.2. The semi-local analysis of (13.2) is presented in Section 13.3, followed by the examples used to validate the convergence conditions in Section 13.4. The article ends with concluding remarks in Section 13.5.

13.2 Scalar sequences

A real sequence $\{l_m\}$ is defined which is majorizing for $\{x_m\}$ [2, 37] (see Theorem 13.1). It is convenient for the convergence analysis to introduce some parameters and polynomials.

Let $\tau, R, R_0, R_1, R_2, R_3, R_4, R_5$ be given nonnegative parameters. By $|\ |$ we mean the standard norm on S. Set $R_6 = R_1R_3|c| + |c-a|R_1R_4$, $R_7 = \frac{RR_5}{2}|d| + \frac{RR_4}{2}|d-b|$, $R_8 = \frac{R}{2}|c|R_3$ and $R_9 = \frac{|d|R_2R}{4}$.

Define sequence $\{l_m\}$ by

$$
\begin{aligned}
l_0 &= 0, \quad h_0 = \tau, \\
l_1 &= \tau + (R_8 + R_9\tau)\tau^3, \\
h_{m+1} &= l_{m+1} + \frac{\xi_{m+1}(l_{m+1} - h_m)}{1 - R_0 l_{m+1}}, \\
l_{m+1} &= h_m + \frac{(R_8 + R_9(h_m - l_m))(h_m - l_m)^3}{(1 - a_m)(1 - R_0 l_m)},
\end{aligned}
\tag{13.3}
$$

where

$$b_m = R_4(|c|(l_m - l_0) + |d|(h_m - h_0)),$$

$$\gamma_m = \frac{1}{1-b_m},$$
$$\xi_{m+1} = \frac{R}{2}(l_{m+1}-h_m) + \gamma_m(R_6 + R_7(h_m - l_m))(h_m - l_m)^2,$$
$$\bar{\xi}_{m+1} = \frac{R}{2}(l_{m+1}-h_m) + 2(R_6 + R_7(h_m - l_m))(h_m - l_m)^2$$

and

$$a_m = R_3(|a|(l_m - l_0) + |b|(h_m - h_0)).$$

Next, two convergence results are presented for sequence $\{l_m\}$ starting from a very general one.

Lemma 13.1. *Assume that for all* $m = 0, 1, 2, \ldots$

$$R_0 l_m < 1, \quad a_m < 1, \quad \xi_{m+1} < 1 \quad \textit{and} \quad b_m < 1. \tag{13.4}$$

Then, the sequence $\{l_m\}$ *is nondecreasing and convergent to* $l^* \in [0, l^{**}]$, *where* $l^{**} = \frac{1}{R_0}$.

Proof. Using (13.3) and (13.4), we see that the sequence $\{l_m\}$ is bounded from above by $\frac{1}{R_0}$ and nondecreasing. Thus it is convergent to l^*. □

Stronger assumption can be introduced to show convergence of sequence $\{l_m\}$. These assumptions imply (13.4), but not necessarily vice versa. Moreover, they are easier to verify than (13.4). Let us define some recurrent polynomials on the interval $[0, 1)$ by

$$g_m^{(1)}(l) = \frac{R}{2}l^{2m}\tau + 2R_6 l^{2m+1}\tau^2 + 2R_7 l^{2m+3}\tau^3 + R_0(1 + l + \cdots + l^{2m+1})\tau - 1,$$
$$\bar{q}_m^{(1)}(l) = \frac{R}{2}l^{2m} + 2R_6 l^{2m+1}\tau^2 + 2R_7 l^{2m+3}\tau^3 + R_0(1 + l + \cdots + l^{2m+1})\tau - 1,$$
$$g_m^{(2)}(l) = 2R_8 l^{2m+1}\tau^2 + 2R_9 l^{2m+2}\tau^3 + R_0(1 + l + \cdots + l^{2m+1})\tau - 1,$$
$$\bar{q}_m^{(2)}(l) = 2R_8 l^{2m+1}\tau + 2R_9 l^{2m+2}\tau + R_0(1 + l + \cdots + l^{2m+1})\tau - 1,$$
$$g_1(l) = \frac{R}{2}l^2 - \frac{R}{2} + 2R_6 l^3\tau - 2R_6 l\tau + 2R_7 l^5\tau^2 - 2R_7 l^3\tau^2 + R_0(l^2 + l^3)$$
$$\bar{g}_1(l) = \frac{R}{2}l^2 - \frac{R}{2} + 2R_6 l^3\tau - 2R_6 l\tau + 2R_7 l^5 - 2R_7 l^3 + R_0(l^2 + l^3),$$
$$g_2(l) = 2R_8 l^2\tau - 2R_8\tau + 2R_9 l^3\tau^2 - 2R_9 l\tau^2 + R_0(l + l^2),$$

and

$$\bar{g}_2(l) = 2R_8 l^2 - 2R_8 + 2R_9 l^3 - 2R_9 l + R_0(l + l^2).$$

The polynomials g_1, $\bar{g}_1$, g_2, $\bar{g}_2$ have zeros in $(0,1)$, since they vanish at $l = 0$ and are positive at $l = 1$. Denote by δ_1, $\bar{\delta}_1$, δ_2, $\bar{\delta}_2$ the minimal such zeros, respectively. Notice that δ_1 and δ_2 depend on m, but not $\bar{\delta}_1$ and $\bar{\delta}_2$. Moreover, set

$$\alpha = \frac{a_1}{1 - R_0 l_1}, \quad \beta = (R_8 + R_9\tau)\tau^2, \quad \gamma = \max\{\alpha, \beta\},$$
$$\delta_3 = \min\{\delta_1, \delta_2\}, \quad \delta = \max\{\delta_1, \delta_2\}, \quad \bar{\delta}_3 = \min\{\bar{\delta}_1, \bar{\delta}_2\},$$
$$\lambda_1 = 1 - R_0\tau, \quad \lambda_2 = \frac{1 - 2R_3|a|\tau}{1 + 2R_3|a|\tau},$$
$$\lambda_3 = \frac{1 - 2R_4|c|\tau}{1 + 2R_4|c|\tau}, \quad \mu = \min\{\lambda_1, \lambda_2, \lambda_3\}.$$

Assume:

$$R_0 l_1 < 1, \quad 2R_3|a|\tau < 1, \quad 2R_4|c|\tau < 1, \tag{13.5}$$

and

$$\alpha \le \beta \le \gamma \le \bar{\delta}_1 \le \delta < \mu \le 1 \tag{13.6}$$

or

$$\alpha \le \beta \le \gamma \le \bar{\delta} \le \bar{\delta}_1 < \mu \le 1. \tag{13.7}$$

This way, we can develop another result for the method (13.2).

Lemma 13.2. *Under assumptions* (13.3) *and* (13.4) *(or* (13.3) *and* (13.5)*) the result of Lemma* 13.1 *holds for* $l^{**} = \frac{\tau}{1-\delta}$. *Moreover, the following estimates hold*

$$0 \le l_{m+1} - h_m \le \delta(h_m - l_m) \le \delta^{2m+1}\tau \tag{13.8}$$

and

$$0 \le h_m - l_m \le \delta(l_m - h_{m-1}) \le \delta^{2m}\tau. \tag{13.9}$$

Proof. Estimates

$$0 \le \frac{\xi_p}{1 - R_0 l_p} \le \delta, \tag{13.10}$$
$$0 \le \frac{(R_8 + R_9(h_p - l_{p+1}))(h_p - l_p)}{(1 - a_p)(1 - R_0 l_p)} \le \delta, \tag{13.11}$$
$$0 \le R_0 l_p < 1, \tag{13.12}$$
$$0 \le g a_p < 1, \tag{13.13}$$
$$0 \le 2b_p \le 1, \tag{13.14}$$

and

$$0 \le l_p \le h_p \le l_{p+1} \tag{13.15}$$

are shown by mathematical induction.

These estimates are true for $p = 0$ by the initial data and assumptions (13.5)–(13.7). Then, we have

$$0 \le l_1 - h_0 \le \delta(h_0 - l_0) = \delta\tau$$

and

$$0 \le h_1 - l_1 \le \delta(l_1 - h_0) \le \delta^2\tau.$$

So, (13.8) and (13.9) also hold for $p = 0, p = 1$, respectively. Assume estimates (13.10)–(13.15) hold for all integer smaller than p. Then, by these hypotheses, we get

$$\begin{aligned} 0 \le h_p &\le l_p + \delta^{2p}\tau \le h_{p-1} + \delta^{2p-1}\tau + \delta^{2p}\tau \le \cdots \\ &\le \tau + \delta\tau + \cdots + \delta^{2p}\tau = \frac{1-\delta^{2p+1}}{1-\delta}\tau < \frac{\tau}{1-\delta} = l^{**}, \end{aligned} \tag{13.16}$$

and

$$\begin{aligned} l_{p+1} &\le h_p + \delta^{2p+1}\tau \le l_p + \delta^{2p}\tau + \delta^{2p+1}\tau \le \cdots \\ &\le \tau + \delta\tau + \cdots + \delta^{2k+1}\tau = \frac{1-\delta^{2p+2}}{1-\delta}\tau < l^{**}. \end{aligned} \tag{13.17}$$

Then, evidently (13.10) holds if

$$\frac{R}{2}\delta^{2p+1}\tau + 2(R_6 + R_7\delta^{2p}\tau)(\delta^{2p}\tau)^2 + \delta R_0\frac{1-\delta^{2p+2}}{1-\delta}\tau - \delta \le 0. \tag{13.18}$$

Case 1. Assume (13.3) and (13.4) hold. Then, (13.18) holds if

$$g_p^{(1)}(l) \le 0 \quad \text{at } l = \delta_1, \tag{13.19}$$

(since $\delta^{2p} < \delta^{p+1}$ for $p = 1, 2, \ldots$). We can have:

$$\begin{aligned} g_{p+1}^{(1)}(l) &= g_{p+1}^{(1)}(l) - g_p^{(1)}(l) + g_p^{(1)}(l) \\ &= \frac{R}{2}l^{2p+2}\tau + 2R_6l^{2p+3}\tau^2 + 2R_7l^{2p+5}\tau^3 + R_0(1 + l + \cdots + l^{2p+3})\tau - 1 \\ &\quad + g_p^{(1)}(l) - \frac{R}{2}l^{2p}\tau - 2R_6l^{2p+1}\tau^2 - 2R_7l^{2p+3}\tau^3 \\ &\quad - R_0(1 + l + \cdots + l^{2p+1})\tau + 1 = g_p^{(1)}(l) + g_1(l)l^{2p}\tau, \end{aligned}$$

so

$$g_{p+1}^{(1)}(l) = g_p^{(1)}(l) + g_1(l)l^{2p}\tau \tag{13.20}$$

In particular, we have

$$g_{p+1}^{(1)}(l) = g_p^{(1)}(l) \quad \text{at } l = \delta_1, \tag{13.21}$$

since $g^{(1)}(\delta_1) = 0$.

Define function

$$g_\infty^{(1)}(l) = \lim_{p\to\infty} g_p^{(1)}(l). \tag{13.22}$$

It then follows from the definition of $g_p^{(1)}$ and (13.22) that

$$g_\infty^{(1)}(l) = \frac{R_0\tau}{1-l} - 1. \tag{13.23}$$

Thus, (13.19) holds if

$$g_\infty^{(1)}(l) \le 0 \quad \text{at } l = \delta_1, \tag{13.24}$$

which is true by (13.4). Similarly, (13.11) holds if

$$2R_8\delta^{4m}\tau^2 + 2R_9\delta^{6m}\tau^3 + \delta R_0(1 + \delta + \cdots + \delta^{2p+1})\tau - \delta \le 0 \tag{13.25}$$

or

$$g_p^{(2)} \le 0 \quad \text{at } l = \delta_2. \tag{13.26}$$

As before, a relationship between two consecutive polynomials $g_p^{(2)}$ is needed:

$$\begin{aligned}
g_{p+1}^{(2)}(l) &= g_{p+1}^{(2)}(l) - g_p^{(2)}(l) + g_p^{(2)}(l) \\
&= g_p^{(2)}(l) + 2R_8l^{2p+3}\tau^2 + 2R_9l^{2p+4}\tau^3 + R_0(1 + l + \cdots + l^{2p+3})\tau - 1 \\
&\quad - 2R_9l^{2p+1}\tau^2 - 2R_9l^{2p+2}\tau^3 - R_0(1 + l + \cdots + l^{2p+1})\tau + 1 \\
&= g_p^{(2)}(l) + g_p^{(2)}(l)l^{2p+1}\tau,
\end{aligned}$$

i. e.,

$$g_{p+1}^{(2)}(l) = g_p^{(2)}(l) + g_p^{(2)}(l)l^{2p+1}\tau, \tag{13.27}$$

and

$$g_{p+1}^{(2)}(l) = g_p^{(2)}(l) \quad \text{at } l = \delta_2. \tag{13.28}$$

Define function

$$g_\infty^{(1)}(l) = \lim_{p\to\infty} g_p^{(2)}(l). \tag{13.29}$$

Then, it follows from the definition of $g_p^{(2)}$ and (13.29) that

$$g_\infty^{(2)}(l) = \frac{R_0\tau}{1-l} - 1. \tag{13.30}$$

Hence, (13.26) holds if

$$g_p^{(2)}(\infty) \le 0 \quad \text{at } l = \delta_2, \tag{13.31}$$

which is true by (13.4).

Case 2. Assume (13.3) and (13.5). Simply replace $g_p^{(1)}, g_1, g_p^{(2)}, g_2, \delta_1$ and δ_2 by $\bar{g}_p^{(1)}, \bar{g}_1, \bar{g}_p^{(2)}, \bar{g}_2, \bar{\delta}_1$ and $\bar{\delta}_2$, respectively. Concerning estimates (13.12)–(13.14), they shall be true if

$$R_0\frac{\tau}{1-\delta} < 1, \quad 2R_3\left(|a|\frac{\tau}{1-\delta} + |b|\left(\frac{\tau}{1-\delta} - \tau\right)\right) \le 1$$

and

$$2R_4\left(|c|\frac{\tau}{1-\delta} + |d|\left(\frac{\tau}{1-\delta} - \tau\right)\right) \le 1.$$

However, these estimates hold by the choice of $\lambda_1, \lambda_2, \lambda_3$, respectively. Finally, estimate (13.15) holds by (13.10)–(13.14) and the definition of sequence $\{l_p\}$. It follows that in either Case 1 or Case 2, sequence $\{l_p\}$ is nondecreasing and bounded from above by $l^* \in [0, l^{**}]$. □

13.3 Semi-local analysis of convergence for method (13.2)

Majorizing sequence $\{l_m\}$ and the following assumptions are required in the convergence analysis. These assumptions are:

(C_1) There exist $x_0 \in D$, $\tau > 0$ so that

$$G'(x_0) \ne 0, \quad V_0 \ne 0, \quad Q_0 \ne 0 \quad \text{and} \quad |G'(x_0)^{-1}G(x_0)| \le \tau.$$

(C_2) There exists $R_0 > 0$ so that for each $u \in D$

$$|G'(x_0)^{-1}(G'(u) - G'(u_0))| \le R_0|u - x_0|.$$

Define $D_0 = U(x_0, \frac{1}{R_0}) \cap D$.

(C_3) There exist positive parameters R, R_1, R_2, R_3, R_4 and R_5 so that for each $z, u \in D_0$

$$\begin{aligned} \left|G'(x_0)^{-1}(G'(z) - G'(u))\right| &\le R|z - u|, \\ \left|G'(x_0)^{-1}G'(u)\right| &\le R_1, \\ \left|V_0^{-1}(G'(z) - G'(u))\right| &\le R_2|z - u|, \\ \left|V_0^{-1}G'(z)\right| &\le R_3, \\ \left|Q_0^{-1}G'(u)\right| &\le R_4 \end{aligned}$$

and

$$\left|Q_0^{-1}(G'(z) - G'(u))\right| \le R_5|z - u|.$$

(C_4) Assumptions of Lemma 13.1 hold.
The assumptions of the Lemma 13.2 can also be used in the condition (C_4).
The notation $U(x_0, \lambda) = \{z \in S : |z - x_0| < \lambda\}$ for some $\lambda > 0$, where as $U[x_0, \lambda]$ stands for the complement of $U(x_0, \lambda)$.

(C_5) $U[x_0, l^*] \subset D$ (or $U[x_0, l^{**}] \subset D$), where $l^{**} = \frac{1}{R_0}$ (or $l^{**} = \frac{\tau}{1-\delta}$), in the case of Lemma 13.1 (or Lemma 13.2).

Then, we can provide the semi-local analysis of convergence of method (13.2).

Theorem 13.1. *Assume that the conditions (C_1)–(C_5) hold. Then, the iteration $\{x_m\}$ produced by* (13.2) *exists in $U(x_0, l^*)$, stays in $U(x_0, l^*)$ for each $m = 0, 1, 2, \ldots$, and converges to some $x^* \in U[x_0, l^*]$ solving equation $G(x) = 0$, and so that*

$$\left|x^* - x_m\right| \le l^* - l_m. \tag{13.32}$$

Proof. Estimates

$$|y_p - x_p| \le h_p - l_p \tag{13.33}$$

and

$$|x_{p+1} - y_p| \le l_{p+1} - h_p \tag{13.34}$$

shall be shown using induction. It follows from (C_1) and (13.2) for $p = 0$ that

$$|y_0 - x_0| = \left|G'(x_0)^{-1}G(x_0)\right| \le \tau = h_0 - l_0 = h_0 < l^*,$$

so (13.33) holds if $p = 0$ and $y_0 \in U(x_0, l^*)$. We use the identity

$$G(y_p) = G(y_p) - G(x_p) - G'(y_p)(y_p - x_p),$$

since by method (13.2)

$$G(x_p) = -G'(x_p)(y_p - x_p).$$

Then, by (C_3), we have

$$\begin{aligned}|G'(x_0)^{-1}G(y_p)| &= \left|\int_0^1 G'(x_0)^{-1}(G'(x_p+\theta(y_p-x_p)) - G'(x_p))d\theta(y_p-x_p)\right| \\ &\leq \frac{R}{2}|y_p - x_p|^2. \end{aligned} \tag{13.35}$$

Suppose (13.33) holds for all positive integers smaller or equal to p. Then, by (13.35)

$$|G'(x_0)^{-1}G(y_p)| \leq \frac{R}{2}(h_p - l_p)^2. \tag{13.36}$$

Moreover, we need some estimates:

$$|G'(x_0)^{-1}(G'(u) - G(x_0))| \leq R_0|u - x_0| \leq R_0 l^* < 1, \tag{13.37}$$

for all $u \in U(x_0, l^*)$, and

$$\begin{aligned}|V_0^{-1}(V_p - V_0)| &= |V_0^{-1}(aG(x_p) + bG(y_p) - aG(x_0) - bG(y_0))| \\ &\leq |a|\left|\int_0^1 V_0^{-1}G'(x_0+\theta(x_p-x_0))d\theta(x_p-x_0)\right| \\ &\quad + |b|\left|\int_0^1 V_0^{-1}G'(y_0+\theta(y_p-y_0))d\theta(y_p-y_0)\right| \\ &\leq |a|R_3(x_p - x_0) + |b|R_3|y_p - y_0| \\ &\leq |a|R_3(l_p - l_0) + |b|R_3(h_p - h_0) = \alpha_p < 1.\end{aligned}$$

Thus, $G'(x_0) \neq 0$, $V_p \neq 0$

$$|G'(u)^{-1}G'(x_0)| \leq \frac{1}{1 - R_0|u - x_0|} \tag{13.38}$$

and

$$|V_p^{-1}V_0| \leq \frac{1}{1 - \alpha_p} \tag{13.39}$$

by the Banach standard lemma on nonzero functions [23, 37].

It follows for $u = x_0$ in (13.38) that y_0, x_1 exist by the first and second substeps of (13.2) for $p = 0$, respectively. We also have by (C_3), (13.35) and (13.36) that

$$\begin{aligned}|V_0^{-1}Q_p| &= |V^{-1}(cG(x_p)+dG(y_p))|\\ &\le |c||V_0^{-1}G(x_p)| + |d||V_0^{-1}G(y_p)|\\ &\le |c||V_0^{-1}G'(x_p)(y_p-x_p)| + |d||V_0^{-1}G(y_p)|\\ &\le |c|R_3|y_p-x_p| + \frac{|d|R_2}{2}|y_p-x_p|^2\\ &\le |c|R_3(h_p-l_p) + \frac{|d|R_2}{2}(s_p-l_p)^2 \end{aligned} \tag{13.40}$$

leading together with (13.3), (13.38)–(13.40) to

$$\begin{aligned}|x_{p+1}-y_p| &\le |V_p^{-1}V_0||V_0^{-1}Q_p||G'(x_p)^{-1}G'(x_0)||G'(x_0)^{-1}G(y_p)|\\ &\le \frac{(|c|R_3(h_p-l_p)+\frac{|d|R_2}{2}(h_p-l_p)^2)\frac{R}{2}(h_p-l_p)^2}{(1-\alpha_p)(1-R_0l_p)} = l_{p+1}-h_p \end{aligned} \tag{13.41}$$

and

$$|x_{p+1}-x_0| \le |x_{p+1}-y_p| + |y_p-x_0| \le l_{p+1}-h_p+h_p-l_0 = l_{p+1} < l^*. \tag{13.42}$$

Thus, (13.33) holds and $x_{p+1} \in U(x_0, l^*)$.

We also need the estimates:

$$\begin{aligned}|Q_0^{-1}(Q_p-Q_0)| &\le |c||Q_0^{-1}(G(x_p)-G(x_0))| + |d||Q_0^{-1}(G(y_p)-G(y_0))|\\ &\le |c|R_4(l_p-l_0) + |d|R_4(h_p-h_0) = b_p < 1.\end{aligned}$$

Hence, $Q_p \ne 0$,

$$|Q_p^{-1}Q_0| \le \frac{1}{1-b_p} \tag{13.43}$$

and

$$G(y_p) = -G'(x_p)Q_p^{-1}V_p(x_{p+1}-y_p) \tag{13.44}$$

by the second substep of method (13.2). Then, we can have

$$\begin{aligned}G(x_{p+1}) &= G(x_{p+1}) - G(y_p) + G(y_p) = G(x_{p+1}) - G(y_p) - G'(y_p)(x_{p+1}-y_p)\\ &\quad + G'(y_p)(x_{p+1}-y_p) - G'(x_p)Q_p^{-1}V_p(x_{p+1}-y_p)\\ &= \int_0^1 (G'(y_p+\theta(x_{p+1}-y_p)) - G'(y_p))d\theta(x_{p+1}-y_p) + M_p(x_{p+1}-y_p),\end{aligned} \tag{13.45}$$

where

$$M_p = G'(y_p) - Q_p^{-1}V_pG'(x_p) = Q_p^{-1}(Q_pG'(x_p) - V_pG'(x_p)).$$

However, we get

$$\begin{aligned} M_p &= (cG(x_p) + dG(y_p))G'(y_p) - (aG(x_p) + bG(y_p))G'(x_p) \\ &= G(x_p)(c(G'(y_p) - G'(x_p)) + (c-a)G'(x_p)) \\ &\quad + G(y_p)(d(G'(y_p) - G'(x_p)) + (d-b)G'(x_p)). \end{aligned}$$

Thus,

$$\begin{aligned} |G'(x_0)^{-1}Q_0^{-1}M_p| &\le |G'(x_0)^{-1}G(x_p)||c||Q_0^{-1}(G'(y_p) - G'(x_p))| \\ &\quad + |c-a||G'(x_0)^{-1}G(x_p)||Q_0^{-1}G(x_p)| \\ &\quad + |d||Q_0^{-1}(G'(y_p) - G'(x_p))||G'(x_0)^{-1}G(y_p)| \\ &\quad + |d-b||G'(x_0)^{-1}G(y_p)||Q_0^{-1}G'(x_p)| \\ &\le |c|R_5|y_p - x_p|R_1|y_p - x_p| + |c-a|R_4|y_p - x_p|R_1|y_p - x_p| \\ &\quad + |d|R_5|y_p - x_p|\frac{R}{2}|y_p - x_p|^2 + |d-b|\frac{R}{2}|y_p - x_p|^2 R_4|y_p - x_p| \\ &\le (R_1R_5|c| + |c-a|R_1R_4)|y_p - x_p|^2 \\ &\quad + \left(\frac{RR_5}{2}|d| + \frac{RR_4}{2}|d-b|\right)|y_p - x_p|^3 \\ &\le (R_6 + R_7(h_p - l_p))(h_p - l_p)^2. \end{aligned} \tag{13.46}$$

It then follows from (13.38) for $u = x_{p+1}$, (13.43), (13.45), (13.46) and (13.2) that

$$\begin{aligned} |G'(x_0)^{-1}G(x_{p+1})| &\le \frac{R}{2}|x_{p+1} - y_p|^2 + |Q_p^{-1}Q_0||G'(x_0)^{-1}Q_0^{-1}M_p||x_{p+1} - y_p| \\ &\le \frac{R}{2}(l_{p+1} - h_p)^2 + \gamma_p(R_6 + R_7(h_p - l_p))(h_p - l_p)^2(l_{p+1} - h_p). \end{aligned} \tag{13.47}$$

So,

$$|y_{p+1} - x_{p+1}| \le h_{p+1} - l_{p+1}$$

and

$$|y_{p+1} - x_0| \le |y_{p+1} - x_{p+1}| + |x_{p+1} - x_0| \le h_{p+1} - l_{p+1} + l_{p+1} - l_0 = h_{p+1} < l^*. \tag{13.48}$$

Therefore, the induction for (13.34) is completed too and $h_{p+1} \in U(x_0, l^*)$. Sequence $\{l_p\}$ is fundamental too, so it is convergent to $x^* \in U[x_0, l^*]$. By letting $p \to \infty$ in (13.35), we obtain $G(x^*) = 0$. The estimate (13.32) is obtained from

$$|x_{i+p} - x_p| \le |x_{i+p} - x_{i+p-1}| + |x_{i+p-1} - x_{i+p-2}| + \cdots + |x_{p+1} - x_p| \le l_{i+p} - l_p \tag{13.49}$$

by letting $i \to \infty$ in (13.49). □

Remark 13.1. It following as in the proof of Theorem 13.1 (see (13.8) and (13.25)) that the conclusions hold if we replace (13.6) or (13.7) by

$$\alpha \leq \beta \leq \gamma \leq \delta < \mu \leq 1,$$

provided that

$$\frac{R}{2}\delta^2\tau + 2(R_6 + R_7\delta^2\tau)\delta^3\tau^2 + \frac{R_0\tau}{1-\delta} - 1 \leq 0$$

and

$$2R_8\delta^3\tau^2 + 2L_9\delta^5\tau^3 + \frac{R_0\tau}{1-\delta} - 1 \leq 0,$$

for some pair of values (δ, τ) also hold for $\delta \in (0,1)$. The last two inequalities provide an estimate of the smallness of τ. These assumptions may hold (or not) in case those of Lemma 13.2. In both cases, they imply the assumptions of Lemma 13.1. In practice, we will test all these assumptions and pick those that hold, if any.

Next, a uniqueness result in given for a root of equation $G(x) = 0$.

Proposition 13.1. *Assume the conditions (C_1)–(C_5) hold, and there exists $\rho_1 \geq l^*$ so that*

$$R_0(l^* + \rho_1) < 2. \tag{13.50}$$

Set $D_1 = U[x_0, \rho_1] \cap D$. Then, the number x^* is the only root of $G(x) = 0$ in the set D_1.

Proof. Define $T = \int_0^1 G'(x^* + \theta(\bar{x} - x^*))d\theta$ for some $\bar{x} \in D_1$ with $G(\bar{x}) = 0$. Using (C_2) and (13.50), we get in turn that

$$|G'(x_0)^{-1}(T - G'(x_0))| \leq R_0 \int_0^1 ((1-\theta)|\overline{x} - x_0| + \theta|x^* - x_0|)d\theta \leq \frac{R_0}{2}(l^* + \rho_1) < 1.$$

Hence, $\overline{x} = x^*$ follows by the identity

$$x^* - \bar{x} = T^{-1}(G(x^*) - G(\overline{x})) = 0 - 0 = 0, \quad \text{since } T \neq 0. \tag{13.51}$$

□

13.4 Examples

We present numerical examples to validate the convergence results. The parameters a, b, c, d appearing on (13.2) are chosen to satisfy the convergence conditions of (13.2).

Example 13.1. Let $D = (-1.5, 1.5)$ and consider scalar function $G : D \to \mathbb{R}$

$$G(u) = e^u + 2u - 1.$$

The zero of this function is $u^* = 0$ and the derivative $G'(u) = e^u + 2$. Then, we get

$$|G'(u) - G'(v)| \le e^{\varkappa}|u - v|, \quad \varkappa \in D.$$

So, $R_0 = \frac{e^{1.5}}{e^{u_0}+2}$, $R = \frac{e^u}{e^{u_0}+2}$, $R_1 = \frac{e^u+2}{e^{u_0}+2}$, $R_2 = \frac{e^u}{|V_0|}$, $R_3 = \frac{e^u+2}{|V_0|}$, $R_4 = \frac{e^u+2}{|Q_0|}$ and $R_5 = \frac{e^u}{|Q_0|}$. Here, $u = d_2$ and

$$D_0 = (d_1, d_2) = \left(\max\left\{-1, u_0 - \frac{1}{R_0}\right\}, \min\left\{1, u_0 + \frac{1}{R_0}\right\}\right).$$

Let $u_0 = 0.2$. Then, for coefficients $a = 1, b = 0, c = 1, d = 0$, we get $l^* \approx 0.2242$, $U[u_0, l^*] = [-0.024227, 0.424227]$. For coefficients $a = 1, b = -1.5$ and $c = d = 1$, we get $l^* \approx 0.2291$, $U[u_0, l^*] = [-0.029146, 0.429146]$.

Table 13.1 shows that the condition (13.32) is satisfied.

Table 13.1: Error estimate (13.32) at each iteration for Example 13.1.

m	$a = 1, b = 0, c = 1, d = 0$		$a = 1, b = -1.5, c = 1, d = 1$	
	$\lvert u^* - u_m\rvert$	$l^* - l_m$	$\lvert u^* - u_m\rvert$	$l^* - l_m$
0	2.0000e-01	2.2423e-01	2.0000e-01	2.2915e-01
1	4.8025e-04	1.1081e-02	9.4993e-06	1.4155e-02
2	6.1549e-12	3.3380e-08	2.9734e-17	5.5171e-08

The graphs (see Figures 13.1 and 13.2) show the values of errors ($|u_m - u^*|$, $|u_m - u_{m-1}|$, $|f(u_m)|$) at each iteration. The results are given for KTLM ($a = 1, b = 0, c = 1, d = 0$) and Newton's method (NM) for initial approximations u_0 far from the exact solution. The iterative process was stopped under the condition $|f(u_m)| < 10^{-10}$. For $u_0 = 5$ KTLM converges under 6 iteration and $|f(u_6)| = 0$.

Example 13.2. Let $D = (0, 2)$ and consider scalar function $G : D \to \mathbb{R}$

$$G(u) = u^3 + 3u - 1.$$

The zero of this function is $u^* \approx 0.3222$ and the derivative $G'(u) = 3(u^2 + 1)$. We can write

$$|G'(u) - G'(v)| \le 3|u + v||u - v|, \quad u, v \in D.$$

Hence, $R_0 = \frac{2+u_0}{u_0^2+1}$, $R = \frac{2u}{u_0^2+1}$, $R_1 = \frac{u^2+1}{u_0^2+1}$, $R_2 = \frac{u^2+1}{|V_0|}$, $R_3 = \frac{2u}{|V_0|}$, $R_4 = \frac{u^2+1}{|Q_0|}$ and $R_5 = \frac{2u}{|Q_0|}$. Here $u = d_2$ and

$$D_0 = (d_1, d_2) = \left(\max\left\{0, u_0 - \frac{1}{R_0}\right\}, \min\left\{2, u_0 + \frac{1}{R_0}\right\}\right).$$

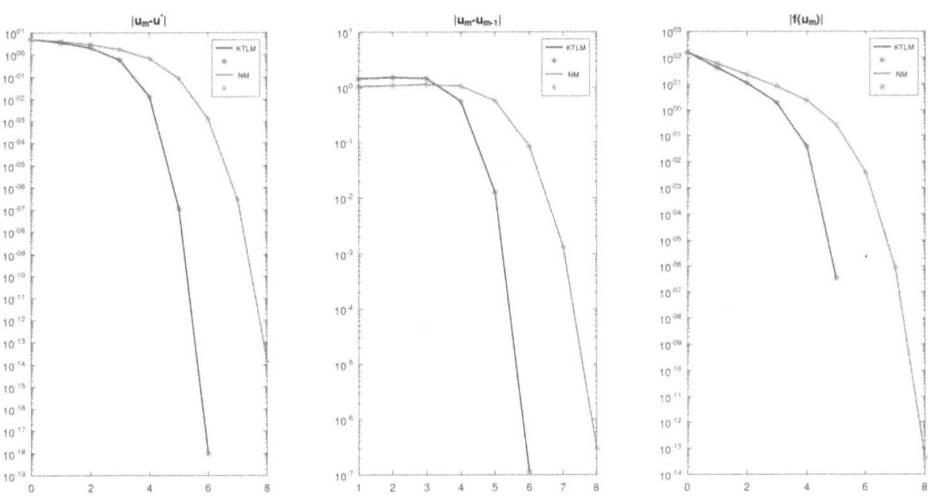

Figure 13.1: Results for $u_0 = 5$.

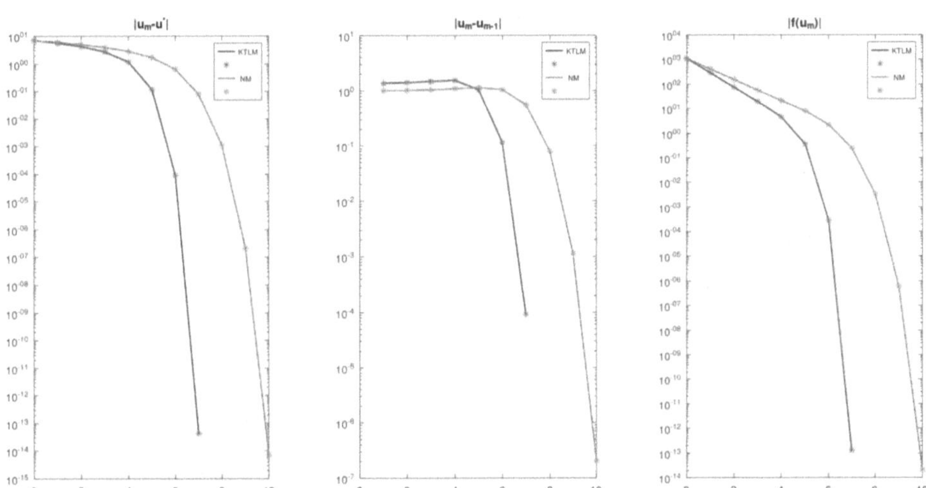

Figure 13.2: Results for $u_0 = 7$.

Let $u_0 = 0.45$. Then, for coefficients $a = 1$, $b = 0$, $c = 1$, $d = 0$, we get $l^* \approx 0.1487$, $U[u_0, l^*] = [0.3013, 0.5987]$. For coefficients $a = 1$, $b = -1.5$ and $c = d = 1$, we get $l^* \approx 0.1535$, $U[u_0, l^*] = [0.2965, 0.6035]$.

Table 13.2 shows that the condition (13.32) is satisfied.

Table 13.2: Error estimate (13.32) at each iteration for Example 13.2.

m	$a = 1, b = 0, c = 1, d = 0$		$a = 1, b = -1.5, c = 1, d = 1$	
	$\|u^* - u_m\|$	$l^* - l_m$	$\|u^* - u_m\|$	$l^* - l_m$
0	1.2781e-01	1.4870e-01	1.2781e-01	1.5351e-01
1	4.4601e-04	8.0840e-03	4.0922e-06	1.0924e-02
2	1.5134e-11	2.1935e-08	5.5511e-17	4.7885e-08

13.5 Conclusion

The class of functions for which KTLM converges is extended for solving nonlinear equations in cases not possible before. Earlier convergence approaches using Taylor series assume the existence of derivatives not on the method. The convergence methodology in this article uses only conditions on the functions which appear on KTLM. The numerical applications further, illuminate the theory. In the future, this methodology shall be appeared on other high convergence order multi-point and multi-step methods along the same lines [1, 4–6, 31, 32, 36, 37, 39].

Bibliography

[1] Argyros, I. K., *Unified convergence for iterative Banach space valued methods with applications*, Mathematics 9(16) (2021), 1942. https://doi.org/10.3390/math9161942

[2] Argyros, I. K., *The Theory and Applications of Iterations Methods*, 2nd ed., CRC Press, Boca Raton, USA, 2022.

[3] Argyros, I. K. and Hilout, S., *Weaker convergence conditions for the convergence of Newton's method*, J. Complex. 28 (2012), 364–387.

[4] Argyros, I. K. and Shakhno, S., *Extending the applicability of two-step solvers for solving equations*, Mathematics 7(1) (2019), 62. https://doi.org/10.3390/math7010062.

[5] Argyros, I. K., Shakhno, S. and Yarmola, H., *Two-step solver for nonlinear equations*, Symmetry 11(2) (2019), 128. https://doi.org/10.3390/sym11020128.

[6] Argyros, I. K., Sharma, D., Argyros, C. I., Parhi, S. K. and Sunanda, S. K., *A family of fifth and sixth convergence order methods for nonlinear models*, Symmetry 13(4) (2021), 715. https://doi.org/10.3390/sym13040715.

[7] Behl, R., Sarría, Í., González, R. and Magreñán, Á. A., *Highly efficient family of iterative methods for solving nonlinear models*, J. Comput. Appl. Math. 346 (2019), 110–132.

[8] Chun, C. and Neta, B., *Comparative study of eighth order methods for finding simple roots of nonlinear equations*, Numer. Algorithms 74 (2017), 1169–1201.

[9] Chun, C. and Neta, B., *Developing high order methods for the solution of systems of nonlinear equations*, Appl. Math. Comput. 342 (2019), 178–190.

[10] Cordero, A. and Torregrosa, J. R., *Variants of Newton's method for functions of several variables*, Appl. Math. Comput. 183 (2006), 199–208.

[11] Cordero, A. and Torregrosa, J. R., *Variants of Newton's method using fifth-order quadrature formulas*, Appl. Math. Comput. 190 (2007), 686–698.

[12] Cordero, A., Hueso, J. L., Martínez, E. and Torregrosa, J. R., *A modified Newton-Jarratt's composition*, Numer. Algorithms 55 (2010), 87–99.

[13] Cordero, A., Garrido, N., Torregrosa, J. R. and Triguero-Navarro, J. R., *Memory in the iterative processes for nonlinear problems*, Math. Methods Appl. Sci. 46(4) (2023), 4145–4158.

[14] Cordero, A., Garrido, N., Torregrosa, J. R. and Triguero-Navarro, P., *Design of iterative methods with memory for solving nonlinear systems*, Math. Methods Appl. Sci. 46(12) (2023), 12361–12377.

[15] Darvishi, M. T. and Barati, A., *A third-order Newton-type method to solve systems of nonlinear equations*, Appl. Math. Comput. 187 (2007), 630–635.

[16] Darvishi, M. T. and Barati, A., *A fourth-order method from quadrature formulae to solve systems of nonlinear equations*, Appl. Math. Comput. 188 (2007), 257–261.

[17] Frontini, M. and Sormani, E., *Third-order methods from quadrature formulae for solving systems of nonlinear equations*, Appl. Math. Comput. 149 (2004), 771–782.

[18] Grau-Sánchez, M., Grau, Á. and Noguera, M., *On the computational efficiency index and some iterative methods for solving systems of nonlinear equations*, J. Comput. Appl. Math. 236 (2011), 1259–1266.

[19] Grau-Sánchez, M., Noguera, M. and Amat, S., *On the approximation of derivatives using divided difference operators preserving the local convergence order of iterative methods*, J. Comput. Appl. Math. 237(1) (2013), 363–372.

[20] Hernández-Veron, M. A. and Romero, N., *On the local convergence of a third order family of iterative processes*, Algorithms 8(4) (2015), 1121–1128.

[21] Homeier, H., *A modified Newton method with cubic convergence: the multivariate case*, J. Comput. Appl. Math. 169 (2004), 161–169.

[22] Hueso, J. L., Martínez, E. and Teruel, C., *Convergence, efficiency and dynamics of new fourth and sixth order families of iterative methods for nonlinear systems*, J. Comput. Appl. Math. 275 (2015), 412–420.

[23] Jarratt, P., *Some efficient fourth-order multipoint methods for solving equations*, BIT 9 (1969), 119–124.

[24] Jarratt, P., *Some fourth order multipoint iterative methods for solving equations*, Math. Comput. 20 (1966), 434–437.

[25] Kansal, M., Cordero, A., Bhalla, S. and Torregrosa, J. R., *New fourth-and sixth-order classes of iterative methods for solving systems of nonlinear equations and their stability analysis*, Numer. Algorithms 87 (2021), 1017–1060.

[26] King, R. F., *A family of fourth order methods for nonlinear equations*, SIAM J. Numer. Anal. 10(5) (1973), 876–879.

[27] Kou, J. and Li, Y., *An improvement of the Jarratt method*, Appl. Math. Comput. 189(2) (2007), 1816–1821.

[28] Kung, H. and Traub, J. F., *Optimal order of one-point and multipoint iteration*, J. ACM 21(4) (1974), 643–651.

[29] Lotfi, T. and Assari, P., *New three-and four-parametric iterative with memory methods with efficiency index near 2*, Appl. Math. Comput. 270 (2015), 1004–1010.

[30] Lotfi, T., Soleymani, F., Ghorbanzadeh, M. and Assari, P., *On the construction of some tri-parametric iterative methods with memory*, Numer. Algorithms 70(4) (2015), 835–845.

[31] Magreñán, Á. A., *Different anomalies in a Jarratt family of iterative root-finding methods*, Appl. Math. Comput. 233 (2014), 29–38.

[32] Magreñán, Á. A. and Guttierrez, S. M., *Real dynamics for damped Newton's method applied to cubic polynomials*, J. Comput. Appl. Math. 275 (2015), 527–538.

[33] Neta, B., Chun, C. and Scott, M., *Basins of attraction for optimal eight order methods to find simple roots of nonlinear equations*, Appl. Math. Comput. 277 (2014), 67–92.

[34] Noor, M. A. and Waseem, M., *Some iterative methods for solving a system of nonlinear equations*, Comput. Math. Appl. 57 (2009), 101–106.

[35] Petkovic, M. S., Neta, B., Petkovic, L. D. and Dzunic, J., *Multipoint Methods for Solving Nonlinear Equations*, Elsevier, New York, USA, 2013.

[36] Proinov, P. D., *New general convergence theory for iterative processes and its applications to Newton–Kantorovich-type theorems*, J. Complex. 25 (2009), 38–62.

[37] Rheinboldt, W. C., An adaptive continuation process for solving systems of nonlinear equations, in: Tikhonov, A. N. et al. (eds.), *Mathematical Models and Numerical Methods*, vol. 3, pp. 129–142, Banach Center, Warsaw, Poland, 1978.

[38] Sayevand, K., Erfanifar, R. and Esmaeili, H., *On computational efficiency and dynamical analysis for a class of novel multi-step iterative schemes*, Int. J. Comput. Math. 6(6) (2020), 1–23.

[39] Shakhno, S. M., *On an iterative algorithm with superquadratic convergence for solving nonlinear operator equations*, J. Comput. Appl. Math. 231(1) (2009), 222–235.

[40] Sharma, D. and Parhi, S. K., *On the local convergence of higher-order methods*, Fixed Point Theory 22(2) (2021), 855–870, https://doi.org/10.24193/fpt-ro.2021.2.55.

[41] Sharma, J. R. and Arora, H., *Efficient Jarratt-like methods for solving systems of nonlinear equations*, Calcolo 51(1) (2014), 193–210.

[42] Sharma, J. R., Guha, R. K. and Sharma, R., *An efficient fourth order weighted-Newton method for systems of nonlinear equations*, Numer. Algorithms 62 (2013), 307–323.

[43] Sharma, J. R., Kumar, S. and Argyros, I. K., *Generalized Kung–Traub method and its multi-step iteration in Banach spaces*, J. Complex. 54 (2019), 101400.

[44] Traub, J. F., *Iterative Methods for Solution of Equations*, Prentice Hall, USA, 1964.

Juan E. Nápoles Valdés and Bahtiyar Bayraktar

14 New integral inequalities for Atangana–Baleanu fractional integrals via (h, m)-convex functions

Abstract: In this article, some integral inequalities of the Hermite–Hadamard type are obtained for functions with second derivative in the class of functions modified (h, m)-convex of the second type, via integrals of the Atangana–Baleanu type. Throughout the manuscript, we show that several results known from the literature are particular cases of ours.

Keywords: Convexity, Hermite–Hadamard inequality, Hölder inequality, Atangana–Baleanu integral, Caputo–Fabrizio fractional integral, (h, m)-convex modified functions

MSC 2020: 26A51, 26D15, 26D10

14.1 Introduction

An important fundamental mathematical concept that plays a vital role in other areas of pure and applied sciences is inequalities and their applications.

In mathematical analysis, inequalities are powerful instruments for comparing and scrutinizing functions, particularly when establishing boundaries for integrals. Consequently, this research delves into integral inequalities, commonly employed in contemporary mathematical analysis.

Within the realm of mathematical sciences, the concept of convex functions assumes a significant role owing to its myriad applications and its theoretical intersections with diverse mathematical domains. Furthermore, many theories of inequalities often stem from convex functions, with their discoveries being reported through various generalizations and extensions of convexity principles. Readers interested in exploring this concept further can refer to [32], which provides an extensive overview of these branches.

Definition 14.1. Let $\upsilon_1, \upsilon_2 \in \mathbb{R}$ with $\upsilon_1 < \upsilon_2$. A function $\mathbf{\Phi} : [\upsilon_1, \upsilon_2] \to \mathbb{R}$ is termed convex if it satisfies the condition $\mathbf{\Phi}(\tau\boldsymbol{\xi} + (1-\tau)\boldsymbol{\zeta}) \le \tau\mathbf{\Phi}(\boldsymbol{\xi}) + (1-\tau)\mathbf{\Phi}(\boldsymbol{\zeta})$, $\forall \boldsymbol{\xi}, \boldsymbol{\zeta} \in [\upsilon_1, \upsilon_2]$ and $\tau \in [0, 1]$.

Juan E. Nápoles Valdés, UNNE, FaCENA, Ave. Libertad 5450, Corrientes 3400, Argentina; and UTN-FRRE, French 414, Resistencia, Chaco 3500, Argentina, e-mail: jnapoles@exa.unne.edu.ar
Bahtiyar Bayraktar, Faculty of Education, Bursa Uludag University, Gorukle Capus, Bursa, Turkey, e-mail: bbayraktar@uludag.edu.tr

https://doi.org/10.1515/9783112215760-014

The celebrated Hermite–Hadamard inequality stands out as one of the most prominent inequalities for convex functions.

For any function Φ satisfying the conditions of Definition 14.1, the inequality

$$\Phi\left(\frac{\upsilon_1+\upsilon_2}{2}\right) \le \frac{1}{\upsilon_2-\upsilon_1}\int_{\upsilon_1}^{\upsilon_2}\Phi(\xi)d\xi \le \frac{\Phi(\upsilon_1)+\Phi(\upsilon_2)}{2}$$

holds. Hermite [25] first published this inequality in 1883, while independently, Hadamard introduced it in 1893 [24].

Various extensions and generalizations can be consulted in [2, 3, 9, 19, 21, 23, 29, 33, 39] and references therein.

In [7, 8], the concepts of first and second type modified (h,m)-convex function in the interval $I=[0,+\infty)$ are given (classes $N_{h,m}^{s,1}(I)$ and $N_{h,m}^{s,2}(I)$, respectively).

Definition 14.2. Let $\Phi: I \to I$ and $h:[0,1]\to(0,1]$. If

$$\Phi(\varsigma\upsilon_1+m(1-\varsigma)\upsilon_2) \le \Phi(\upsilon_1)h^s(\varsigma)+m\Phi(\upsilon_2)(1-h^s(\varsigma)) \tag{14.1}$$

is true $\forall \upsilon_1,\upsilon_2\in I$, $\varsigma\in[0,1]$, where $s\in[-1,1]$, $m\in[0,1]$, then $\Phi\in N_{h,m}^{s,1}(I)$.

Definition 14.3. Let $\Phi: I \to I$ and $h:[0,1]\to(0,1]$. If

$$\Phi(\varsigma\upsilon_1+m(1-\varsigma)\upsilon_2) \le \Phi(\upsilon_1)h^s(\varsigma)+m\Phi(\upsilon_2)(1-h(\tau))^s \tag{14.2}$$

is true $\forall \upsilon_1,\upsilon_2\in I$, $\varsigma\in[0,1]$, where $s\in[-1,1]$, $m\in[0,1]$, then $\Phi\in N_{h,m}^{s,2}(I)$.

Remark 14.1. In [7, 8, 22] one can see a number of convex classes-special cases of the triple $(h(\tau),m,s)$, obtained from Definitions 14.2 and 14.3,

Fractional calculus is a field of recent consolidation in applied mathematics, which arises as a result of the open problem related to how to solve some differential equations that contain derivatives of non-integer order. The answer to this problem has motivated mathematicians increasingly in the last 50 years, mainly.

In our work, we often use Euler gamma functions $\Gamma(\alpha)$ and $\Gamma_k(\alpha)$ (see [36, 14]):

$$\Gamma(z)=\int_0^\infty \varsigma^{z-1}e^{-\varsigma}\,d\varsigma, \quad \Re(z)>0,$$

$$\Gamma_k(z)=\int_0^\infty \varsigma^{z-1}e^{-\varsigma^k/k}\,d\varsigma, \quad k>0.$$

Moreover, $\lim_{k\to 1}\Gamma_k(z)=\Gamma(z)$, $\Gamma_k(K+z)=z\Gamma_k(z)$ and $\Gamma_k(z)=(k)^{\frac{z}{k}-1}\Gamma(\frac{z}{k})$.

We consider fractional integrals under the conditions $0\le\upsilon_1<\varsigma<\upsilon_2\le\infty$. First, we define the classical Riemann–Liouville fractional integral [30].

Definition 14.4. Let $\Phi \in L_1[\upsilon_1, \upsilon_2]$. Then the Riemann–Liouville fractional integrals of order $\alpha \in \mathbb{C}$, $\Re(\alpha) > 0$ are defined by (left and right-side respectively):

$$I^{\alpha}_{\upsilon_2-}\Phi(\xi) = \frac{1}{\Gamma(\alpha)} \int_{\xi}^{\upsilon_2} (\varsigma - \xi)^{\alpha-1}\Phi(\varsigma)\, d\varsigma, \quad \xi < \upsilon_2$$

$$I^{\alpha}_{\upsilon_1+}\Phi(\xi) = \frac{1}{\Gamma(\alpha)} \int_{\upsilon_1}^{\xi} (\xi - \varsigma)^{\alpha-1}\Phi(\varsigma)\, d\varsigma, \quad \xi > \upsilon_1.$$

Below, we use the $AC^1[\upsilon_1, \upsilon_2]$-space of functions that have absolutely continuous first-order derivatives on a given interval.

Definition 14.5 ([11]). Let $0 < \alpha \leq 1$ and let $\Phi \in AC^1[\upsilon_1, \upsilon_2]$. The Caputo fractional derivatives of order α, both left- and right-sided, are defined as follows

$$({}^{C}D^{\alpha}_{\upsilon_2-}\Phi)(\varsigma) = \frac{-1}{\Gamma(1-\alpha)} \int_{\varsigma}^{\upsilon_2} \frac{\Phi'(\varsigma)d\varsigma}{(\xi - \varsigma)^{\alpha}}, \quad \upsilon_2 > \varsigma,$$

$$({}^{C}D^{\alpha}_{\upsilon_1+}\Phi)(\varsigma) = \frac{1}{\Gamma(1-\alpha)} \int_{\upsilon_1}^{\varsigma} \frac{\Phi'(\xi)d\xi}{(\varsigma - \xi)^{\alpha}}, \quad \varsigma > \upsilon_1.$$

In 2015, Caputo and Fabrizio proposed the following operator (see [12, 37]):

Definition 14.6. Let $0 < \alpha \leq 1$ and let $\Phi \in AC^1[\upsilon_1, \upsilon_2]$. The Caputo–Fabrizio fractional derivatives of order α, both left- and right-sided, are defined as follows

$$({}^{C}D^{\alpha}_{\upsilon_2-}\Phi)(\varsigma) = -\frac{B(\alpha)}{1-\alpha} \int_{\varsigma}^{\upsilon_2} \Phi'(\xi)e^{-\frac{\alpha(\xi-\varsigma)^{\alpha}}{1-\alpha}}\, d\xi, \quad \varsigma < \upsilon_2,$$

$$({}^{C}D^{\alpha}_{\upsilon_1+}\Phi)(\varsigma) = \frac{B(\alpha)}{1-\alpha} \int_{\upsilon_1}^{\varsigma} \Phi'(\xi)e^{-\frac{\alpha(\varsigma-\xi)^{\alpha}}{1-\alpha}}\, d\xi, \quad \varsigma > \upsilon_1,$$

where $B(\alpha)$ is a normalization function and $B(0) = B(1) = 1$.

Their corresponding integral operators given by:

Definition 14.7. Let $0 < \alpha \leq 1$ and let $\Phi \in AC^1[\upsilon_1, \upsilon_2]$. The left and right-sided Caputo–Fabrizio integral of order α are defined as follows:

$$({}^{CF}I^{\alpha}_{\upsilon_2-}\Phi)(\varsigma) = \frac{1-\alpha}{B(\alpha)}\Phi(\varsigma) + \frac{\alpha}{B(\alpha)} \int_{\varsigma}^{\upsilon_2} \Phi'(y)dy, \quad \varsigma < \upsilon_2,$$

$$\left({}^{CF}I^{\alpha}_{\upsilon_1+}\Phi\right)(\varsigma) = \frac{1-\alpha}{B(\alpha)}\Phi(\varsigma) + \frac{\alpha}{B(\alpha)}\int\limits_{\upsilon_1}^{\varsigma}\Phi'(y)dy, \quad \varsigma > \upsilon_1,$$

where $B(\alpha)$ is a normalization function and $B(0) = B(1) = 1$.

A number of recent studies in which integral inequalities were obtained through Caputo operators (see for example [15, 18, 20, 26, 41]).

Although the Caputo–Fabrizio operator has proven to be an effective tool to solve many problems associated with systems of differential equations, one difficulty has persisted: not reducing to the original function in special cases, which was resolved with the Atangana–Baleanu differential operator. The normalization function described in the following definitions shares similar characteristics with the function used in the Caputo–Fabrizio operator.

Definition 14.8 ([5]). Let $0 < \alpha \le 1$ and let $\Phi \in AC^1[\upsilon_1, \upsilon_2]$. The new Atangana–Baleanu fractional derivative of order α, with $\alpha \in [0,1]$, is defined as follows:

$$\left({}^{ABC}_{\upsilon_1}D^{\alpha}_{\varsigma}\Phi\right)(\varsigma) = \frac{B(\alpha)}{1-\alpha}\int\limits_{\upsilon_1}^{\varsigma}\Phi'(\xi)E_{\alpha}\left(-\alpha\frac{(\varsigma-\xi)^{\alpha}}{1-\alpha}\right)d\xi, \quad \varsigma > \upsilon_1,$$

where $B(\alpha)$ is the normalization function and $B(0) = B(1) = 1$, and $E_{\alpha}(z)$ is the monoparametric Mittag–Leffler function (see [28]).

The above differential operators have a non-local kernel.

The corresponding integral operators to Atangana–Baleanu derivative is given by:

Definition 14.9 ([5] and [4]). Let $0 < \alpha \le 1$ and let $\Phi \in AC^1[\upsilon_1, \upsilon_2]$. The left- and right-sided Atangana–Baleanu integral of order α are defined as follows:

$$\left({}^{ABC}I^{\alpha}_{b-}\Phi\right)(\varsigma) = \frac{1-\alpha}{B(\alpha)}\Phi(\varsigma) + \frac{\alpha}{B(\alpha)}\int\limits_{\varsigma}^{\upsilon_2}\Phi(y)(y-\varsigma)^{\alpha-1}dy, \quad \varsigma < \upsilon_2,$$

$$\left({}^{ABC}I^{\alpha}_{\upsilon_1+}\Phi\right)(\varsigma) = \frac{1-\alpha}{B(\alpha)}\Phi(\varsigma) + \frac{\alpha}{B(\alpha)}\int\limits_{\upsilon_1}^{\varsigma}\Phi(y)(\varsigma-y)^{\alpha-1}dy, \quad \varsigma > \upsilon_1,$$

where $\alpha \in [0,1]$, $B(\alpha)$ is the normalization function and $B(0) = B(1) = 1$.

The following operators will be useful in our paper [7, 8].

Definition 14.10. Let $\Phi \in L[\upsilon_1, \upsilon_2]$, and let $w : [0,1] \to [0, +\infty)$ be a continuous and positive function whose first-order derivative is piecewise continuous on $[\upsilon_1, \upsilon_2]$. Under these conditions, the right- and left-sided weighted fractional integrals are defined as follows:

$$J^{w}_{\upsilon_1+}\Phi(\upsilon_2) = \int_{\upsilon_1}^{\upsilon_2} w''\left(\frac{\varsigma - \upsilon_1}{\upsilon_2 - \upsilon_1}\right)\Phi(\varsigma)d\varsigma$$

and

$$J^{w}_{\upsilon_2-}\Phi(\upsilon_1) = \int_{\upsilon_1}^{\upsilon_2} w''\left(\frac{\upsilon_2 - \varsigma}{\upsilon_2 - \upsilon_1}\right)\Phi(\varsigma)d\varsigma.$$

This paper introduces novel adaptations of the Hermite–Hadamard inequality within the context of modified (h, m)-convex functions, utilizing integral operators as defined in Definition 14.9.

14.2 Complementary result

The following result will simplify obtaining our results, first, we will give a general version in the framework of weighted integrals.

Lemma 14.1. *Let Φ: $(0, \upsilon_2^*] \longrightarrow \mathbb{R}$ and $\kappa \in \mathbb{N}$. If $\Phi \in C^2(0, \upsilon_2^*)$, $\Phi'' \in L_1(0, \upsilon_2^*)$, $\xi \in (\frac{\upsilon_1}{1+\kappa}, \upsilon_2)$ with $0 \leq \frac{\upsilon_1}{1+\kappa} \leq \upsilon_1 \leq \upsilon_2 < \upsilon_2^* < +\infty$, then valid the identity:*

$$\begin{aligned}
&J^{w}_{\frac{\upsilon_1}{1+\kappa}+}\Phi\left(\frac{\xi}{1+\kappa}\right) + J^{w}_{\frac{\upsilon_2}{1+\kappa}-}\Phi\left(\frac{\xi}{1+\kappa}\right)\\
&= \left(\frac{\xi - \upsilon_1}{1+\kappa}\right)^2 \left[w(0)\Phi'\left(\frac{\xi}{1+\kappa}\right) - w(1)\Phi'\left(\frac{\upsilon_1}{1+\kappa}\right)\right]\\
&\quad + \frac{\xi - \upsilon_1}{1+\kappa}\left[w'(0)\Phi\left(\frac{\xi}{1+\kappa}\right) - w'(1)\Phi\left(\frac{\upsilon_1}{1+\kappa}\right)\right]\\
&\quad + \left(\frac{\upsilon_2 - \xi}{1+\kappa}\right)^2 \left[w(1)\Phi'\left(\frac{\upsilon_2}{1+\kappa}\right) - w(0)\Phi'\left(\frac{\xi}{1+\kappa}\right)\right]\\
&\quad + \frac{\upsilon_2 - \xi}{1+\kappa}\left[w'(1)\Phi\left(\frac{\upsilon_2}{1+\kappa}\right) - w'(0)\Phi\left(\frac{\xi}{1+\kappa}\right)\right]\\
&\quad + \left(\frac{\xi - \upsilon_1}{1+\kappa}\right)^3 \int_0^1 w(1-\varsigma)\Phi''\left(\frac{\varsigma\xi}{1+\kappa} + \frac{(1-\varsigma)\upsilon_1}{1+\kappa}\right)d\varsigma\\
&\quad + \left(\frac{\upsilon_2 - \xi}{1+\kappa}\right)^3 \int_0^1 w(\varsigma)\Phi''\left(\frac{\varsigma\upsilon_2}{1+\kappa} + \frac{(1-\varsigma)\xi}{1+\kappa}\right)d\varsigma. \qquad (14.3)
\end{aligned}$$

Proof. Denoting $I_1 = \int_0^1 w(1-\varsigma)\Phi''(\frac{\varsigma\xi}{1+\kappa} + \frac{(1-\varsigma)\upsilon_1}{1+\kappa})d\varsigma$ and $I_2 = \int_0^1 w(\varsigma)\Phi''(\frac{\varsigma\upsilon_2}{1+\kappa} + \frac{(1-\varsigma)\xi}{1+\kappa})d\varsigma$, we have after integrating by parts twice the following results

$$I_1 = \int_0^1 w(1-\varsigma)\Phi''\left(\frac{\varsigma\xi}{1+\kappa} + \frac{(1-\varsigma)\upsilon_1}{1+\kappa}\right)d\varsigma$$

$$= \frac{1+\kappa}{\xi - \upsilon_1}\left[w(0)\Phi'\left(\frac{\xi}{1+\kappa}\right) - w(1)\Phi'\left(\frac{\upsilon_1}{1+\kappa}\right)\right] + \frac{(1+\kappa)^2}{(\xi-\upsilon_1)^2}\left[w'(0)\Phi\left(\frac{\xi}{1+\kappa}\right) - w'(1)\Phi\left(\frac{\upsilon_1}{1+\kappa}\right)\right] + \frac{(1+\kappa)^2}{(\xi-\upsilon_1)^2}\int_0^1 w''(1-\varsigma)\Phi\left(\frac{\varsigma\xi}{1+\kappa} + \frac{(1-\varsigma)\upsilon_1}{1+\kappa}\right)d\varsigma. \tag{14.4}$$

By changing the variables $u = \frac{\varsigma\xi}{1+\kappa} + \frac{(1-\varsigma)\upsilon_1}{1+\kappa}$ in the last integral, we obtain

$$\int_0^1 w''(1-\varsigma)\Phi\left(\frac{\varsigma\xi}{1+\kappa} + \frac{(1-\varsigma)\upsilon_1}{1+\kappa}\right)d\varsigma$$

$$= \frac{1+\kappa}{\xi-\upsilon_1}\int_{\frac{\upsilon_1}{1+\kappa}}^{\frac{\xi}{1+\kappa}} w''\left(\frac{u - \frac{\upsilon_1}{1+\kappa}}{\frac{\xi}{1+\kappa} - \frac{\upsilon_1}{1+\kappa}}\right)\Phi(u)du = \frac{1+\kappa}{\xi-\upsilon_1} J^{w}_{(\frac{\upsilon_1}{1+\kappa})+}\Phi\left(\frac{\xi}{1+\kappa}\right).$$

Analogously, we have for I_2:

$$I_2 = \int_0^1 w(\varsigma)\Phi''\left(\frac{\varsigma\upsilon_2}{1+\kappa} + \frac{(1-\varsigma)\xi}{1+\kappa}\right)d\varsigma = \frac{1+\kappa}{\upsilon_2-\xi}\left[w(1)\Phi'\left(\frac{\upsilon_2}{1+\kappa}\right) - w(0)\Phi'\left(\frac{\xi}{1+\kappa}\right)\right] + \left(\frac{1+\kappa}{\upsilon_2-\xi}\right)^2\left[w'(1)\Phi\left(\frac{\upsilon_2}{1+\kappa}\right) - w'(0)\Phi\left(\frac{\xi}{1+\kappa}\right)\right] + \left(\frac{1+\kappa}{\upsilon_2-\xi}\right)^3 J^{w}_{(\frac{\upsilon_2}{1+\kappa})-}\Phi\left(\frac{\xi}{1+\kappa}\right). \tag{14.5}$$

From (14.4) and (14.5), we obtain the desired result. □

Remark 14.2. We would like to present the generality of the previous result, showing how several known results from the literature can be obtained as particular cases of ours. From the previous result, if $w(\varsigma) = \varsigma(1-\varsigma)$, $\kappa = 0$ and $\xi = \frac{\upsilon_1+\upsilon_2}{2}$ we obtain the Lemma 2.1 of [40], under the same assumptions on κ and ξ if we put $w(\varsigma) = z^2$ we derive the Lemma 1 of [35], taking $w(\varsigma) = \varsigma(1-\varsigma)$, $\kappa = 0$, $\xi = \upsilon_2$ and working only with I_1 we obtain an equivalent result to Lemma 1 of [1].

Remark 14.3. With slight modifications and adaptations, following the idea presented in the Lemma 14.1, we obtain the Lemma 2 of [34], the Lemma 2.1 of [6] and Lemma 2.1 of [13].

14.3 Main results for Atangana–Baleanu integral operators

With the particular case $w(z) = z^{\alpha+2}$, from Lemma 14.1, we obtain the following result in terms of the Atangana–Baleanu integral operator.

Lemma 14.2. *Let Φ: $(0, \upsilon_2^*] \longrightarrow \mathbb{R}$ and $\kappa \in \mathbb{N}$. If $\Phi \in C^2(0, \upsilon_2^*)$, $\Phi'' \in L_1(0, \upsilon_2^*)$, $\xi \in (\frac{\upsilon_1}{1+\kappa}, \upsilon_2)$ with $0 \le \frac{\upsilon_1}{1+\kappa} \le \upsilon_1 \le \upsilon_2 < \upsilon_2^* < +\infty$, then valid the identity:*

$$
\begin{aligned}
&{}^{ABC}J^{\alpha}_{\frac{\upsilon_1}{1+\kappa}+}\Phi\left(\frac{\xi}{1+\kappa}\right) + {}^{ABC}J^{w}_{\frac{\upsilon_2}{1+\kappa}-}\Phi\left(\frac{\xi}{1+\kappa}\right) - \frac{2(1-\alpha)}{B(\alpha)} - \frac{(\xi-\upsilon_1)^{\alpha}\Phi(\frac{\upsilon_1}{1+\kappa}) + (\upsilon_2-\xi)^{\alpha}\Phi(\frac{\upsilon_2}{1+\kappa})}{(1+\kappa)^{\alpha}B(\alpha)\Gamma(\alpha)} \\
&= \frac{1}{(\alpha+1)(1+\kappa)^2B(\alpha)\Gamma(\alpha)}\left\{(\xi-\upsilon_1)^2\Phi'\left(\frac{\upsilon_1}{1+\kappa}\right) + (\upsilon_2-\xi)^2\Phi'\left(\frac{\upsilon_2}{1+\kappa}\right)\right. \\
&\quad + (\xi-\upsilon_1)^{\alpha+2}\int_0^1 (1-\varsigma)^{\alpha+1}\Phi''\left(\frac{\varsigma\xi}{1+\kappa} + \frac{(1-\varsigma)\upsilon_1}{1+\kappa}\right)d\varsigma \\
&\quad \left. + (\upsilon_2-\xi)^{\alpha+2}\int_0^1 \varsigma^{\alpha+1}\Phi''\left(\frac{\varsigma\upsilon_2}{1+\kappa} + \frac{(1-\varsigma)\xi}{1+\kappa}\right)d\varsigma\right\}.
\end{aligned} \tag{14.6}
$$

Proof. After integrating by parts on the right side, completing the differential operator and changing variables, the desired result is obtained. □

Remark 14.4. If $\kappa = 0$, we obtain the Lemma 2.1 of [38].

Denoting

$$
\begin{aligned}
L = {}&{}^{ABC}J^{\alpha}_{\frac{\upsilon_1}{1+\kappa}+}\Phi\left(\frac{\xi}{1+\kappa}\right) + {}^{ABC}J^{w}_{\frac{\upsilon_2}{1+\kappa}-}\Phi\left(\frac{\xi}{1+\kappa}\right) \\
&- \frac{2(1-\alpha)}{B(\alpha)} - \frac{(\xi-\upsilon_1)^{\alpha}\Phi(\frac{\upsilon_1}{1+\kappa}) + (\upsilon_2-\xi)^{\alpha}\Phi(\frac{\upsilon_2}{1+\kappa})}{(1+\kappa)^{\alpha}B(\alpha)\Gamma(\alpha)}
\end{aligned}
$$

throughout the paper, we can obtain a new variant of the Hadamard-type inequality using the Atangana–Baleanu fractional integrals, as follows:

Theorem 14.1. *Let $m \in (0,1]$, $\Phi : [0, \upsilon_2^*] \to \mathbb{R}$, and $\Phi \in C^2(0, \upsilon_2^*)$, and $\xi \in [\frac{\upsilon_1}{1+\kappa}, \upsilon_2]$ with $0 \le \frac{\upsilon_1}{1+\kappa} \le \upsilon_1 < \upsilon_2 \le \upsilon_2^* < +\infty$. Suppose $\Phi'' \in L_1[\frac{\upsilon_1}{1+\kappa}, \upsilon_2]$ and $|\Phi''| \in N^{s,2}_{h,m}[\frac{\upsilon_1}{1+\kappa}, \upsilon_2]$, then valid the identity:*

$$
\begin{aligned}
|L| \le {}&\frac{(\xi-\upsilon_1)^2\Phi'(\frac{\upsilon_1}{1+\kappa}) + (\upsilon_2-\xi)^2\Phi'(\frac{\upsilon_2}{1+\kappa})}{(\alpha+1)(1+\kappa)^2B(\alpha)\Gamma(\alpha)} \\
&+ \frac{(\xi-\upsilon_1)^{\alpha+2}}{(\alpha+1)(1+\kappa)^{\alpha+2}B(\alpha)\Gamma(\alpha)}\left[\left|\Phi''\left(\frac{\xi}{1+\kappa}\right)\right|\int_0^1 (1-\varsigma)^{\alpha+1}h^s(\varsigma)d\varsigma\right.
\end{aligned}
$$

$$+ m\left|\Phi''\left(\frac{\upsilon_1}{m(1+\kappa)}\right)\right|\int_0^1 (1-\varsigma)^{\alpha+1}(1-h(\varsigma))^s d\varsigma\Bigg]$$

$$+ \frac{(\upsilon_2-\xi)^{\alpha+2}}{(\alpha+1)(1+\kappa)^{\alpha+2}B(\alpha)\Gamma(\alpha)}\left[\left|\Phi''\left(\frac{\upsilon_2}{1+\kappa}\right)\right|\int_0^1 (1-\varsigma)^{\alpha+1}h^s(\varsigma)d\varsigma\right.$$

$$\left.+ m\left|\Phi''\left(\frac{\xi}{m(1+\kappa)}\right)\right|\int_0^1 (1-\varsigma)^{\alpha+1}(1-h(\varsigma))^s d\varsigma\right]. \tag{14.7}$$

Proof. Let us denote

$$I_1 = \int_0^1 (1-\varsigma)^{\alpha+1}\Phi''\left(\frac{\varsigma\xi}{1+\kappa} + \frac{(1-\varsigma)\upsilon_1}{1+\kappa}\right)d\varsigma$$

and

$$I_2 = \int_0^1 \varsigma^{\alpha+1}\Phi''\left(\frac{\varsigma\upsilon_2}{1+\kappa} + \frac{(1-\varsigma)\xi}{1+\kappa}\right)d\varsigma$$

so we have:

$$|L| \leq \frac{(\xi-\upsilon_1)^2\Phi'(\frac{\upsilon_1}{1+\kappa}) + (\upsilon_2-\xi)^2\Phi'(\frac{\upsilon_2}{1+\kappa})}{(\alpha+1)(1+\kappa)^2B(\alpha)\Gamma(\alpha)} + \frac{|(\xi-\upsilon_1)^{\alpha+2}I_1 + (\upsilon_2-\xi)^{\alpha+2}I_2|}{(\alpha+1)(1+\kappa)^{\alpha+2}B(\alpha)\Gamma(\alpha)}. \tag{14.8}$$

Considering that $|\Phi''| \in N_{h,m}^{s,2}[\frac{\upsilon_1}{1+\kappa}, \upsilon_2]$, we have

$$\left|\Phi''\left(\frac{\varsigma\xi}{1+\kappa} + \frac{(1-\varsigma)\upsilon_1}{1+\kappa}\right)\right| \leq \left|\Phi''\left(\frac{\xi}{1+\kappa}\right)\right|h^s(\varsigma) + m\left|\Phi''\left(\frac{\upsilon_1}{m(1+\kappa)}\right)\right|(1-h(\varsigma))^s.$$

Analogously, we have

$$\left|\Phi''\left(\frac{\varsigma\upsilon_2}{1+\kappa} + \frac{(1-\varsigma)\xi}{1+\kappa}\right)\right| \leq \left|\Phi''\left(\frac{\upsilon_2}{1+\kappa}\right)\right|h^s(\varsigma) + m\left|\Phi''\left(\frac{\xi}{m(1+\kappa)}\right)\right|(1-h(\varsigma))^s.$$

Substituting these results into I_1 and I_2 leads us to the following:

$$|I_1| \leq \int_0^1 (1-\varsigma)^{\alpha+1}\left|\Phi''\left(\frac{\varsigma\xi}{1+\kappa} + \frac{(1-\varsigma)\upsilon_1}{1+\kappa}\right)\right|d\varsigma$$

$$\leq \left|\Phi''\left(\frac{\xi}{1+\kappa}\right)\right|\int_0^1 (1-\varsigma)^{\alpha+1}h^s(\varsigma)d\varsigma + m\left|\Phi''\left(\frac{\upsilon_1}{m(1+\kappa)}\right)\right|\int_0^1 (1-\varsigma)^{\alpha+1}(1-h(\varsigma))^s d\varsigma,$$

and

$$|I_2| \le \int_0^1 \varsigma^{\alpha+1}\left|\Phi''\left(\frac{\varsigma \upsilon_2}{1+\kappa} + \frac{(1-\varsigma)\xi}{1+\kappa}\right)\right| d\varsigma$$
$$\le \left|\Phi''\left(\frac{\upsilon_2}{1+\kappa}\right)\right| \int_0^1 \varsigma^{\alpha+1} h^s(\varsigma)d\varsigma + m\left|\Phi''\left(\frac{\xi}{m(1+\kappa)}\right)\right| \int_0^1 \varsigma^{\alpha+1}(1-h(\varsigma))^s d\varsigma.$$

Substituting these results into (14.8), after grouping and sorting, we obtain the desired result. □

Remark 14.5. If $\kappa = 0$, and $s = m = 1$, we get Theorem 2.1 of [38] and Corollary 2.1 of above paper is still valid.

The following results provides more precision to the previous theorem.

Theorem 14.2. *Let $m \in (0,1]$, $\Phi : [0, \upsilon_2^*] \to \mathbb{R}$, and $\Phi \in C^2(0, \upsilon_2^*)$, and $\xi \in [\frac{\upsilon_1}{1+\kappa}, \upsilon_2]$ with $0 \le \frac{\upsilon_1}{1+\kappa} \le \upsilon_1 < \upsilon_2 \le \upsilon_2^* < +\infty$. Suppose $\Phi'' \in L_1[\frac{\upsilon_1}{1+\kappa}, \upsilon_2]$ and $|\Phi''|^q \in N_{h,m}^{s,2}[\frac{\upsilon_1}{1+\kappa}, \upsilon_2]$, then valid the identity:*

$$|L| \le \frac{(\xi-\upsilon_1)^2\Phi'(\frac{\upsilon_1}{1+\kappa}) + (\upsilon_2-\xi)^2\Phi'(\frac{\upsilon_2}{1+\kappa})}{(\alpha+1)(1+\kappa)^2 B(\alpha)\Gamma(\alpha)} + \left(\frac{1}{(\alpha+1)p+1}\right)^{\frac{1}{p}}$$
$$\times \frac{(\xi-\upsilon_1)^{\alpha+2}(A_1)^{\frac{1}{q}} + (\upsilon_2-\xi)^{\alpha+2}(A_2)^{\frac{1}{q}}}{(\alpha+1)(1+\kappa)^{\alpha+2}B(\alpha)\Gamma(\alpha)}, \tag{14.9}$$

where $\frac{1}{p} + \frac{1}{q} = 1$, A_1 and A_2 will be specified later.

Proof. Taking into account the fact that $|\Phi''|^q \in N_{h,m}^{s,2}[\frac{\upsilon_1}{1+\kappa}, \upsilon_2]$, we have

$$\left|\Phi''\left(\frac{\varsigma\xi}{1+\kappa} + \frac{(1-\varsigma)\upsilon_1}{1+\kappa}\right)\right|^q \le \left|\Phi''\left(\frac{\xi}{1+\kappa}\right)\right|^q h^s(\varsigma) + m\left|\Phi''\left(\frac{\upsilon_1}{m(1+\kappa)}\right)\right|^q (1-h(\varsigma))^s.$$

Analogously, we have

$$\left|\Phi''\left(\frac{\varsigma\upsilon_2}{1+\kappa} + \frac{(1-\varsigma)\xi}{1+\kappa}\right)\right|^q \le \left|\Phi''\left(\frac{\upsilon_2}{1+\kappa}\right)\right|^q h^s(\varsigma) + m\left|\Phi''\left(\frac{\xi}{m(1+\kappa)}\right)\right|^q (1-h(\varsigma))^s.$$

Substituting these results into I_1 and I_2 and using the Hölder inequality leads us to the following:

$$|I_1| \le \int_0^1 (1-\varsigma)^{\alpha+1}\left|\Phi''\left(\frac{\varsigma\xi}{1+\kappa} + \frac{(1-\varsigma)\upsilon_1}{1+\kappa}\right)\right| d\varsigma$$
$$\le \left(\int_0^1 (1-\varsigma)^{(\alpha+1)p} d\varsigma\right)^{\frac{1}{p}} \left(\int_0^1 \left|\Phi''\left(\frac{\varsigma\xi}{1+\kappa} + \frac{(1-\varsigma)\upsilon_1}{1+\kappa}\right)\right|^q d\varsigma\right)^{\frac{1}{q}}$$

$$\leq \left(\frac{1}{(\alpha+1)p+1}\right)^{\frac{1}{p}} \left\{\int_0^1 \left[\left|\Phi''\left(\frac{\xi}{1+\kappa}\right)\right|^q h^s(\varsigma) + m\left|\Phi''\left(\frac{\upsilon_1}{m(1+\kappa)}\right)\right|^q (1-h(\varsigma))^s\right] d\varsigma\right\}^{\frac{1}{q}},$$

and

$$|I_2| \leq \int_0^1 \varsigma^{\alpha+1}\left|\Phi''\left(\frac{\varsigma\upsilon_2}{1+\kappa} + \frac{(1-\varsigma)\xi}{1+\kappa}\right)\right| d\varsigma \leq \left(\frac{1}{(\alpha+1)p+1}\right)^{\frac{1}{p}}$$
$$\times \left\{\int_0^1 \left[\left|\Phi''\left(\frac{\upsilon_2}{1+\kappa}\right)\right|^q h^s(\varsigma) + m\left|\Phi''\left(\frac{\xi}{m(1+\kappa)}\right)\right|^q (1-h(\varsigma))^s\right] d\varsigma\right\}^{\frac{1}{q}}.$$

In this way, we have

$$\begin{aligned}|L| \leq{} & \frac{(\xi-\upsilon_1)^2\Phi'(\frac{\upsilon_1}{1+\kappa}) + (\upsilon_2-\xi)^2\Phi'(\frac{\upsilon_2}{1+\kappa})}{(\alpha+1)(1+\kappa)^2 B(\alpha)\Gamma(\alpha)} \\ & + \frac{(\xi-\upsilon_1)^{\alpha+2}}{(\alpha+1)(1+\kappa)^{\alpha+2}B(\alpha)\Gamma(\alpha)}\left(\frac{1}{(\alpha+1)p+1}\right)^{\frac{1}{p}}(A_1)^{\frac{1}{q}} \\ & + \frac{(\upsilon_2-\xi)^{\alpha+2}}{(\alpha+1)(1+\kappa)^{\alpha+2}B(\alpha)\Gamma(\alpha)}\left(\frac{1}{(\alpha+1)p+1}\right)^{\frac{1}{p}}(A_2)^{\frac{1}{q}}, \end{aligned} \tag{14.10}$$

with

$$A_1 = \left|\Phi''\left(\frac{\xi}{1+\kappa}\right)\right|^q \int_0^1 h^s(\varsigma)d\varsigma + m\left|\Phi''\left(\frac{\upsilon_1}{m(1+\kappa)}\right)\right|^q \int_0^1 (1-h(\varsigma))^s d\varsigma$$

and

$$A_2 = \left|\Phi''\left(\frac{\upsilon_2}{1+\kappa}\right)\right|^q \int_0^1 h^s(\varsigma)d\varsigma + m\left|\Phi''\left(\frac{\xi}{m(1+\kappa)}\right)\right|^q \int_0^1 (1-h(\varsigma))^s d\varsigma.$$

From (14.10), we have the required inequality (14.9). □

Remark 14.6. As before, if $\kappa = 0$, and $s = m = 1$, we get Theorem 2.2 and Corollary 2.2 of [38].

Theorem 14.3. *Let $m \in (0,1]$, $\Phi : [0, \upsilon_2^*] \to \mathbb{R}$, and $\Phi \in C^2(0, \upsilon_2^*)$, and $\xi \in [\frac{\upsilon_1}{1+\kappa}, \upsilon_2]$ with $0 \leq \frac{\upsilon_1}{1+\kappa} \leq \upsilon_1 < \upsilon_2 \leq \upsilon_2^* < +\infty$. Suppose $\Phi'' \in L_1[\frac{\upsilon_1}{1+\kappa}, \upsilon_2]$ and $|\Phi''|^q \in N_{h,m}^{s,2}[\frac{\upsilon_1}{1+\kappa}, \upsilon_2]$, then valid the identity:*

$$|L| \leq \frac{(\xi-\upsilon_1)^2\Phi'(\frac{\upsilon_1}{1+\kappa}) + (\upsilon_2-\xi)^2\Phi'(\frac{\upsilon_2}{1+\kappa})}{(\alpha+1)(1+\kappa)^2 B(\alpha)\Gamma(\alpha)}$$

$$+\left(\frac{1}{(\alpha+1)p+1}\right)^{\frac{1}{p}}\frac{(\xi-\upsilon_1)^{\alpha+2}(A_1)^{\frac{1}{q}}+(\upsilon_2-\xi)^{\alpha+2}(A_2)^{\frac{1}{q}}}{(\alpha+1)(1+\kappa)^{\alpha+2}B(\alpha)\Gamma(\alpha)},\tag{14.11}$$

where A_1 and A_2 will be specified later.

Proof. Considering that $|\Phi''|^q \in N^{s,2}_{h,m}[\frac{\upsilon_1}{1+\kappa},\upsilon_2]$, we can write:

$$\left|\Phi''\left(\frac{\varsigma\xi}{1+\kappa}+\frac{(1-\varsigma)\upsilon_1}{1+\kappa}\right)\right|^q \le \left|\Phi''\left(\frac{\xi}{1+\kappa}\right)\right|^q h^s(\varsigma)+m\left|\Phi''\left(\frac{\upsilon_1}{m(1+\kappa)}\right)\right|^q(1-h(\varsigma))^s.$$

Analogously, we have

$$\left|\Phi''\left(\frac{\varsigma\upsilon_2}{1+\kappa}+\frac{(1-\varsigma)\xi}{1+\kappa}\right)\right|^q \le \left|\Phi''\left(\frac{\upsilon_2}{1+\kappa}\right)\right|^q h^s(\varsigma)+m\left|\Phi''\left(\frac{\xi}{m(1+\kappa)}\right)\right|^q(1-h(\varsigma))^s.$$

Substituting these results into I_1 and I_2 and using the Hölder Inequality leads us to the following:

$$\begin{aligned}|I_1| &\le \int_0^1 (1-\varsigma)^{\alpha+1}\left|\Phi''\left(\frac{\varsigma\xi}{1+\kappa}+\frac{(1-\varsigma)\upsilon_1}{1+\kappa}\right)\right|d\varsigma\\ &\le \left(\int_0^1(1-\varsigma)^{(\alpha+1)p}d\varsigma\right)^{\frac{1}{p}}\left(\int_0^1\left|\Phi''\left(\frac{\varsigma\xi}{1+\kappa}+\frac{(1-\varsigma)\upsilon_1}{1+\kappa}\right)\right|^q d\varsigma\right)^{\frac{1}{q}}\\ &\le \left(\frac{1}{(\alpha+1)p+1}\right)^{\frac{1}{p}}\left(\left|\Phi''\left(\frac{\xi}{1+\kappa}\right)\right|^q\int_0^1 h^s(\varsigma)d\varsigma+m\left|\Phi''\left(\frac{\upsilon_1}{m(1+\kappa)}\right)\right|^q\int_0^1(1-h(\varsigma))^s d\varsigma\right)^{\frac{1}{q}},\end{aligned}$$

and

$$\begin{aligned}|I_2| &\le \int_0^1 \varsigma^{\alpha+1}\left|\Phi''\left(\frac{\varsigma\upsilon_2}{1+\kappa}+\frac{(1-\varsigma)\xi}{1+\kappa}\right)\right|d\varsigma\\ &\le \left(\frac{1}{(\alpha+1)p+1}\right)^{\frac{1}{p}}\left(\left|\Phi''\left(\frac{\upsilon_2}{1+\kappa}\right)\right|^q\int_0^1 h^s(\varsigma)d\varsigma+m\left|\Phi''\left(\frac{\xi}{m(1+\kappa)}\right)\right|^q\int_0^1(1-h(\varsigma))^s d\varsigma\right)^{\frac{1}{q}}.\end{aligned}$$

In this way, we have

$$\begin{aligned}|L| \le\ &\frac{(\xi-\upsilon_1)^2\Phi'(\frac{\upsilon_1}{1+\kappa})+(\upsilon_2-\xi)^2\Phi'(\frac{\upsilon_2}{1+\kappa})}{(\alpha+1)(1+\kappa)^2B(\alpha)\Gamma(\alpha)}\\ &+\frac{(\xi-\upsilon_1)^{\alpha+2}}{(\alpha+1)(1+\kappa)^{\alpha+2}B(\alpha)\Gamma(\alpha)}\left[\left(\frac{1}{(\alpha+1)p+1}\right)^{\frac{1}{p}}(A_1)^{\frac{1}{q}}\right]\\ &+\frac{(\upsilon_2-\xi)^{\alpha+2}}{(\alpha+1)(1+\kappa)^{\alpha+2}B(\alpha)\Gamma(\alpha)}\left[\left(\frac{1}{(\alpha+1)p+1}\right)^{\frac{1}{p}}(A_2)^{\frac{1}{q}}\right],\end{aligned}\tag{14.12}$$

with

$$A_1 = \left|\Phi''\left(\frac{\xi}{1+\kappa}\right)\right|^q \int_0^1 h^s(\varsigma) + m\left|\Phi''\left(\frac{\upsilon_1}{m(1+\kappa)}\right)\right|^q \int_0^1 (1-h(\varsigma))^s d\varsigma$$

and

$$A_2 = \left|\Phi''\left(\frac{\upsilon_2}{1+\kappa}\right)\right|^q \int_0^1 h^s(\varsigma)d\varsigma + m\left|\Phi''\left(\frac{\xi}{m(1+\kappa)}\right)\right|^q \int_0^1 (1-h(\varsigma))^s d\varsigma.$$

From (14.12), we have the required inequality (14.11). □

Remark 14.7. As before, if $\kappa = 0$, and $s = m = 1$, we get Theorem 2.2 and Corollary 2.2 of [38].

Lemma 14.3. *Let $\Phi\colon [0, \upsilon_2^*] \longrightarrow \mathbb{R}^+$, such that $\Phi^{(\varrho+1)} \in L_1[0, \upsilon_2^*]$ with $\varrho \in \mathbb{N}$ and $0 < \upsilon_1 < \upsilon_2 \le \upsilon_2^* < +\infty$, then valid the identity:*

$$\begin{aligned}
&-\left(\Phi^{(\varrho)}\left(\frac{\upsilon_1+\kappa\upsilon_2}{1+\kappa}\right) + \Phi^{(\varrho)}\left(\frac{\kappa\upsilon_1+\upsilon_2}{1+\kappa}\right)\right)\\
&+\frac{(1+\kappa)^{\varrho-\frac{\alpha}{k}} k\Gamma_k(\varrho-\frac{\alpha}{k}+1)}{(\upsilon_2-\upsilon_1)^{\varrho-\frac{\alpha}{k}}}\left({}^{C}D^{\alpha,k}_{(\frac{\kappa\upsilon_1+\upsilon_2}{1+\kappa})-}\Phi^{(\varrho)}(\upsilon_1) + (-1)^{\varrho}\,{}^{C}D^{\alpha,k}_{(\frac{\kappa\upsilon_1+\upsilon_2}{1+\kappa})+}\Phi^{(\varrho)}(\upsilon_2)\right)\\
&=\frac{\upsilon_2-\upsilon_1}{1+\kappa}\left[\int_0^1 \varsigma^{\varrho-\frac{\alpha}{k}}\Phi^{(\varrho+1)}\left(\frac{\varsigma}{1+\kappa}\upsilon_1 + \frac{1+\kappa-\varsigma}{1+\kappa}\upsilon_2\right)d\varsigma\right.\\
&\left.-\int_0^1 \varsigma^{\varrho-\frac{\alpha}{k}}\Phi^{(\varrho+1)}\left(\frac{\varsigma}{1+\kappa}\upsilon_2 + \frac{1+\kappa-\varsigma}{1+\kappa}\upsilon_1\right)d\varsigma\right].
\end{aligned} \tag{14.13}$$

Proof. Integrating by parts both integrals on the right-hand side and taking into account that

$$\begin{aligned}
&\frac{(1+\kappa)^{\varrho-\frac{\alpha}{k}}(\kappa-\frac{\alpha}{k})}{(\upsilon_2-\upsilon_1)^{\varrho-\frac{\alpha}{k}}}\left(\int_{\upsilon_1}^{\frac{\kappa\upsilon_1+\upsilon_2}{1+\kappa}} (\xi-\upsilon_1)^{\varrho-\frac{\alpha}{k}-1}\Phi^{(\varrho)}(\xi)d\xi + \int_{\frac{\upsilon_1+\kappa\upsilon_2}{1+\kappa}}^{\upsilon_2} (\upsilon_2-\xi)^{\varrho-\frac{\alpha}{k}-1}\Phi^{(\varrho)}(\xi)d\xi\right)\\
&=\frac{(1+\kappa)^{\varrho-\frac{\alpha}{k}} k\Gamma_k(\varrho-\frac{\alpha}{k}+1)}{(\upsilon_2-\upsilon_1)^{\varrho-\frac{\alpha}{k}}}\left({}^{C}D^{\alpha,k}_{(\frac{\kappa\upsilon_1+\upsilon_2}{1+\kappa})-}\Phi^{(\varrho)}(\upsilon_1) + (-1)^{\varrho}\,{}^{C}D^{\alpha,k}_{(\frac{\kappa\upsilon_1+\upsilon_2}{1+\kappa})+}\Phi^{(\varrho)}(\upsilon_2)\right),
\end{aligned}$$

we obtain the desired result. □

Remark 14.8. One can readily observe that deriving Lemma 2.1 of [10] from the preceding outcome involves setting $\kappa = 0$, examining convex functions, and exclusively relying on the second integral pertaining to the function $\Phi^{(\varrho+1)}(\varsigma u + (1-\varsigma)v)$, where $\upsilon_1 \le u < v \le \upsilon_2$.

Remark 14.9. Lemma 3.1 from [31] for the case of $\kappa = 0$ and (h, m)-convex functions; Lemma 1 from [16], Lemma 2 from [17], and Lemma 3.1 from [27] for $h(\varsigma) = \varsigma$, $\kappa = 1$ and $s = m = 1$ can be derived from the previous lemma as special cases.

Theorem 14.4. *Let $m \in (0,1]$, $\Phi : [0, \upsilon_2^*] \to \mathbb{R}$, and $\Phi \in C^{(\varrho+1)}(0, \upsilon_2^*)$ with $0 \le \upsilon_1 \le \frac{\upsilon_1}{m} < \upsilon_2 \le \frac{\upsilon_2}{m} \le \upsilon_2^* < +\infty$. If $\Phi^{(\varrho+1)} \in L_1(\frac{\upsilon_1}{m}, \frac{\upsilon_2}{m})$ and $|\Phi^{(\varrho+1)}| \in N_{h,m}^{s,2}[\frac{\upsilon_1}{m}, \frac{\upsilon_2}{m}]$, then valid the inequality:*

$$\begin{aligned}
&\Bigg|\mathbb{A}\Big[{}^{C}D^{\alpha,k}_{(\frac{\kappa\upsilon_1+\upsilon_2}{1+\kappa})-}\Phi^{(\varrho)}(\upsilon_1) + (-1)^{\varrho}\,{}^{C}D^{\alpha,k}_{(\frac{\kappa\upsilon_1+\upsilon_2}{1+\kappa})+}\Phi^{(\varrho)}(\upsilon_2)\Big] \\
&\quad - \left[\Phi^{(\varrho)}\left(\frac{\upsilon_1+\kappa\upsilon_2}{1+\kappa}\right) + \Phi^{(\varrho)}\left(\frac{\kappa\upsilon_1+\upsilon_2}{1+\kappa}\right)\right]\Bigg| \\
&\le \frac{\upsilon_2-\upsilon_1}{1+\kappa}\big[|\Phi^{(\varrho+1)}(\upsilon_1)| + |\Phi^{(\varrho+1)}(\upsilon_2)|\big]\mathbb{B} + \frac{\upsilon_2-\upsilon_1}{1+\kappa}m\left[\left|\Phi^{(\varrho+1)}\left(\frac{\upsilon_1}{m}\right)\right| + \left|\Phi^{(\varrho+1)}\left(\frac{\upsilon_2}{m}\right)\right|\right]\mathbb{C}
\end{aligned} \tag{14.14}$$

with

$$\mathbb{A} = \frac{(1+\kappa)^{\varrho-\frac{\alpha}{k}}k\Gamma_k(\varrho-\frac{\alpha}{k}+1)}{(\upsilon_2-\upsilon_1)^{\varrho-\frac{\alpha}{k}}}, \quad \mathbb{B} = \int_0^1 \varsigma^{\varrho-\frac{\alpha}{k}} h^s\left(\frac{\varsigma}{1+\kappa}\right)d\varsigma,$$

and

$$\mathbb{C} = \int_0^1 \varsigma^{\varrho-\frac{\alpha}{k}}\left(1-h\left(\frac{\varsigma}{1+\kappa}\right)\right)^s d\varsigma.$$

Proof. From (14.13), we get

$$\begin{aligned}
&\left|\int_0^1 \varsigma^{\varrho-\frac{\alpha}{k}}\left[\Phi^{(\varrho+1)}\left(\frac{\varsigma}{1+\kappa}\upsilon_1 + \frac{1+\kappa-\varsigma}{1+\kappa}\upsilon_2\right) - \Phi^{(\varrho+1)}\left(\frac{\varsigma}{1+\kappa}\upsilon_2 + \frac{1+\kappa-\varsigma}{1+\kappa}\upsilon_1\right)\right]d\varsigma\right| \\
&\quad \le \int_0^1 \varsigma^{\varrho-\frac{\alpha}{k}}\left|\Phi^{(\varrho+1)}\left(\frac{\varsigma}{1+\kappa}\upsilon_1 + \frac{1+\kappa-\varsigma}{1+\kappa}\upsilon_2\right)\right|d\varsigma \\
&\quad\quad + \int_0^1 \varsigma^{\varrho-\frac{\alpha}{k}}\left|\Phi^{(\varrho+1)}\left(\frac{\varsigma}{1+\kappa}\upsilon_2 + \frac{1+\kappa-\varsigma}{1+\kappa}\upsilon_1\right)\right|d\varsigma.
\end{aligned}$$

Taking into account the fact that $|\Phi^{(\varrho+1)}| \in N^{s,2}_{h,m}[\frac{\upsilon_1}{m}, \frac{\upsilon_2}{m}]$, we can write

$$\begin{aligned}
&\int_0^1 \varsigma^{\varrho-\frac{\alpha}{k}} \left|\Phi^{(\varrho+1)}\left(\frac{\varsigma}{1+\kappa}\upsilon_1 + \frac{1+\kappa-\varsigma}{1+\kappa}\upsilon_2\right)\right| d\varsigma \\
&\quad \le \int_0^1 \varsigma^{\varrho-\frac{\alpha}{k}} \left[h^s\left(\frac{\varsigma}{1+\kappa}\right)|\Phi^{(\varrho+1)}(\upsilon_1)| + m\left(1-h\left(\frac{\varsigma}{1+\kappa}\right)\right)^s \left|\Phi^{(\varrho+1)}\left(\frac{\upsilon_2}{m}\right)\right|\right] d\varsigma \\
&\quad = |\Phi^{(\varrho+1)}(\upsilon_1)| \int_0^1 \varsigma^{\varrho-\frac{\alpha}{k}} h^s\left(\frac{\varsigma}{1+\kappa}\right) d\varsigma + m\left|\Phi'\left(\frac{\upsilon_2}{m}\right)\right| \int_0^1 \varsigma^{\varrho-\frac{\alpha}{k}}\left(1-h\left(\frac{\varsigma}{1+\kappa}\right)\right)^s d\varsigma. \quad (14.15)
\end{aligned}$$

In the same way,

$$\begin{aligned}
&\int_0^1 \varsigma^{\varrho-\frac{\alpha}{k}} \left|\Phi^{(\varrho+1)}\left(\frac{\varsigma}{1+\kappa}\upsilon_2 + \frac{1+\kappa-\varsigma}{1+\kappa}\upsilon_1\right)\right| d\varsigma \\
&\quad \le |\Phi^{(\varrho+1)}(\upsilon_2)| \int_0^1 \varsigma^{\varrho-\frac{\alpha}{k}} h^s\left(\frac{\varsigma}{1+\kappa}\right) d\varsigma + m\left|\Phi'\left(\frac{\upsilon_1}{m}\right)\right| \int_0^1 \varsigma^{\varrho-\frac{\alpha}{k}}\left(1-h\left(\frac{\varsigma}{1+\kappa}\right)\right)^s d\varsigma. \quad (14.16)
\end{aligned}$$

Using (14.15) and (14.16), we readily derive inequality (14.14). Thus, the theorem is established. □

Remark 14.10. For $\kappa = 0$, one can readily verify the first part of Theorem 2.7 in [31] pertaining to (h, m)-convex functions. When $\kappa = 1$ and function is convex Theorem 7 (with $q = 1$) in [16], along with Theorem 5 in [17] and Theorem 3.2 (with $q = 1$) in [27] these are all specific instances of the preceding theorem.

The preceding result could be improved, if we impose some conditions on $|\Phi^{(\varrho+1)}|^q$.

Theorem 14.5. *Let $m \in (0,1]$, $\Phi : [0, \upsilon_2^*] \to \mathbb{R}^+$, and $\Phi \in C^{(\varrho+1)}(0, \upsilon_2^*)$ with $0 \le \upsilon_1 \le \frac{\upsilon_1}{m} < \upsilon_2 \le \frac{\upsilon_2}{m} \le \upsilon_2^* < +\infty$. If $\Phi^{(\varrho+1)} \in L_1(\frac{\upsilon_1}{m}, \frac{\upsilon_2}{m})$ and $|\Phi^{(\varrho+1)}|^q \in N^{s,2}_{h,m}[\frac{\upsilon_1}{m}, \frac{\upsilon_2}{m}]$, then valid the inequality:*

$$\begin{aligned}
&\Bigg|\mathbb{A}\left[{}^C D^{\alpha,k}_{(\frac{\kappa\upsilon_1+\upsilon_2}{1+\kappa})-}\Phi^{(\varrho)}(\upsilon_1) + (-1)^{\varrho}\, {}^C D^{\alpha,k}_{(\frac{\kappa\upsilon_1+\upsilon_2}{1+\kappa})+}\Phi^{(\varrho)}(\upsilon_2)\right] \\
&\qquad - \left[\Phi^{(\varrho)}\left(\frac{\upsilon_1+\kappa\upsilon_2}{1+\kappa}\right) + \Phi^{(\varrho)}\left(\frac{\kappa\upsilon_1+\upsilon_2}{1+\kappa}\right)\right]\Bigg| \\
&\quad \le \frac{(\upsilon_2-\upsilon_1)B_q}{1+\kappa}\{(p_1C_{11} + mp_{12}C_{12})^{\frac{1}{q}} + (p_2C_{11} + mp_{21}C_{12})^{\frac{1}{q}}\},
\end{aligned}$$

where $\mathbb{A}$ as before,

$$B_p = \frac{1}{(p\varrho - \frac{p\alpha}{k} + 1)^{\frac{1}{p}}},$$

$$p_1 = \left|\Phi^{(\varrho+1)}(\upsilon_1)\right|^q, \quad p_{12} = \left|\Phi^{(\varrho+1)}\left(\frac{\upsilon_2}{m}\right)\right|^q,$$
$$p_2 = \left|\Phi^{(\varrho+1)}(\upsilon_2)\right|^q, \quad p_{21} = \left|\Phi^{(\varrho+1)}\left(\frac{\upsilon_1}{m}\right)\right|^q,$$
$$C_{12} = \int_0^1 \left(1 - h\left(\frac{\varsigma}{1+\kappa}\right)\right)^s d\varsigma \quad and \quad C_{11} = \int_0^1 h^s\left(\frac{\varsigma}{1+\kappa}\right) d\varsigma.$$

Proof. From (14.13), we get

$$\left|\int_0^1 \varsigma^{\varrho-\frac{\alpha}{k}} \Phi^{(\varrho+1)}\left(\frac{\varsigma}{1+\kappa}\upsilon_1 + \frac{1+\kappa-\varsigma}{1+\kappa}\upsilon_2\right)d\varsigma - \int_0^1 \varsigma^{\varrho-\frac{\alpha}{k}} \Phi^{(\varrho+1)}\left(\frac{\varsigma}{1+\kappa}\upsilon_2 + \frac{1+\kappa-\varsigma}{1+\kappa}\upsilon_1\right)d\varsigma\right|$$
$$\leq \int_0^1 \varsigma^{\varrho-\frac{\alpha}{k}} \left|\Phi^{(\varrho+1)}\left(\frac{\varsigma}{1+\kappa}\upsilon_1 + \frac{1+\kappa-\varsigma}{1+\kappa}\upsilon_2\right)\right| d\varsigma$$
$$+ \int_0^1 \varsigma^{\varrho-\frac{\alpha}{k}} \left|\Phi^{(\varrho+1)}\left(\frac{\varsigma}{1+\kappa}\upsilon_2 + \frac{1+\kappa-\varsigma}{1+\kappa}\upsilon_1\right)\right| d\varsigma.$$

By using Hölder's inequality, we can write

$$\int_0^1 \varsigma^{\varrho-\frac{\alpha}{k}} \left|\Phi^{(\varrho+1)}\left(\frac{\varsigma}{1+\kappa}\upsilon_1 + \frac{1+\kappa-\varsigma}{1+\kappa}\upsilon_2\right)\right| d\varsigma$$
$$\leq \left(\int_0^1 \varsigma^{p(\varrho-\frac{\alpha}{k})} d\varsigma\right)^{\frac{1}{p}} \left(\int_0^1 \left|\Phi^{(\varrho+1)}\left(\frac{\varsigma}{1+\kappa}\upsilon_1 + \frac{1+\kappa-\varsigma}{1+\kappa}\upsilon_2\right)\right|^q d\varsigma\right)^{\frac{1}{q}} \tag{14.17}$$

and

$$\int_0^1 \varsigma^{\varrho-\frac{\alpha}{k}} \left|\Phi^{(\varrho+1)}\left(\frac{\varsigma}{1+\kappa}\upsilon_2 + \frac{1+\kappa-\varsigma}{1+\kappa}\upsilon_1\right)\right| d\varsigma$$
$$\leq \left(\int_0^1 \varsigma^{p(\varrho-\frac{\alpha}{k})} d\varsigma\right)^{\frac{1}{p}} \left(\int_0^1 \left|\Phi^{(\varrho+1)}\left(\frac{\varsigma}{1+\kappa}\upsilon_2 + \frac{1+\kappa-\varsigma}{1+\kappa}\upsilon_1\right)\right|^q d\varsigma\right)^{\frac{1}{q}} \tag{14.18}$$

Taking into account the fact that $|\Phi^{(\varrho+1)}|^q \in N^{s,2}_{h,m}[\frac{\upsilon_1}{m}, \frac{\upsilon_2}{m}]$ from (14.17) and (14.18), we have

$$\int_0^1 \left|\Phi^{(\varrho+1)}\left(\frac{\varsigma}{1+\kappa}\upsilon_1 + \frac{1+\kappa-\varsigma}{1+\kappa}\upsilon_2\right)\right|^q d\varsigma$$

$$\leq |\Phi^{(\varrho+1)}(\upsilon_1)|^q \int_0^1 h^s\left(\frac{\varsigma}{1+\kappa}\right)d\varsigma + m\left|\Phi^{(\varrho+1)}\left(\frac{\upsilon_2}{m}\right)\right|^q \int_0^1 \left(1-h\left(\frac{\varsigma}{1+\kappa}\right)\right)^s d\varsigma, \quad (14.19)$$

$$\int_0^1 \left|\Phi^{(\varrho+1)}\left(\frac{\varsigma}{1+\kappa}\upsilon_2 + \frac{1+\kappa-\varsigma}{1+\kappa}\upsilon_1\right)\right|^q d\varsigma$$

$$\leq |\Phi^{(\varrho+1)}(\upsilon_2)|^q \int_0^1 h^s\left(\frac{\varsigma}{1+\kappa}\right)d\varsigma + m\left|\Phi^{(\varrho+1)}\left(\frac{\upsilon_1}{m}\right)\right|^q \int_0^1 \left(1-h\left(\frac{\varsigma}{1+\kappa}\right)\right)^s d\varsigma. \quad (14.20)$$

Denoting, for brevity $B_p = (\int_0^1 \varsigma^{p(\varrho-\frac{\alpha}{k})}\,d\varsigma)^{\frac{1}{p}} = \frac{1}{(p(\varrho-\frac{\alpha}{k})+1)^{\frac{1}{p}}}$, substituting (14.19), (14.20) in (14.17) and (14.18), we obtain the required inequality. □

Remark 14.11. From this result for $q > 1$ follows Theorem 8 of [16] and Theorem 3.2 of [27]. Theorem 6 of [17] and he second part of Theorem 2.7 of [31], also we can be derived for different values of κ and different notions of convexity.

Theorem 14.6. *Let $m \in (0,1]$, $\Phi : [0, \upsilon_2^*] \to \mathbb{R}$, and $\Phi \in C^{(\varrho+1)}(0, \upsilon_2^*)$ with $0 \leq \upsilon_1 \leq \frac{\upsilon_1}{m} < \upsilon_2 \leq \frac{\upsilon_2}{m} \leq \upsilon_2^* < +\infty$. If $\Phi^{(\varrho+1)} \in L_1(\frac{\upsilon_1}{m}, \frac{\upsilon_2}{m})$ and $|\Phi^{(\varrho+1)}|^q \in N_{h,m}^{s,2}[\frac{\upsilon_1}{m}, \frac{\upsilon_2}{m}]$, then valid the inequality:*

$$\begin{aligned}&\Bigg|\mathbb{A}\left[{}^C D^{\alpha,k}_{(\frac{\kappa\upsilon_1+\upsilon_2}{1+\kappa})-}\Phi^{(\varrho)}(\upsilon_1) + (-1)^{\varrho}\,{}^C D^{\alpha,k}_{(\frac{\kappa\upsilon_1+\upsilon_2}{1+\kappa})+}\Phi^{(\varrho)}(\upsilon_2)\right] \\ &\quad - \left[\Phi^{(\varrho)}\left(\frac{\upsilon_1+\kappa\upsilon_2}{1+\kappa}\right) + \Phi^{(\varrho)}\left(\frac{\kappa\upsilon_1+\upsilon_2}{1+\kappa}\right)\right]\Bigg| \\ &\leq \frac{(\upsilon_2-\upsilon_1)B_q}{1+\kappa}\{(p_1C_1 + mp_{12}C_2)^{\frac{1}{q}} + (p_2C_1 + mp_{21}C_2)^{\frac{1}{q}}\}\end{aligned} \quad (14.21)$$

with $q \geq 1, p_1, p_{12}, p_2, p_{21}$ and $\mathbb{A}$ as before,

$$C_1 = \int_0^1 \varsigma^{\varrho-\frac{\alpha}{k}} h^s\left(\frac{\varsigma}{1+\kappa}\right)d\varsigma, \quad C_2 = \int_0^1 \varsigma^{\varrho-\frac{\alpha}{k}}\left(1-h\left(\frac{\varsigma}{1+\kappa}\right)\right)^s d\varsigma,$$

$$B_q = \frac{1}{(\varrho-\frac{\alpha}{k}+1)^{1-\frac{1}{q}}}.$$

Proof. Similarly, we deduce from identity (14.13):

$$\left|\int_0^1 \varsigma^{\varrho-\frac{\alpha}{k}}\Phi^{(\varrho+1)}\left(\frac{\varsigma}{1+\kappa}\upsilon_1 + \frac{1+\kappa-\varsigma}{1+\kappa}\upsilon_2\right)d\varsigma - \int_0^1 \varsigma^{\varrho-\frac{\alpha}{k}}\Phi^{(\varrho+1)}\left(\frac{\varsigma}{1+\kappa}\upsilon_2 + \frac{1+\kappa-\varsigma}{1+\kappa}\upsilon_1\right)d\varsigma\right|$$

$$\leq \int_0^1 \varsigma^{\varrho-\frac{\alpha}{k}}\left|\Phi^{(\varrho+1)}\left(\frac{\varsigma}{1+\kappa}\upsilon_1 + \frac{1+\kappa-\varsigma}{1+\kappa}\upsilon_2\right)\right|d\varsigma$$

$$+\int_0^1 \varsigma^{\varrho-\frac{\alpha}{k}}\left|\Phi^{(\varrho+1)}\left(\frac{\varsigma}{1+\kappa}\upsilon_2+\frac{1+\kappa-\varsigma}{1+\kappa}\upsilon_1\right)\right|d\varsigma.$$

Utilizing the widely recognized power mean inequality, we obtain:

$$\int_0^1 \varsigma^{\varrho-\frac{\alpha}{k}}\left|\Phi^{(\varrho+1)}\left(\frac{\varsigma}{1+\kappa}\upsilon_1+\frac{1+\kappa-\varsigma}{1+\kappa}\upsilon_2\right)\right|d\varsigma$$
$$\leq\left(\int_0^1 \varsigma^{\varrho-\frac{\alpha}{k}}d\varsigma\right)^{1-\frac{1}{q}}\left(\int_0^1 \varsigma^{\varrho-\frac{\alpha}{k}}\left|\Phi^{(\varrho+1)}\left(\frac{\varsigma}{1+\kappa}\upsilon_1+\frac{1+\kappa-\varsigma}{1+\kappa}\upsilon_2\right)\right|^q d\varsigma\right)^{\frac{1}{q}} \tag{14.22}$$

and

$$\int_0^1 \varsigma^{\varrho-\frac{\alpha}{k}}\left|\Phi^{(\varrho+1)}\left(\frac{\varsigma}{1+\kappa}\upsilon_2+\frac{1+\kappa-\varsigma}{1+\kappa}\upsilon_1\right)\right|d\varsigma$$
$$\leq\left(\int_0^1 \varsigma^{\varrho-\frac{\alpha}{k}}d\varsigma\right)^{1-\frac{1}{q}}\left(\int_0^1 \varsigma^{\varrho-\frac{\alpha}{k}}\left|\Phi^{(\varrho+1)}\left(\frac{\varsigma}{1+\kappa}\upsilon_2+\frac{1+\kappa-\varsigma}{1+\kappa}\upsilon_1\right)\right|^q d\varsigma\right)^{\frac{1}{q}}. \tag{14.23}$$

Taking into account the fact that $|\Phi^{(\varrho+1)}|^q\in N^{s,2}_{h,m}[\frac{\upsilon_1}{m},\frac{\upsilon_2}{m}]$, we get

$$\int_0^1 \varsigma^{\varrho-\frac{\alpha}{k}}\left|\Phi^{(\varrho+1)}\left(\frac{\varsigma}{1+\kappa}\upsilon_1+\frac{1+\kappa-\varsigma}{1+\kappa}\upsilon_2\right)\right|^q d\varsigma$$
$$\leq|\Phi^{(1+\kappa)}(\upsilon_1)|^q\int_0^1 \varsigma^{\varrho-\frac{\alpha}{k}}h^s\left(\frac{\varsigma}{\varrho+1}\right)d\varsigma+m\left|\Phi^{(\varrho+1)}\left(\frac{\upsilon_2}{m}\right)\right|^q\int_0^1 \varsigma^{\varrho-\frac{\alpha}{k}}\left(1-h\left(\frac{\varsigma}{1+\kappa}\right)\right)^s d\varsigma. \tag{14.24}$$

Similarly,

$$\int_0^1 \varsigma^{\varrho-\frac{\alpha}{k}}\left|\Phi^{(\varrho+1)}\left(\frac{\varsigma}{1+\kappa}\upsilon_2+\frac{1+\kappa-\varsigma}{1+\kappa}\upsilon_1\right)\right|^q d\varsigma$$
$$\leq|\Phi^{(\varrho+1)}(\upsilon_2)|^q\int_0^1 \varsigma^{\varrho-\frac{\alpha}{k}}h^s\left(\frac{\varsigma}{1+\kappa}\right)d\varsigma+m\left|\Phi^{(\varrho+1)}\left(\frac{\upsilon_1}{m}\right)\right|^q\int_0^1 \varsigma^{\varrho-\frac{\alpha}{k}}\left(1-h\left(\frac{\varsigma}{1+\kappa}\right)\right)^s d\varsigma. \tag{14.25}$$

If we put (14.24) and (14.25), in (14.22) and in (14.23), it allows us to obtain the inequality (14.21). In this way, the proof is completed. □

Remark 14.12. From this result for $q > 1$ follows Theorem 7 from [16] and Theorem 5 from [17] for different values of κ and different definitions of convexity, as well as Theorem 3.3 from [27].

14.4 Conclusions

Within this paper, various adaptations of the renowned Hermite–Hadamard Inequality have been derived within the context of the Atangana–Baleanu fractional derivative. The breadth of our findings has been validated by showcasing their encapsulation of numerous previously established instances documented in the literature.

Bibliography

[1] Alomari, M., Darus, M. and Dragomir, S. S., *New inequalities of Hermite–Hadamard type for functions whose second derivates absolute values are quasi-convex*, RGMIA Res. Rep. Collect. 12 (2009), 14, Supplement. http://www.staff.vu.edu.au/RGMIA/v12

[2] Amlashi, M. E. and Hassani, M., *More on the Hermite–Hadamard inequality*, Int. J. Nonlinear Anal. Appl. 12(2) (2021), 2153–2159, https://doi.org/10.22075/ijnaa.2021.21753.2293.

[3] Ali, M. A., Nápoles, J. E., Kashuri, A. and Zhang, Z., *Fractional non conformable Hermite–Hadamard inequalities for generalized-convex functions*, Fasc. Math. 64 (2020), 5–16, https://doi.org/10.21008/j.0044-4413.2020.0007.

[4] Abdeljawada, T. and Baleanu, D., *Integration by parts and its applications of a new nonlocal fractional derivative with Mittag-Leffler nonsingular kernel*, J. Nonlinear Sci. Appl. 10 (2017), 1098–1107, https://doi.org/10.22436/jnsa.010.03.20.

[5] Atangana, A. and Baleanu, D., *New fractional derivatives with non-local and non-singular kernel: Theory and application to Heat transfer model*, Therm. Sci. 20 (2016), 763–769.

[6] Bayraktar, B., *Some integral inequalities of Hermite–Hadamard type for differentiable (s, m)-convex functions via fractional integrals*, TWMS J. Appl. Eng. Math. 10(3) (2020), 625–637.

[7] Bayraktar, B. and Nápoles Valdes, J. E., *New generalized integral inequalities via (h, m)-convex modified functions*, Izv. Inst. Mat. Inform. Udmurt. Gos. Univ. 60 (2022), 3–15, https://doi.org/10.35634/2226-3594-2022-60-01.

[8] Bayraktar, B. and Nápoles Valdes, J. E., *Hermite–Hadamard Weighted integral inequalities for (h, m)-convex modified functions*, Fract. Differ. Calc. 12(2) (2022), 235–248, https://doi.org/10.7153/fdc-2022-12-15.

[9] Bermudo, S., Kórus, P. and Nápoles, J. E., *On q-Hermite–Hadamard inequalities for general convex functions*, Acta Math. Hung. 162 (2020), 364–374, https://doi.org/10.1007/s10474-020-01025-6.

[10] Butt, S. I., Ozdemir, M. E., Umar, M. and Celik, B., *Several new integral inequalities via Caputo k-fractional derivative operators*, Asian-Eur. J. Math. 14(9) (2021), 2150150. https://doi.org/10.1142/S1793557121501503.

[11] Caputo, M., *Elasticite Dissipazione*, Zanichelli, Bologna, 1969.

[12] Caputo, M. and Fabrizio, M., *A new definition of fractional derivative without singular kernel*, Prog. Fract. Differ. Appl. 1(2) (2015), 73–85.

[13] Chen, F. X., *On the generalization of some Hermite–Hadamard inequalities for functions with convex absolute values of the second derivatives via fractional integrals*, Ukr. Math. J. 70(12) (2019), 1953–1965, https://doi.org/10.1007/s11253-019-01618-7.

[14] Díaz, R. and Pariguan, E., *On hypergeometric functions and Pochhammer k-symbol*, Divulg. Mat. 15(2) (2007), 179–192.

[15] Dong, Y., Zeb, M., Farid, G. and Bibi, S., *Hadamard inequalities for strongly (α, m)-convex functions via Caputo fractional derivatives*, J. Math. 2021 (2021), 6691151, 16 pages. https://doi.org/10.1155/2021/6691151.

[16] Farid, G., Javed, A., Rehman, A. U. and Qureshi, M. I., *On Hadamard-type inequalities for differentiable functions via Caputo k-fractional derivatives*, Cogent Math. 4 (2017), 1355429, https://doi.org/10.1080/23311835.2017.1355429.

[17] Farid, G., Naqvi, S. and Rehman, A. U., *A version of the Hadamard inequality for Caputo fractional derivatives and related results*, RGMIA Res. Rep. Collect. 20 (2017), 59.

[18] Farid, G., Rehman, A. U., Bibi, S. and Chu, Y. M., *Refinements of two fractional versions of Hadamard inequalities for Caputo fractional derivatives and related results*, Open J. Math. Sci. 5 (2021), 1–10, https://doi.org/10.30538/oms2021.0139.

[19] Farissi, A. E., *Simple proof and refinement of Hermite–Hadamard inequality*, J. Math. Inequal. 4(3) (2010), 365–369.

[20] Feng, X., Feng, B., Farid, G., Bibi, S., Xiaoyan, Q. and Wu, Z., *Caputo fractional derivative Hadamard inequalities for strongly m-convex functions*, J. Funct. Spaces (2021), 6642655, https://doi.org/10.1155/2021/6642655.

[21] Gasimov, Y. S., Nachaoui, A. and Niftiyev, A. A., *Non-linear eigenvalue problems for p-Laplacian with variable domain*, Optim. Lett. 4(1) (2010), 67–84.

[22] Gasimov, Y. S. and Nápoles Valdés, J. E., *Some refinements of Hermite–Hadamard inequality using k-fractional Caputo derivatives*, Fract. Differ. Calc. 12(2) (2022), 209–221.

[23] Guzmán, P. M., Nápoles, V. J. E. and Gasimov, Y., *Integral inequalities within the framework of generalized fractional integrals*, Fract. Differ. Calc. 11(1) (2021), 69–84.

[24] Hadamard, J., *Étude sur les propriétés des fonctions entières et en particulier d'une fonction considérée par Riemann*, J. Math. Pures Appl. 9 (1893), 171–216.

[25] Hermite, C., *Sur deux limites d'une intégrale définie*, Mathesis 3 (1883), 82.

[26] Hussain, R., Ali, A., Ayub, A. and Latif, A., *Some new fractional integral inequalities for harmonically h-convex via Caputo k-fractional derivatives*, Bull. Int. Math. Virtual Inst. 11(1) (2021), 99–110.

[27] Kang, S. M., Farid, G., Nazeer, W. and Naqvi, S., *A version of the Hadamard inequality for Caputo fractional derivatives and related results*, J. Comput. Anal. Appl. 27(6) (2019), 962–972.

[28] Mainardi, F., *On some properties of the Mittag-Leffler function $E_\alpha(-\varsigma\alpha)$, completely monotone for $\varsigma > 0$ with $0 < \alpha < 1$*, Discrete Contin. Dyn. Syst., Ser. B 19(7) (2014), 2267–2278.

[29] Marinescu, D. S. and Monea, M., *A very short proof of the Hermite–Hadamard inequalities*, Am. Math. Mon. 127(9) (2020), 850–851.

[30] Miller, K. and Ross, B., *An Introduction to the Fractional Calculus and Fractional Differential Equations*, Wiley, 1993.

[31] Mishra, L. N., Ain, Q. U., Farid, G. and Rehman, A. U., *k-fractional integral inequalities for (h, m)-convex functions via Caputo k-fractional derivatives*, Korean J. Math. 27(2) (2019), 357–374.

[32] Nápoles Valdés, J. E., Rabossi, F. and Samaniego, A. D., *Convex functions: Ariadne's thread or Charlotte's spiderweb?* Adv. Math. Model. Appl. 5(2) (2020), 176–191.

[33] Nápoles Valdés, J. E., Rodríguez, J. M. and Sigarreta, J. M., *On Hermite–Hadamard type inequalities for non-conformable integral operators*, Symmetry 11 (2019), 1108.

[34] Özdemir, M. E., Ekinci, A. and Akdemir, A. O., *Generalizations of integral inequalities for functions whose second derivatives are convex and m-convex*, Miskolc Math. Notes 13(2) (2012), 441–457.

[35] Özdemir, M. E., Yıldız, Ç., Akdemir, A. O. and Set, E., *On some inequalities for s-convex functions and applications*, J. Inequal. Appl. 2013, 333 (2013).

[36] Rainville, E. D., *Special Functions*, Macmillan, New York, 1960.

[37] Sahoo, S. K., Mohammed, P. O., Kodamasingh, B., Tariq, M. and Hamed, Y. S., *New fractional integral inequalities for convex functions pertaining to Caputo–Fabrizio operator*, Fractal Fract. 6(3) (2022), 171.

[38] Set, E., Butt, S. I., Akdemir, A. O., Karaoglan, A. and Abdeljawad, T., *New integral inequalities for differentiable convex functions via Atangana–Baleanu fractional integral operators*, Chaos Solitons Fractals 143 (2021), 110554.

[39] Simić, S., *Some refinements of Hermite–Hadamard inequality and an open problem*, Kragujev. J. Math. 42(3) (2018), 349–356.

[40] Xi, B. Y., Gao, D. D. and Qi, F., *Integral inequalities of Hermite–Hadamard type for* (α, s)*-convex and* (α, s, m)*-convex functions*, Ital. J. Pure Appl. Math. 44 (2020), 499–510.

[41] Zhao, J., Butt, S. I., Nasir, J., Wang, Z. and Tlili, I., *Hermite–Jensen–Mercer type inequalities for Caputo fractional derivatives*, J. Funct. Spaces 2020 (2020), 7061549.

Nabanita Konwar

15 Analysis of fixed-point theory in the setting of non-Archimedean fuzzy-*b*-metric spaces

Abstract: In this work, we established some new definitions of fuzzy *b*-contractions on non-Archimedean fuzzy-*b*-metric spaces. We developed fuzzy interpolative Reich–Rus–Ciric (shortly R–R–C) type *b* contraction fixed-point theorem and fuzzy $\mathfrak{Y}^\circ$ interpolative R–R–C-type *b* contraction fixed-point theorem on non-Archimedean fuzzy-*b*-metric spaces.

Keywords: Non-Archimedean fuzzy-*b*-metric spaces, fuzzy interpolative Reich–Rus–Ciric-type *b* contraction, fuzzy $\mathfrak{Y}^\circ$ interpolative Reich–Rus–Ciric-type *b* contraction, fixed point

MSC 2020: 47H10, 54H25, 54E50

15.1 Introduction

The idea of *b*-metric space is highly significant in establishing the foundation of fixed-point theory. Since the existence of fixed points often leads to the identification of solutions in graph theory, studying fixed-point theorem within various generalized spaces holds considerable importance in system design and modeling. However, accurately modeling systems can sometimes be challenging due to the lack of precise measurements between elements or points. In such situations, fuzzy sets and fuzzy logic offer an effective framework for supporting modeling and design processes.

The Banach fixed-point theorem, established in 1922, stands as one of the most fundamental results in fixed-point theory. Following its introduction, the field experienced significant growth and development. In 1965, Zadeh [30] introduced fuzzy set theory—a novel framework designed to address situations involving vague or imprecise data. The notion of a fuzzy metric space was initially proposed by Kaleva and Seikkala [24], and later refined by researchers such as Kramosil and Michalek [26], as well as George and Veeramani [21]. Around the same time, Bakhtin [3] and Czerwik [8] put forward the notion of *b*-metric space by weakening the traditional triangle inequality condition.

Heilpern [23] was among the first to explore the theory of fixed point within fuzzy metric spaces, extending Banach's contraction principle into the fuzzy setting. Since then, many researchers have further generalized and expanded this principle in fuzzy

Competing interests: The authors state that there are no conflicts of interest.

Nabanita Konwar, Department of Mathematics, Birjhora Mahavidyalaya, Bongaigaon, Assam-783380, India, e-mail: nabnitakonwar@gmail.com

https://doi.org/10.1515/9783112215760-015

environments from various perspectives. Simultaneously, parallel developments of Banach's theorem and alternative contraction conditions have been explored in works such as [1, 2, 4, 6, 7, 18, 25, 27]. For more relevant literature, we refer to the recent works of Debnath [9–17, 19] and others [20, 28, 29].

In this work, we establish the concept of fuzzy interpolative Reich–Rus–Ciric-type *b* contraction in non-Archimedean fuzzy-*b*-metric spaces and investigate some fixed-point theorems related to this newly established framework.

15.2 Basic definitions

This section contains some preliminary definitions that are useful for the work.

Definition 15.1 ([3, 8]). Let $\mathfrak{X} \neq \phi$ and $\mathfrak{D}_{\mathfrak{b}} : \mathfrak{X} \times \mathfrak{X} \to [0, \infty)$ be a function. If $\mathfrak{D}_{\mathfrak{b}}$ satisfy the following conditions:
(1) $\mathfrak{D}_{\mathfrak{b}}(\mathfrak{A}_1, \mathfrak{A}_2) = 0$ if and only if $\mathfrak{A}_1 = \mathfrak{A}_2$;
(2) $\mathfrak{D}_{\mathfrak{b}}(\mathfrak{A}_1, \mathfrak{A}_2) = \mathfrak{D}_{\mathfrak{b}}(\mathfrak{A}_2, \mathfrak{A}_1)$ for all $\mathfrak{A}_1, \mathfrak{A}_2 \in \mathfrak{X}$;
(3) $\exists$ a $\mathbb{R} \ni b \geq 1$ such that $\mathfrak{D}_{\mathfrak{b}}(\mathfrak{A}_1, \mathfrak{x}_3) \leq [\mathfrak{D}_{\mathfrak{b}}(\mathfrak{A}_1, \mathfrak{A}_2) + \mathfrak{D}_{\mathfrak{b}}(\mathfrak{A}_2, \mathfrak{x}_3)]$ for all $\mathfrak{A}_1, \mathfrak{A}_2, \mathfrak{A}_3 \in \mathfrak{X}$,

then it is a *b*-metric on $\mathfrak{X}$. And the pair $(\mathfrak{X}, \mathfrak{D}_{\mathfrak{b}})$ is called *b*-metric space.

Definition 15.2 ([21]). Let $\mathfrak{X} \neq \phi$. Consider a continuous t-norm $*$ and a fuzzy set $\mathfrak{H}$ on $\mathfrak{X}^2 \times (0, \infty)$. Then $(\mathfrak{X}, \mathfrak{H}, *)$ is called a fuzzy *b*-metric space if $\exists\ \mathfrak{b}, \mathfrak{J}, z \in \mathfrak{X}$ for a given number $b \geq 1$ and $s, t > 0$ such that the conditions stated below are satisfied:
(i) $\mathfrak{H}(\mathfrak{b}, \mathfrak{J}, t) > 0$.
(ii) $\mathfrak{H}(\mathfrak{b}, \mathfrak{J}, t) = 1$ iff $\mathfrak{b} = \mathfrak{J}$.
(iii) $\mathfrak{H}(\mathfrak{b}, \mathfrak{J}, t) = \mathfrak{H}(\mathfrak{J}, \mathfrak{b}, t)$.
(iv) $\mathfrak{H}(\mathfrak{b}, z, t + s) \geq (\mathfrak{H}(\mathfrak{b}, \mathfrak{J}, \frac{t}{b})) * (\mathfrak{H}(\mathfrak{J}, z, \frac{s}{b}))$.
(v) $\mathfrak{H}(\mathfrak{b}, \mathfrak{J}, \cdot) : (0, \infty) \to [0, 1]$ is continuous.

Definition 15.3 ([5]). Let $\mathfrak{X} \neq \phi$. Consider a continuous t-norm $*$ and a fuzzy set $\mathfrak{H}$ on $\mathfrak{X}^2 \times (0, \infty)$. Then, $(\mathfrak{X}, \mathfrak{H}, *)$ is called a non-Archimedean fuzzy *b*-metric space if there exist $\mathfrak{b}, \mathfrak{J}, z \in \mathfrak{X}$ for a given number $b \geq 1$ and $s, t > 0$ such that the conditions stated below are satisfied:
(i) $\mathfrak{H}(\mathfrak{b}, \mathfrak{J}, t) > 0$.
(ii) $\mathfrak{H}(\mathfrak{b}, \mathfrak{J}, t) = 1$ iff $\mathfrak{b} = \mathfrak{J}$.
(iii) $\mathfrak{H}(\mathfrak{b}, \mathfrak{J}, t) = \mathfrak{H}(\mathfrak{J}, \mathfrak{b}, t)$.
(iv) $\mathfrak{H}(\mathfrak{b}, z, \max\{t, s\}) \geq (\mathfrak{H}(\mathfrak{b}, \mathfrak{J}, \frac{t}{b})) * (\mathfrak{H}(\mathfrak{J}, z, \frac{s}{b}))$ or $(\mathfrak{H}(\mathfrak{b}, \mathfrak{J}, \frac{t}{b})) * (\mathfrak{H}(\mathfrak{J}, z, \frac{t}{b})) \leq \mathfrak{H}(\mathfrak{b}, z, t)$.
(v) $\mathfrak{H}(\mathfrak{b}, \mathfrak{J}, \cdot) : (0, \infty) \to [0, 1]$ is continuous.

Definition 15.4 ([22]). Let $\mathfrak{X} \neq \phi$. Consider a continuous t-norm $*$ and a fuzzy set $\mathfrak{H}$ on $\mathfrak{X}^2 \times (0, \infty)$. Then, $(\mathfrak{X}, \mathfrak{H}, *)$ is called an extended fuzzy *b*-metric space if there exist $\mathfrak{b}, \mathfrak{J}, z \in \mathfrak{X}$ for $\mathbb{R} \ni b \geq 1$ and $s, t > 0$ such that the conditions stated below are satisfied:

(i) $\mathfrak{H}(\mathfrak{b}, \mathfrak{I}, t) > 0$.
(ii) $\mathfrak{H}(\mathfrak{b}, \mathfrak{I}, t) = 1$ iff $\mathfrak{A}_1 = \mathfrak{I}$.
(iii) $\mathfrak{H}(\mathfrak{b}, \mathfrak{I}, t) = \mathfrak{H}(\mathfrak{I}, \mathfrak{b}, t)$.
(iv) $\mathfrak{H}(\mathfrak{b}, z, t + s) \geq (\mathfrak{H}(\mathfrak{b}, \mathfrak{I}, \frac{t}{b})) * (\mathfrak{H}(\mathfrak{I}, z, \frac{s}{b}))$.
(v) $\mathfrak{H}(\mathfrak{b}, \mathfrak{I}, \cdot) : [0, \infty) \to (0, 1]$ is continuous.

Definition 15.5 ([5]). Let $\mathfrak{X} \neq \phi$. Consider a continuous t-norm $*$ and a fuzzy set $\mathfrak{H}$ on $\mathfrak{X}^2 \times (0, \infty)$. Then $(\mathfrak{X}, \mathfrak{H}, *)$ is known as a non-Archimedean extended fuzzy *b*-metric space if for a $\mathbb{R} \ni b \geq 1$ and $s, t > 0$, there exist $\mathfrak{b}, \mathfrak{A}_2, z \in \mathfrak{X}$ such that the conditions stated below are satisfied:
(i) $\mathfrak{H}(\mathfrak{b}, \mathfrak{I}, t) > 0$.
(ii) $\mathfrak{H}(\mathfrak{b}, \mathfrak{I}, t) = 1$ iff $\mathfrak{b} = \mathfrak{I}$.
(iii) $\mathfrak{H}(\mathfrak{b}, \mathfrak{I}, t) = \mathfrak{H}(\mathfrak{I}, \mathfrak{b}, t)$.
(iv) $\mathfrak{H}(\mathfrak{b}, z, \max\{t, s\}) \geq (\mathfrak{H}(\mathfrak{b}, \mathfrak{I}, \frac{t}{b})) * (\mathfrak{H}(\mathfrak{I}, z, \frac{s}{b}))$ or $(\mathfrak{H}(\mathfrak{b}, \mathfrak{I}, \frac{t}{b})) * (\mathfrak{H}(\mathfrak{I}, z, \frac{t}{b})) \leq \mathfrak{H}(\mathfrak{b}, z, t)$.
(v) $\mathfrak{H}(\mathfrak{b}, \mathfrak{I}, \cdot) : [0, \infty) \to (0, 1]$ is continuous.

15.3 Fuzzy *b* contraction

This section contains some definitions related to fuzzy *b* metric space.

Definition 15.6. Consider a fuzzy *b*-MS $(\mathfrak{X}, \mathfrak{Y}, *)$. A function $\mathfrak{J} : \mathfrak{X} \to \mathfrak{X}$ is called a fuzzy-interpolative Reich–Rus–Ciric-type *b*-contraction if for some $b \geq 1$ and $\alpha, \beta \in (0, 1)$ $\exists$ a constant $\lambda \in [0, 1)$ such that

$$\left[1 - \mathfrak{Y}\left(\mathfrak{J}(\mathfrak{A}_1), \mathfrak{J}(\mathfrak{A}_2), \frac{t}{b}\right)\right] \geq \lambda\left[1 - \mathfrak{Y}\left(\mathfrak{A}_1, \mathfrak{A}_2, \frac{t}{b}\right)\right]^{\beta}\left[1 - \mathfrak{Y}\left(\mathfrak{A}_1, \mathfrak{J}(\mathfrak{A}_1), \frac{t}{b}\right)\right]^{\alpha}\left[1 - \mathfrak{Y}\left(\mathfrak{A}_2, \mathfrak{J}(\mathfrak{A}_2), \frac{t}{b}\right)\right]^{1-\alpha-\beta} \quad (15.1)$$

for all $\mathfrak{A}_1, \mathfrak{A}_2 \in \mathfrak{X} - \mathrm{Fix}(\mathfrak{J})$

Definition 15.7. Consider a extended fuzzy *b*-MS $(\mathfrak{X}, \mathfrak{Y}, *)$. A function $\mathfrak{J} : \mathfrak{X} \to \mathfrak{X}$ is called a fuzzy-$\mathfrak{Y}^0$-interpolative Reich–Rus–Ciric-type *b*-contraction if it satisfy equation (15.1) for all $t \geq 0$. Again $\mathfrak{J}$ is known as fuzzy-0-interpolative Reich–Rus–Ciric-type *b*-contraction if it satisfy equation (15.1) for $t = 0$.

15.4 Theorems on fuzzy *b* contraction

Theorem 15.1. *Consider that $(\mathfrak{X}, \mathfrak{Y}, *)$ is a complete non-archimedean fuzzy b-metric space and $\mathfrak{J} : \mathfrak{X} \to \mathfrak{X}$ is a fuzzy interpolative Reich–Rus–Ciric-type b contraction. Then, $\mathfrak{J}$ admits a fixed point in $\mathfrak{X}$.*

Proof. Consider $\mathfrak{A}_0 \in \mathfrak{X}$. Let us construct a sequence $(\mathfrak{A}_n)_{n\in\mathbb{N}}$ such that $\mathfrak{A}_{n+1} = \mathfrak{I}(\mathfrak{A}_n)$. Then we have,

$$\left[1-\mathfrak{Y}\left(\mathfrak{I}(\mathfrak{A}_{n-1}),\mathfrak{I}(\mathfrak{A}_n),\frac{t}{b}\right)\right] \geq \lambda\left[1-\mathfrak{Y}\left(\mathfrak{A}_{n-1},\mathfrak{A}_n,\frac{t}{b}\right)\right]^{\beta}\left[1-\mathfrak{Y}\left(\mathfrak{A}_{n-1},\mathfrak{I}(\mathfrak{A}_{n-1}),\frac{t}{b}\right)\right]^{\alpha} \times\left[1-\mathfrak{Y}\left(\mathfrak{A}_n,\mathfrak{I}(\mathfrak{A}_n),\frac{t}{b}\right)\right]^{1-\alpha-\beta},$$

$$\begin{aligned}\left[1-\mathfrak{Y}\left(\mathfrak{A}_n,\mathfrak{I}(\mathfrak{A}_n),\frac{t}{b}\right)\right]^{\alpha+\beta} &\geq \lambda\left[1-\mathfrak{Y}\left(\mathfrak{A}_n,\mathfrak{A}_{n-1},\frac{t}{b}\right)\right]^{\beta}\left[1-\mathfrak{Y}\left(\mathfrak{A}_{n-1},\mathfrak{I}(\mathfrak{A}_{n-1}),\frac{t}{b}\right)\right]^{\alpha}\\ &= \lambda\left[1-\mathfrak{Y}\left(\mathfrak{A}_{n-1},\mathfrak{I}(\mathfrak{A}_{n-1}),\frac{t}{b}\right)\right]^{\alpha+\beta}.\end{aligned}$$

Therefore,

$$\left[1-\mathfrak{Y}\left(\mathfrak{A}_n,\mathfrak{I}(\mathfrak{A}_n),\frac{t}{b}\right)\right]^{\alpha+\beta} \geq \lambda\left[1-\mathfrak{Y}\left(\mathfrak{A}_{n-1},\mathfrak{I}(\mathfrak{A}_{n-1}),\frac{t}{b}\right)\right]^{\alpha+\beta},$$

i. e., $\{\mathfrak{Y}(\mathfrak{A}_{n-1},\mathfrak{I}(\mathfrak{A}_{n-1}),\frac{t}{b})\}$ is non-increasing.

Hence, we have

$$\left[1-\mathfrak{Y}\left(\mathfrak{A}_n,\mathfrak{I}(\mathfrak{A}_n),\frac{t}{b}\right)\right] \geq \lambda\left[1-\mathfrak{Y}\left(\mathfrak{A}_{n-1},\mathfrak{I}(\mathfrak{A}_{n-1}),\frac{t}{b}\right)\right].$$

Continuing in this manner $\forall n \geq 1$, we obtain

$$\left[1-\mathfrak{Y}\left(\mathfrak{A}_n,\mathfrak{I}(\mathfrak{A}_n),\frac{t}{b}\right)\right] \geq \lambda^n\left[1-\mathfrak{Y}\left(\mathfrak{A}_0,\mathfrak{A}_1,\frac{t}{b}\right)\right].$$

Letting the limit approach as $n \to \infty$, we obtain

$$\lim_{n\to\infty}\left[1-\mathfrak{Y}\left(\mathfrak{A}_n,\mathfrak{I}(\mathfrak{A}_n),\frac{t}{b}\right)\right] = 0 \Rightarrow \mathfrak{Y}\left(\mathfrak{A}_n,\mathfrak{I}(\mathfrak{A}_n),\frac{t}{b}\right) = 1.$$

Then from the property of 15.3, we have for $n < m$,

$$\mathfrak{Y}\left(\mathfrak{A}_n,\mathfrak{A}_m,\frac{t}{b}\right) \geq \mathfrak{Y}\left(\mathfrak{A}_n,\mathfrak{A}_{n+1},\frac{t}{b}\right) * \mathfrak{Y}\left(\mathfrak{A}_{n+1},\mathfrak{A}_{n+2},\frac{t}{b}\right) * \cdots * \mathfrak{Y}\left(\mathfrak{A}_{m-1},\mathfrak{A}_m,\frac{t}{b}\right).$$

Letting the limit approach as $n, m \to \infty$, we obtain

$$\begin{aligned}&\lim_{n,m\to\infty}\mathfrak{Y}\left(\mathfrak{A}_n,\mathfrak{A}_m,\frac{t}{b}\right)\\ &\geq \lim_{n\to\infty}\mathfrak{Y}\left(\mathfrak{A}_n,\mathfrak{A}_{n+1},\frac{t}{b}\right) * \lim_{n\to\infty}\mathfrak{Y}\left(\mathfrak{A}_{n+1},\mathfrak{A}_{n+2},\frac{t}{b}\right) * \cdots * \lim_{m\to\infty}\mathfrak{Y}\left(\mathfrak{A}_{m-1},\mathfrak{A}_m,\frac{t}{b}\right)\\ &\geq 1 * 1 * \cdots * 1\\ &\geq 1.\end{aligned}$$

Therefore,

$$\lim_{n,m\to\infty} \mathfrak{Y}\left(\mathfrak{A}_n, \mathfrak{A}_m, \frac{t}{b}\right) = 1.$$

Thus the sequence $(\mathfrak{A}_n)$ is Cauchy in $(\mathfrak{X}, \mathfrak{Y}, *)$.

Since $(\mathfrak{X}, \mathfrak{Y}, *)$ is a complete non-archimedean fuzzy b-metric space, therefore $(\mathfrak{A}_n)$ is a convergent sequence and let it converges to $\mathfrak{A}^*$ as $n \to \infty$.

Next, for all $n \in \mathbb{N}$, we have

$$\begin{aligned}&\left[1 - \mathfrak{Y}\left(\mathfrak{J}(\mathfrak{A}_n), \mathfrak{J}(\mathfrak{A}^*), \frac{t}{b}\right)\right]\\&\quad \geq \lambda\left[1 - \mathfrak{Y}\left(\mathfrak{A}_n, \mathfrak{A}^*, \frac{t}{b}\right)\right]^{\beta}\left[1 - \mathfrak{Y}\left(\mathfrak{A}_n, \mathfrak{J}(\mathfrak{A}_n), \frac{t}{b}\right)\right]^{\alpha}\left[1 - \mathfrak{Y}\left(\mathfrak{A}^*, \mathfrak{J}(\mathfrak{A}^*), \frac{t}{b}\right)\right]^{1-\alpha-\beta}.\end{aligned}$$

Taking limit as $n \to \infty$, we have

$$\begin{aligned}&\left[1 - \mathfrak{Y}\left(\mathfrak{J}(\mathfrak{A}^*), \mathfrak{J}(\mathfrak{A}^*), \frac{t}{b}\right)\right]\\&\quad \geq \lambda\left[1 - \mathfrak{Y}\left(\mathfrak{A}^*, \mathfrak{A}^*, \frac{t}{b}\right)\right]^{\beta}\left[1 - \mathfrak{Y}\left(\mathfrak{A}^*, \mathfrak{J}(\mathfrak{A}^*), \frac{t}{b}\right)\right]^{\alpha}\left[1 - \mathfrak{Y}\left(\mathfrak{A}^*, \mathfrak{J}(\mathfrak{A}^*), \frac{t}{b}\right)\right]^{1-\alpha-\beta}.\end{aligned}$$

Therefore,

$$\left[1 - \mathfrak{Y}\left(\mathfrak{A}^*, \mathfrak{J}(\mathfrak{A}^*), \frac{t}{b}\right)\right] = 0.$$

Implies $\mathfrak{J}(\mathfrak{A}^*) = \mathfrak{A}^*$.

Hence $\mathfrak{J}$ possesses a fixed point $\mathfrak{A}^*$ in $(\mathfrak{X}, \mathfrak{Y}, *)$. □

Theorem 15.2. *Consider that $(\mathfrak{X}, \mathfrak{Y}^{\circ}, *)$ is a complete non-archimedean extended fuzzy b-metric space and $\mathfrak{J} : \mathfrak{X} \to \mathfrak{X}$ is a fuzzy $\mathfrak{Y}^{\circ}$ interpolative Reich–Rus–Ciric-type b contraction. Then, $\mathfrak{J}$ admits a fixed point in $\mathfrak{X}$.*

Proof. If we consider $t > 0$, the proof proceeds along the same lines as Theorem 15.1.

Next, consider $t = 0$.

Let $\mathfrak{A}_0 \in \mathfrak{X}$ and for all $n \in \mathbb{N}$, $(\mathfrak{A}_n)$ be a sequence such that $\mathfrak{A}_{n+1} = \mathfrak{J}(\mathfrak{A}_n)$. Then, we have

$$\begin{aligned}[1 - \mathfrak{Y}^{\circ}(\mathfrak{J}(\mathfrak{A}_{n-1}), \mathfrak{J}(\mathfrak{A}_n), 0)] &\geq \lambda[1 - \mathfrak{Y}^{\circ}(\mathfrak{A}_{n-1}, \mathfrak{A}_n)]^{\beta}[1 - \mathfrak{Y}^{\circ}(\mathfrak{A}_{n-1}, \mathfrak{J}(\mathfrak{A}_{n-1}))]^{\alpha}\\&\quad \times [1 - \mathfrak{Y}^{\circ}(\mathfrak{A}_n, \mathfrak{J}(\mathfrak{A}_n))]^{1-\alpha-\beta},\\ [1 - \mathfrak{Y}^{\circ}(\mathfrak{A}_n, \mathfrak{J}(\mathfrak{A}_n))]^{\alpha+\beta} &\geq \lambda[1 - \mathfrak{Y}^{\circ}(\mathfrak{A}_n, \mathfrak{A}_{n-1})]^{\beta}[1 - \mathfrak{Y}^{\circ}(\mathfrak{A}_{n-1}, \mathfrak{J}(\mathfrak{A}_{n-1}))]^{\alpha}\\&= \lambda[1 - \mathfrak{Y}^{\circ}(\mathfrak{A}_{n-1}, \mathfrak{J}(\mathfrak{A}_{n-1}))]^{\alpha+\beta}.\end{aligned}$$

Therefore,

$$[1 - \mathfrak{Y}^{\circ}(\mathfrak{A}_n, \mathfrak{I}(\mathfrak{A}_n))] \geq \lambda[1 - \mathfrak{Y}^{\circ}(\mathfrak{A}_{n-1}, \mathfrak{I}(\mathfrak{A}_{n-1}))],$$

i. e., $\{\mathfrak{Y}^{\circ}(\mathfrak{A}_n, \mathfrak{A}_{n+1}))\}$ is non-increasing.

Proceeding this way, we have $\forall n \geq 1$,

$$[1 - \mathfrak{Y}^{\circ}(\mathfrak{A}_n, \mathfrak{A}_{n+1})] \geq \lambda^n [1 - \mathfrak{Y}^{\circ}(\mathfrak{x}_0, \mathfrak{A}_1)].$$

By letting the limit approach as $n \to \infty$, we have

$$\lim_{n\to\infty} [1 - \mathfrak{Y}^{\circ}(\mathfrak{A}_n, \mathfrak{A}_{n+1})] = 0 \Rightarrow \mathfrak{Y}^{\circ}(\mathfrak{A}_n, \mathfrak{A}_{n+1}) = 1.$$

Then, from the property of 15.5, we have for $n < m$

$$\mathfrak{Y}^{\circ}(\mathfrak{A}_n, \mathfrak{A}_m) \geq \mathfrak{Y}^{\circ}(\mathfrak{A}_n, \mathfrak{A}_{n+1}) * \mathfrak{Y}^{\circ}(\mathfrak{A}_{n+1}, \mathfrak{A}_{n+2}) * \cdots * \mathfrak{Y}^{\circ}(\mathfrak{A}_{m-1}, \mathfrak{A}_m).$$

Letting the limit approach as $n, m \to \infty$,

$$\begin{aligned}
&\lim_{n,m\to\infty} \mathfrak{Y}^{\circ}(\mathfrak{A}_n, \mathfrak{A}_m) \\
&\quad \geq \lim_{n\to\infty} \mathfrak{Y}^{\circ}(\mathfrak{A}_n, \mathfrak{A}_{n+1}) * \lim_{n\to\infty} \mathfrak{Y}^{\circ}(\mathfrak{A}_{n+1}x, \mathfrak{A}_{n+2}) * \cdots * \lim_{m\to\infty} \mathfrak{Y}^{\circ}(\mathfrak{A}_{m-1}, \mathfrak{A}_m) \\
&\quad \geq 1 * 1 * \cdots * 1 \\
&\quad \geq 1.
\end{aligned}$$

Therefore,

$$\lim_{n,m\to\infty} \mathfrak{Y}^{\circ}(\mathfrak{A}_n, \mathfrak{A}_m) = 1.$$

Hence the sequence $(\mathfrak{A}_n)$ is Cauchy in $(\mathfrak{X}, \mathfrak{Y}^{\circ}, *)$

Since $(\mathfrak{X}, \mathfrak{Y}^{\circ}, *)$ is a complete non-Archimedean extended fuzzy b-metric space, therefore $(\mathfrak{A}_n)$ is a convergent sequence and let it converges to $\mathfrak{A}^*$ as $n \to \infty$.

Next, we proceed to show that $\mathfrak{A}^*$ satisfies the fixed-point condition for $\mathfrak{I}$.

For all $n \in \mathbb{N}$,

$$\begin{aligned}
&[1 - \mathfrak{Y}^{\circ}(\mathfrak{I}(\mathfrak{A}_n), \mathfrak{I}(\mathfrak{A}^*))] \\
&\quad \geq \lambda[1 - \mathfrak{Y}^{\circ}(\mathfrak{A}_n, \mathfrak{A}^*)]^{\beta} [1 - \mathfrak{Y}^{\circ}(\mathfrak{A}_n, \mathfrak{I}(\mathfrak{A}_n))]^{\alpha} [1 - \mathfrak{Y}^{\circ}(\mathfrak{A}^*, \mathfrak{I}(\mathfrak{A}^*))]^{1-\alpha-\beta}.
\end{aligned}$$

Letting the limit approach as $n \to \infty$, we have

$$\begin{aligned}
&[1 - \mathfrak{Y}^{\circ}(\mathfrak{I}(\mathfrak{A}^*), \mathfrak{I}(\mathfrak{A}^*))] \\
&\quad \geq \lambda[1 - \mathfrak{Y}^{\circ}(\mathfrak{A}^*, \mathfrak{A}^*)]^{\beta} [1 - \mathfrak{Y}^{\circ}(\mathfrak{A}^*, \mathfrak{I}(\mathfrak{A}^*))]^{\alpha} [1 - \mathfrak{Y}^{\circ}(\mathfrak{A}^*, \mathfrak{I}(\mathfrak{A}^*))]^{1-\alpha-\beta}.
\end{aligned}$$

Therefore,

$$[1 - \mathfrak{Y}^{\circ}(\mathfrak{A}^*, \mathfrak{J}(\mathfrak{A}^*))] = 0.$$

Implies $\mathfrak{J}(\mathfrak{A}^*) = \mathfrak{A}^*$.
Hence, $\mathfrak{J}$ possesses a fixed point $\mathfrak{A}^*$ in $(\mathfrak{X}, \mathfrak{Y}^{\circ}, *)$. □

15.5 Conclusion

Within this chapter, we explored the definitions for fuzzy-interpolative Reich–Rus–Ciric-type b-contraction and fuzzy-0-interpolative Reich–Rus–Ciric-type b-contraction. To support these definitions, we establish extended versions of fixed-point theorems within the framework of non-Archimedean fuzzy-b-metric spaces, specifically tailored to these types of fuzzy interpolative contractions. This study lays the groundwork for future research, which may explore new fixed-point results involving other forms of contractions in fuzzy b-metric space.

Bibliography

[1] Alharbi, N., Aydi, H., Felhi, A., Ozel, C. and Sahmim, S., *α-contractive mappings on rectangular b-metric spaces and an application to integral equations*, J. Math. Anal. 9(3) (2018), 47–60.

[2] Aydi, H., Chen, C. M. and Karapinar, E., *Interpolative Ćirić–Reich–Rus type contractions via the Branciari distance*, Mathematics 7 (2019), 84.

[3] Bakhtin, I. A., *The contraction mapping principle in almost metric spaces*, Funct. Anal. 30 (1989), 26–37.

[4] Boyd, D. W. and Wong, J. S., *On nonlinear contractions*, Proc. Am. Math. Soc. 20 (1969), 458–464.

[5] Bozdemir, M., *Fixed-point theorems in some Fuzzy Metric Spaces via interpolative contractions*, J. Univers. Math. 8(1) (2025), 33–39.

[6] Chatterjea, S. K., *Fixed-point theorems*, C. R. Acad. Bulgare Sci. 25(6) (1972), 727–730.

[7] Ćirić, L., *Fixed-point theorems for multivalued contractions in complete metric spaces*, J. Math. Anal. Appl. 348 (2008), 499–507.

[8] Czerwik, S., *Contraction mappings in b-metric spaces*, Acta Math. Univ. Ostrav. 1(1) (1993), 5–11.

[9] Debnath, P., *Optimization through best proximity points for multivalued F-contractions*, Miskolc Math. Notes 22(1) (2021), 143–151.

[10] Debnath, P., *Set-valued Meir–Keeler, Geraghty and Edelstein type fixed point results in b-metric spaces*, Rend. Circ. Mat. Palermo (2) 70(3) (2021), 1389–1398.

[11] Debnath, P., *Banach, Kannan, Chatterjea, and Reich-type contractive inequalities for multivalued mappings and their common fixed points*, Math. Methods Appl. Sci. 45(3) (2022), 1587–1596.

[12] Debnath, P., *New common fixed-point theorems for Gornicki-type mappings and enriched contractions*, São Paulo J. Math. Sci. 16(2) (2022), 1401–1408.

[13] Debnath, P., *Common fixed-point and fixed-circle results for a class of discontinuous f-contractive mappings*, Mathematics 10(9) (2022), 1605.

[14] Debnath, P., *A new extension of Kannan's fixed-point theorem via F-contraction with application to integral equations*, Asian-Eur. J. Math. 15(07) (2022), 2250123.

[15] Debnath, P., *Results on discontinuity at fixed point for a new class of contractive mappings*, Sahand Commun. Math. Anal. 20(4) (2023), 21–32.

[16] Debnath, P., *Best proximity points of multivalued Geraghty contractions*, Miskolc Math. Notes 24(1) (2023), 119–127.

[17] Debnath, P., *On some open questions about multivalued enriched interpolative Kannan and Ciric–Reich–Rus type contractions in a Banach space*, Fixed Point Theory 26(2) (2025), 465–476.

[18] Debnath, P., Konwar, N. and Radenović, S., *Metric Fixed Point Theory*, Springer, Singapore, Germany, 2021.

[19] Debnath, P., Mitrović, Z. and Radenović, S., *Interpolative Hardy–Rogers and Reich–Rus–Ćirić type contractions in b-metric spaces and rectangular b-metric spaces*, Mat. Vesn. 72 (2020), 368–374.

[20] Ding, H. S., Imdad, M., Radenovic, S. and Vujakovic, J., *On some fixed point results in b-metric, rectangular and b-rectangular metric spaces*, Arab J. Math. Sci. 22(2) (2016), 151–164.

[21] George, A. and Veeramani, P., *On some result in fuzzy metric spaces*, Fuzzy Sets Syst. 64 (1994), 359–399.

[22] Gregori, V., Minana, J. J. and Miravet, D., *Extended fuzzy metrics and fixed-point theorems*, Math. J. 7 (2019), 303.

[23] Heilpern, S., *Fuzzy mappings and fixed-point theorem*, J. Math. Anal. Appl. 83(2) (1981), 566–569.

[24] Kaleva, O. and Seikkala, S., *A Banach contraction theorem in fuzzy metric space*, Fuzzy Sets Syst. 12 (1984), 215–229.

[25] Kannan, R., *Some results on fixed points*, Bull. Calcutta Math. Soc. 60(1) (1968), 71–77.

[26] Kramosil, I. and Michalek, J., *Fuzzy metric and statistical metric spaces*, Kybernetika 11 (1975), 326–334.

[27] Meir, A. and Keeler, E., *A theorem on contraction mappings*, J. Math. Anal. Appl. 28 (1969), 326–329.

[28] Rhoades, B. E., *A comparison of various definitions of contractive mappings*, Trans. Am. Math. Soc. 226 (1977), 257–290.

[29] Wardowski, D., *Fixed points of a new type of contractive mappings in complete metric spaces*, Fixed Point Theory Appl. 94(1) (2012), 1–6.

[30] Zadeh, L. A., *Fuzzy sets*, Inf. Control 8 (1965), 338–353.

Talat Nazir and Mujahid Abbas

16 Fixed points of set-valued generalized rational graphic ϕ-contractive operators in semi-metric spaces

Abstract: In this chapter, we study the fixed point results of set-valued generalized rational graphic ϕ-contractive operators in the setup of semi-metric spaces equipped with a directed graph. Examples are presented for support the conclusions developed herein. The results presented their in improve and extend several comparable results in the current literature.

Keywords: Set-valued operator, fixed-point, semi-metric space, generalized graphic contraction

MSC 2020: 47H10, 47H09, 54H25, 54C60

16.1 Introduction

Metric fixed-point theory is useful and valuable way for addressing the wide range of scientific problems. It has extensive applications across various branches of mathematics, including the problems dealing with the presence solutions of differential and integral equations, equilibrium feasibility and optimization-based problems.

Fréchet [7] initially studied the notion of a distance between two abstract objects of a non-empty set Y that makes Hausdorff topology on Y. Subsequently, several generalizations of metric functions were developed. One of the significant generalizations concept is that of semi-metric (also known as symmetric metric) space, which generates a topology that is not necessarily Hausdorff. Zhu et al. [21] explored equal level contraction condition of maps in the symmetric space, while Imdad [9] established fixed points of maps satisfy the strict contraction conditions in such spaces. For more important results in the way, we mention the work in [2–4, 6, 19].

Note that there are various extensions and restrictions of the definition of a metric. We, in this chapter, are the Hausdorff semi-metric d_s on a non-empty set Y. The primary restriction of this semi-metric is that it does not satisfy the triangle inequality. We obtained fixed points of generalized rational graphic ϕ-contractive maps on the bounded and closed sets in the semi-metric space (Y, d_s).

Talat Nazir, Department of Mathematical Sciences, University of South Africa, Florida 0003, South Africa, e-mail: talatn@unisa.ac.za
Mujahid Abbas, Department of Mechanical Engineering Science, Faculty of Engineering and the Built Environment, University of Johannesburg, Johannesburg 2092, South Africa, e-mail: mujahida@uj.ac.za

https://doi.org/10.1515/9783112215760-016

Throughout this work, the collection of positive real numbers, non-negative real numbers, and natural numbers will be denoted by $\mathbb{R}^+$, $\mathbb{R}_+$, and $\mathbb{N}$, respectively. First, we give preliminaries results needed in the sequel.

16.2 Preliminaries

We recall the basic notions and results related to semi metric spaces.

Definition 16.1 (Semi-metric space). [20] For nonempty set Y, a map $\mathrm{d}_s : Y \times Y \to \mathbb{R}_+$ is called (Hausdorff) semi-metric on Y if for all $\chi_1, \chi_2 \in Y$, the following hold:
(1) $\mathrm{d}_s(\chi_1, \chi_2) \geq 0$ and equal to zero if and only if $\chi_1 = \chi_2$.
(2) $\mathrm{d}_s(\chi_2, \chi_1) = \mathrm{d}_s(\chi_1, \chi_2)$.

Pair (Y, d_s) is known as (Hausdorff) semi-metric space.

Example 16.1 ([12]). For $Y = \mathbb{R}_+$, a semi-metric $d_s : Y \times Y \to \mathbb{R}_+$ is given as

$$\begin{aligned} d_s(\chi_1, \chi_2) &= |\chi_1 - \chi_2| \quad \text{for all } \chi_1, \chi_2 \in Y \backslash 2, \\ d_s(2, \chi_1) &= d_s(\chi_1, 2) = \frac{\chi_1 + 2}{4} \quad \text{if } \chi_1 \neq 2, \quad \text{and} \\ d_s(2, 2) &= 0. \end{aligned}$$

Clearly, d_s is not metric. Indeed, we have $d_s(0, 3) \nleq d_s(0, 2) + d_s(3, 2)$.

Consider a semi-metric space (Y, d_s), $\chi \in Y$ and $\varrho > 0$. An open ball, centered at χ with radius ϱ is defined as $\mathrm{B}(\chi, \varrho) = \{x \in Y : \mathrm{d}_s(\chi, x) < \varrho\}$. A topology τ_{d_s} on Y is given by the family of sets as

$$\tau_{\mathrm{d}_s} = \{\mathcal{U} \subseteq Y : \forall \chi \in \mathcal{U}\ \exists\, \varrho \in \mathbb{R}_+ \text{ such that } B(\chi, \varrho) \subseteq \mathcal{U}\}.$$

Definition 16.2 ([20]). If $\{\chi_n\}$ a sequence in semi-metric space, then for $\chi \in Y$ and $\varrho > 0$, open ball $\mathrm{B}(\chi, \varrho)$ is the neighborhood of χ having topology τ_{d_s}. And $\lim_{n\to\infty} \mathrm{d}_s(\chi_n, \chi) = 0$, whenever $\{\chi_n\}$ converges to χ in τ_{d_s} and conversely.

The following are some important properties, which play a vital role in proving the conclusions in semi-metric space. If $\{u_n\}$, $\{v_n\}$ and $\{w_n\}$ are sequences in semi-metric space (Y, d_s) and $u, v \in Y$. Then,
(1) $\lim_{k\to\infty} \mathrm{d}_s(u_k, u) = 0$ and $\lim_{k\to\infty} \mathrm{d}_s(u_k, v) = 0 \Longrightarrow u = v$.
(2) $\lim_{k\to\infty} \mathrm{d}_s(u_k, u) = 0$ and $\lim_{k\to\infty} \mathrm{d}_s(u_k, v_k) = 0 \Longrightarrow \lim_{k\to\infty} \mathrm{d}_s(v_k, u) = 0$.
(3) $\lim_{k\to\infty} \mathrm{d}_s(u_k, v_k) = 0$ and $\lim_{k\to\infty} \mathrm{d}_s(v_k, w_k) = 0 \Longrightarrow \lim_{k\to\infty} d(u_k, w_k) = 0$.
(4) $\lim_{k\to\infty} \mathrm{d}_s(u_k, v_k) = 0$ and $\lim_{k\to\infty} \mathrm{d}_s(v_k, w_k) = 0 \Longrightarrow \lim_{k\to\infty} \mathrm{d}_s(u_k, w_k) < \infty$.
(5) $\lim_{k\to\infty} \mathrm{d}_s(u_k, u) = 0 \Longrightarrow \lim_{k\to\infty} \mathrm{d}_s(u_k, v) = \mathrm{d}_s(u, v)$.

Below, we present the definition of $\{u_n\}$ Cauchy sequence in the semi-metric space.

Definition 16.3 ([10]). Let $\{u_n\}$ be a sequence in the semi-metric space (Y, d_s).

(1) If for any given $\epsilon > 0$, we have $n(\epsilon) \in \mathbb{N}$ such that $\mathrm{d}_s(u_n, u_m) < \epsilon$ for all $n, m \geq n(\epsilon)$, then $\{u_n\}$ is known as d_s-Cauchy sequence.
(2) (Y, d_s) is called S-complete, if for each d_s-Cauchy sequence $\{u_n\}$ in Y, we have an element $u \in Y$ such that $\lim_{n\to\infty} \mathrm{d}_s(u_n, u) = 0$.
(3) (Y, d_s) is called d_s-Cauchy complete, if every d_s-Cauchy sequence $\{u_n\}$ in Y converges to point $u \in Y$.

For non-empty set Y, we then define the following subsets:

$$\begin{aligned} N(Y) &= \{\mathcal{U} : \mathcal{U} \text{is subset of } Y \text{ such that } \mathcal{U} \neq \emptyset\}, \\ B(Y) &= \{\mathcal{U} : \mathcal{U} \text{ is bounded subset of } Y \text{ such that } \mathcal{U} \neq \emptyset\}, \\ Cl(Y) &= \{\mathcal{U} : \mathcal{U} \text{ is closed subset of } Y \text{ such that } \mathcal{U} \neq \emptyset\}, \\ CB(Y) &= \{\mathcal{U} : \mathcal{U} \text{ is bounded closed subset of } Y \text{ such that } \mathcal{U} \neq \emptyset\}. \end{aligned}$$

Definition 16.4. In semi-metric space (Y, d_s), for $\mathcal{A}, \mathcal{B} \in CB(Y)$, define the map $\mathcal{H}_s : CB(Y) \times CB(Y) \to \mathbb{R}_+$ as

$$\mathcal{H}_s(\mathcal{A}, \mathcal{B}) = \max\{\mathcal{D}(\mathcal{A}, \mathcal{B}), \mathcal{D}(\mathcal{B}, \mathcal{A})\},$$

where $\mathcal{D}(\mathcal{A}, \mathcal{B}) = \sup\{\mathrm{d}_s(\mathcal{B}, a) : a \in \mathcal{A}\}$ and $\mathrm{d}_s(x, \mathcal{A}) = \inf\{\mathrm{d}_s(x, a) : a \in \mathcal{A}\}$. The mapping $\mathcal{H}_s$ is called the Pompeiu–Hausdorff (PH)-semi-metric induced by d_s.

Definition 16.5. In semi-metric space (Y, d_s), consider $\emptyset \neq U \subseteq Y$. The set U is known as d_s-closed, if $U = \overline{U}$, where $\overline{U} = \{u \in Y : \mathrm{d}_s(u, U) = 0\}$.

Clearly, whenever (Y, d_s) is the d_s-Cauchy complete semi-metric space, the space $(CB(Y), \mathcal{H}_s)$ inherits this completeness and remains a d_s-Cauchy complete semi-metric space.

Definition 16.6. A semi-metric space (Y, d_s) is known as the ϵ-chainable space for given $\epsilon > 0$, if for $u, v \in Y$, we have $n \in \mathbb{N}$ with a sequence $\{y_n\}$ having

$$y_0 = u, \quad y_n = v \quad \text{and} \quad \mathrm{d}_s(y_{i-1}, y_i) < \epsilon \quad \text{for } i = 1, \dots, n.$$

Analogous results to that of Nadler's lemmas as in [16] are now stated for the semi-metric space.

Lemma 16.1. *Let (Y, d_s) be semi-metric space with $\mathcal{V}, \mathcal{U} \in CB(Y)$. Then $\mathrm{d}_s(\mathcal{V}, u) \leq \mathcal{H}_s(\mathcal{V}, \mathcal{U})$, whenever $u \in \mathcal{V}$.*

Proof. If $u \in \mathcal{U}$, then

$$\mathrm{d}_s(\mathcal{V}, u) = \inf_{v \in \mathcal{V}} \mathrm{d}_s(u, v) \leq \sup_{u \in \mathcal{U}} \mathrm{d}_s(\mathcal{V}, u) \leq \mathcal{H}_s(\mathcal{V}, \mathcal{U}).$$ □

Lemma 16.2 ([12]). *Let (Y, d_s) be a semi-metric space with $\mathcal{V}, \mathcal{U} \in CB(Y)$. Then for any $u \in \mathcal{U}$, we have $v \in \mathcal{V}$ such that for any $\mu > 1$, we have $\mathrm{d}_s(v, u) \leq \mu\mathcal{H}_s(\mathcal{V}, \mathcal{U})$.*

Lemma 16.3 ([10]). *If $\mathcal{U}, \mathcal{V} \in CB(Y)$ with $\mathcal{H}_s(\mathcal{U}, \mathcal{V}) < \epsilon$, then for each $u \in \mathcal{U}$, we have $v \in \mathcal{V}$ such that $\mathrm{d}_s(v, u) < \epsilon$.*

Definition 16.7. Let (Y, d_s) be a semi-metric space. A mapping $f : Y \to Y$ is known as contraction, if $\kappa \in [0, 1)$ exists such that

$$\mathrm{d}_s(fx_1, fx_2) \leq \kappa \mathrm{d}_s(x_1, x_2) \quad \text{for all } x_1, x_2 \in Y.$$

We say the semi-metric space (Y, d_s) is bounded, when a real number $\mathcal{M}$ exists having $\mathcal{M} < \infty$, where $\mathcal{M} = \sup\{\mathrm{d}_s(u_1, u_2) : u_1, u_2 \in Y\}$.

Lemma 16.4 ([10]). *Let (Y, d_s) be a bounded d_s-Cauchy complete semi-metric space and $f : Y \to Y$ a contraction. Then f has at most one fixed point $y^* \in Y$. Also, for each $y_0 \in Y$, the iterative sequence $\{y_0, fy_0, f^2y_0, \ldots\}$ converges to y^*.*

16.3 Graphic contractive maps

Jachymski in [11] studied fixed point of maps by introducing a graph concept on a metric space. Ran and Reurings' result in [18] was the main motivation for such an effort towards generalization of the order or poset construction of a metric space. The usefulness of imposing a graph structure on (Y, d_s) lies in the fact that it generalizes the ordered structure on (Y, d_s). One the significant results in fixed-point theory with a poset structure was obtained by Ran and Reurings in [18]. The latter result was the main thrust in the generalization of equipping spaces with a graph structure which was motivated by Jachymski in [11]. Latif et al. [13] derived fixed point theorems for a class of set-contractive mappings on spaces endowed with a directed graph. In this chapter, we present several fixed-point theorems for generalized rational graphic contractive maps in semi-metric spaces endow with directed graph structure. Our results not only unify but also extend a number of existing findings in [1, 7, 8, 14, 15].

Below, we discuss the mechanism to identify a semi-metric space having graph G.

Definition 16.8. Let Y be non-empty set. The directed graph $G = (V(G), E(G))$ is identify as follows:

(1) $Y = V(G)$, where $V(G)$ is the vertex set.
(2) $\Delta \subseteq E(G)$, where $\Delta \subseteq Y \times Y$ denotes the diagonal and $E(G)$ represents the edge set.
(3) Parallel edges along with different vertices are not allowed.

Given that $V(G) = Y$, we will use the terms $V(G)$ and Y interchangeably throughout the paper.

The semi-metric space (Y, d_s) is associated with the directed graph G, as described in Definition 16.8, where G is a weighted directed graph. Specifically, for each $(u, v) \in E(G)$, we assigning a weight $\mathrm{d}_s(v_1, v_2)$, with loops satisfying $\mathrm{d}_s(u, u) = 0$ for all $u \in V(G)$. Furthermore, the PH-semi-metric induced by d_s assigns a positive PH-weight to every pair $U, V \in CB(Y)$ with $\mathcal{H}_s(U, V) = 0$, if and only if $U = V$. A (directed) path of length $l \in \mathbb{N}$ between vertices, $u, v \in V(G)$ is defined as a finite sequence $\{y_n\}$ for $n = 0, \ldots, l$ of vertices such that $u = y_0$, $v = y_l$, $(y_{i-1}, y_i) \in E(G)$ for all $i \in \{1, \ldots, l\}$. The graph G is known as connected when we have directed path exists in between every two vertices. In case underlying undirected graph (formed by ignoring edge directions) is connected, then G is referred as weakly connected. Inverse graph G^{-1} is defined by reversing the direction of each edge in G, i. e., $(v, u) \in E(G^{-1})$ if $(u, v) \in E(G)$ and conversely. We use notation $Y_f = \{y \in Y : (y, fy) \in E(G)\}$.

Definition 16.9 ([11]). Suppose (Y, d) is a metric space. Map $f : Y \to Y$ is called Banach G-contraction if

(1) f conserves edges of graph G, i. e., if $(u, v) \in E(G)$, then $(fu, fv) \in E(G)$ for all $u, v \in V(G) = Y$.
(2) f shrinks the weights of edges of graph G, i. e., there exits $k \in (0, 1)$ such that for every $u, v \in V(G)$ and $(u_1, v_1) \in E(G)$ implies $d(fu_1, fv_1) \leq kd(u_1, v_1)$.

Jachymski in [11] utilized the property $\mathcal{P}$ given as follows: A sequence $\{y_n\}$ in Y, if $\lim_{n\to\infty} y_n = y$ and also $(y_{n+1}, y_n) \in E(G)$, then $(y_n, y) \in E(G)$.

Theorem 16.1 ([11]). *Suppose (Y, d) is complete metric space having directed graph G with property $\mathcal{P}$ and G-contraction $f : Y \to Y$. Then,*

(1) *$Y_f \neq \emptyset$ iff admits a fixed point and conversely.*
(2) *f is a Picard operator, if only if G is weakly connected and $Y_f \neq \emptyset$.*
(3) *$f|_{[y]_{\tilde{G}}}$ is a Picard operator for every $y \in Y_f$.*
(4) *If f is a weakly Picard operator, then $Y_f \times Y_f \subseteq E(G)$.*

Restricting the topological structure of the domain, extending the range, or generalizing the contractive condition of the mappings have all been used to extend the Banach contraction principle in one way or other. Nadler in [16] used PH-metric to obtain the multi-valued analogue of the Banach contraction by change the map's range by $CB(Y)$, for some arbitrary set Y.

Definition 16.10 ([16]). Consider the two metric spaces (X, d_1) and (Y, d_2). Map $f : X \to CB(Y)$ is known as set-valued Lipschitz mapping of X into Y if $v, z \in X$,

$$H(fz, fv) \leq ad_1(z, v),$$

where $\alpha \geq 0$ and H is the PH-metric. If $\alpha < 1$, then f is referred to as a set-valued contraction.

Theorem 16.2 ([16]). *In a complete metric space (X,d), if $f : X \to CB(X)$ is a set-valued contraction, then f possess a fixed point.*

Definition 16.11. Suppose that $\emptyset \neq V, W \subseteq Y$. Then,

a) When we state that there exists an edge between V and W, we mean that there exists $v \in V$ and $w \in W$ such that $(v,w) \in E(G)$; this relationship may be denoted as $(V,W) \subseteq E(G)$.

b) When we refer to a path between V and W, we are indicating the existence of a path in G from some $w \in W$ to some $v \in V$.

We present a relation R on $CB(Y)$ as follows: for URV for $U, V \in CB(Y)$, we say URV if there exists a path connecting U and V. The relation R is said to be transitive if, whenever there is path from U to V and a path from V to W, there also exists a path from U to W. A set C in $CB(Y)$ is called complete if every pair of elements U, V in C is connected by an edge. An equivalence class of an element $\mathcal{U} \in CB(Y)$ under the relation R is given as follows:

$$[\mathcal{U}]_G = \{X \in CB(Y) : \mathcal{U}RX\}.$$

Instead of taking the mappings $Y \to Y$ or $Y \to CB(Y)$, we are taking $T : CB(Y) \to CB(Y)$.

For map $T : CB(Y) \to CB(Y)$, a set $U^* \in CB(Y)$ is called fixed point of T if $T(U^*) = U^*$. And with $F(T)$, we denote the collection of all fixed point of T.

The set Y_T for map $T : CB(Y) \to CB(Y)$ is described as

$$Y_T = \{\mathcal{X} \in CB(Y) : (\mathcal{X}, T(\mathcal{X})) \subseteq E(G)\}.$$

Definition 16.12 ([1]). A graph G has the property $\overset{*}{\mathcal{P}}$ if, for a sequence $\{U_n\}$ in $CB(Y)$, $\lim_{n\to\infty} U_n = U$ and $(U_n, U_{n+1}) \in E(G)$ for $n \in \mathbb{N}$, implies we have a subsequence $\{U_{n_k}\}$ of $\{U_n\}$ with $(U_{n_k}, U) \in E(G)$ for $n \in \mathbb{N}$.

We present generalized graphic contractive condition that will be central to our results dealing with *generalized rational graphic ϕ-contractive mapping.*

Definition 16.13. In a semi-metric space (Y, d_s), consider $T : CB(Y) \to CB(Y)$ a set-valued mapping. Then T is said to be a generalized rational graphic ϕ-contractive mapping, if the following hold:

(1) An edge (path) between V and U implies the existence of a corresponding path (or edge) between $T(V)$ and $T(U)$ for all $V, U \in CB(Y)$.

(2) We have a non-decreasing upper semi-continuous map $\phi : \mathbb{R}_+ \to \mathbb{R}_+$ having $\phi(s) < s$ for each $s > 0$ along with $\sum_{k=0}^{\infty} \phi^k(s)$ converges such that for all $\chi_1, \chi_2 \in CB(Y)$ having $(\chi_1, \chi_2) \subseteq E(G)$, the following inequality holds:

$$\mathcal{H}_s(T(\chi_1), T(\chi_2)) \leq \phi(M_T(\chi_1, \chi_2)),$$

where

$$M_T(\chi_1,\chi_2) = \max\Bigg\{\mathcal{H}_s(\chi_1,\chi_2), \mathcal{H}_s(T(\chi_1),\chi_1), \mathcal{H}_s(\chi_2, T(\chi_2)), \frac{\mathcal{H}_s(\chi_2, T(\chi_2))[1+\mathcal{H}_s(\chi_1, T(\chi_1))]}{1+\mathcal{H}_s(\chi_1,\chi_2)}, \frac{\mathcal{H}_s(\chi_2, T(\chi_1))[1+\mathcal{H}_s(\chi_1, T(\chi_1))]}{1+\mathcal{H}_s(\chi_1,\chi_2)}, \frac{\mathcal{H}_s(\chi_1, T(\chi_1))[1+\mathcal{H}_s(\chi_1, T(\chi_1))]}{1+\mathcal{H}_s(\chi_1,\chi_2)}\Bigg\}.$$

16.4 Fixed-point results

Theorem 16.3. *Let (Y, d_s) be a d_s-Cauchy complete semi-metric space endowed with directed graph, having $V(G) = Y$ and $\Delta \subseteq E(G)$. If $T : CB(Y) \to CB(Y)$ is the generalized rational graphic ϕ-contractive mapping, then following holds:*

(1) *If $F(T) \neq \emptyset$ is complete, then PH-weight assigned to $U, V \in F(T)$ is 0.*
(2) *If $F(T) \neq \emptyset$, then $Y_T \neq \emptyset$.*
(3) *If $Y_T \neq \emptyset$ with the weakly connected graph $\widetilde{G}$ satisfies $\overset{*}{\mathcal{P}}$, then T has a fixed point.*
(4) *$F(T)$ is a singleton if $F(T) \neq \emptyset$ is complete and conversely.*

Proof. 1) Consider contrary that there exists $\chi_1,\chi_2 \in F(T)$ having $\mathcal{H}_s(\chi_1,\chi_2) > 0$. Now, since T is a generalized rational graphic ϕ-contractive mapping, implies

$$\mathcal{H}_s(\chi_1,\chi_2) = \mathcal{H}_s(T(\chi_1), T(\chi_2)) \leq \phi(M_T(\chi_1,\chi_2)), \tag{16.1}$$

where

$$\begin{aligned} M_T(\chi_1,\chi_2) &= \max\Bigg\{\mathcal{H}_s(\chi_1,\chi_2), \mathcal{H}_s(\chi_1, T(\chi_1)), \mathcal{H}_s(T(\chi_2),\chi_2), \frac{\mathcal{H}_s(\chi_2, T(\chi_2))[1+\mathcal{H}_s(T(\chi_1),\chi_1)]}{1+\mathcal{H}_s(\chi_1,\chi_2)}, \frac{\mathcal{H}_s(T(\chi_1),\chi_2)[1+\mathcal{H}_s(\chi_1, T(\chi_1))]}{1+\mathcal{H}_s(\chi_1,\chi_2)}, \frac{\mathcal{H}_s(\chi_1, T(\chi_1))[1+\mathcal{H}_s(\chi_1, T(\chi_1))]}{1+\mathcal{H}_s(\chi_2,\chi_1)}\Bigg\} \\ &= \mathcal{H}_s(\chi_2,\chi_1). \end{aligned}$$

The above inequality (16.1) implies

$$\begin{aligned} \mathcal{H}_s(\chi_1,\chi_2) &\leq \phi(\mathcal{H}_s(\chi_2,\chi_1)) \\ &< \mathcal{H}_s(\chi_1,\chi_2), \end{aligned}$$

a contradiction. Hence $\mathcal{H}_s(\chi_2,\chi_1) = 0$.

2) Consider $\chi \in F(T)$. As $(\alpha,\alpha) \subseteq E(G)$ for all $\alpha \in V(G)$, that is, $(\chi,\chi) = (\chi, T(\chi)) \subseteq E(G)$. Therefore, we $\chi \in CB(Y)$ such that $(\chi, T(\chi)) \subseteq E(G)$. Thus, $\chi \in Y_T$.

3) Consider $\chi \in Y_T \neq \emptyset$. As $\chi \in CB(Y)$ having $\widetilde{G}$ is weakly connected, implies $CB(Y) \subseteq [\chi]_{\widetilde{G}}$. As map T satisfies equivalence class with transitive property on $CB(Y)$ gives $T(\chi) \in [\chi]_{\widetilde{G}}$. Also, $T(\chi_i) \in [\chi]_{\widetilde{G}}$ for each $\chi_i \in [\chi]_{\widetilde{G}}$. By definition of T, $(T^n(\chi), T^{n+1}(\chi)) \in E(\widetilde{G})$ for all $n \in \{0,1,2,\dots\}$. Now, define

$$\begin{aligned}
\chi &= \chi_0,\\
T(\chi_0) &= \chi_1,\\
T^2(\chi_0) &= T(\chi_1) = \chi_2,\\
&\vdots\\
T^n(\chi_0) &= T(\chi_{n-1}) = \chi_n.
\end{aligned}$$

Consider $\chi_{n+1} \neq \chi_n$ for all $n \in \mathbb{N}\{0\}$. If $\chi_{j+1} = \chi_j$ for some j, then $T(\chi_j) = \chi_{j+1} = \chi_j$, i. e., χ_j is fixed point of T. By weakly connectedness of $\widetilde{G}$, implies a sequence $\{y_i\}$ such that $(y_i, y_{i+1}) \in E(\widetilde{G})$, for $i = 0,\dots,n-1$ and $y_i \in \chi_i$ for $i = 0,\dots,n$. Again, by definition of T, we have

$$\begin{aligned}
\mathcal{H}_s(T^{n+1}(\chi), T^n(\chi)) = \mathcal{H}_s(\chi_n,\chi_{n+1}) &= \mathcal{H}_s(T(\chi_{n-1}), T(\chi_n))\\
&\leq \phi(M_T(\chi_{n-1},\chi_n)),
\end{aligned} \tag{16.2}$$

where

$$\begin{aligned}
M_T(\chi_{n-1},\chi_n) = \max\Bigg\{ &\mathcal{H}_s(\chi_n,\chi_{n-1}), \mathcal{H}_s(\chi_{n-1}, T(\chi_{n-1})), \mathcal{H}_s(T(\chi_n),\chi_n),\\
&\frac{\mathcal{H}_s(T(\chi_n),\chi_n)[1 + \mathcal{H}_s(\chi_{n-1}, T(\chi_{n-1}))])}{1 + \mathcal{H}_s(U_{n-1}, U_n)},\\
&\frac{\mathcal{H}_s(\chi_n, T(\chi_{n-1}))[1 + \mathcal{H}_s(\chi_{n-1}, T(\chi_{n-1}))])}{1 + \mathcal{H}_s(\chi_n,\chi_{n-1})},\\
&\frac{\mathcal{H}_s(\chi_{n-1}, T(\chi_{n-1}))[1 + \mathcal{H}_s(\chi_{n-1}, T(\chi_{n-1}))])}{1 + \mathcal{H}_s(\chi_{n-1},\chi_n)}\Bigg\}.
\end{aligned}$$

Applying the mapping, we get

$$\begin{aligned}
M_T(\chi_{n-1},\chi_n) = \max\Bigg\{ &\mathcal{H}_s(\chi_n,\chi_{n-1}), \mathcal{H}_s(\chi_{n-1},\chi_n), \mathcal{H}_s(\chi_{n+1},\chi_n),\\
&\frac{\mathcal{H}_s(\chi_n,\chi_{n+1})[1 + \mathcal{H}_s(\chi_{n-1},\chi_n)])}{1 + \mathcal{H}_s(\chi_n,\chi_{n-1})},\\
&\frac{\mathcal{H}_s(\chi_n,\chi_n)[1 + \mathcal{H}_s(\chi_n,\chi_{n-1})])}{1 + \mathcal{H}_s(\chi_{n-1},\chi_n)},
\end{aligned}$$

$$\left.\frac{\mathcal{H}_s(\chi_{n-1},\chi_n)[1+\mathcal{H}_s(\chi_n,\chi_{n-1})])}{1+\mathcal{H}_s(\chi_{n-1},\chi_n)}\right\}$$
$$=\max\{\mathcal{H}_s(\chi_{n-1},\chi_n),\mathcal{H}_s(\chi_n,\chi_{n+1})\}.$$

Now, if $M_T(\chi_{n*1},\chi_n) = \mathcal{H}_s(\chi_n,\chi_{n+1})$, then by inequality (16.2), we have a contradiction. Therefore, the only value $M_T(\chi_{n-1},\chi_n)$ can obtain is $\mathcal{H}_s(\chi_{n-1},\chi_n)$. This gives

$$\begin{aligned}\mathcal{H}_s(T^{n+1}(\chi),T^n(\chi)) &= \mathcal{H}_s(\chi_n,\chi_{n+1}) = \mathcal{H}_s(T(\chi_n),T(\chi_{n-1}))\\ &\le \phi(\mathcal{H}_s(\chi_{n-1},\chi_n))\\ &= \phi(\mathcal{H}_s(T(\chi_{n-1}),T(\chi_{n-2})))\\ &\le \phi^2(\mathcal{H}_s(\chi_{n-2},\chi_{n-1}))\\ &\le\cdots\le \phi^n(\mathcal{H}_s(\chi_0,\chi_1)) = \phi^n(\mathcal{H}_s(\chi,T(\chi))).\end{aligned}\tag{16.3}$$

For $n,m\in\mathbb{N}$,

$$\mathcal{H}_s(T^{m+n}(\chi),T^n(\chi))\le\phi^n(\mathcal{H}_s(\chi_0,\chi_m)) = \phi^n(\mathcal{H}_s(\chi,\chi_m)).$$

By employing the upper limits as $n,m\to\infty$ yields $\phi^n(\mathcal{H}_s(\chi,\chi_m))\to 0$. Therefore, $\{T^n(\chi)\}$ is a d_s-Cauchy sequence in $CB(Y)$. As (Y,d) is d_s-Cauchy complete, we can find χ^* in $CB(Y)$ such that $\lim_{n\to\infty}T^n(\chi)=\chi^*$.

Now, let us show that χ^* is a fixed point of T. Since, $\{T^n(\chi)\}$ with $\lim_{n\to\infty}T^n(\chi)=\chi^*$ is a d_s-Cauchy sequence and by our assumption, we have $(T^n(\chi),T^{n+1}(\chi))\in E(\tilde{G})$ for every $n\in\mathbb{N}$. Therefore, by property $\overset{*}{\mathcal{P}}$, we have subsequence $\{T^{n_k}(\chi)\}$ of $\{T^n(\chi)\}$ implies edge between $T^{n_k}(\chi)$ and χ^* for all $n\in\mathbb{N}$. So

$$\mathcal{H}_s(T(\chi^*),T^{n_k}(\chi^*))\le\phi^n(M_T(\chi^*,T^{n_k-1}(\chi^*))).\tag{16.4}$$

Note that $M_T(\chi^*,T^{n_k-1}(\chi^*)) = M_T(T^{n_k-1}(\chi^*),\chi^*)$, where

$$\begin{aligned}&M_T(T^{n_k-1}(\chi^*),\chi^*)\\ &=\max\Bigg\{\mathcal{H}_s(T^{n_k-1}(\chi^*),\chi^*),\mathcal{H}_s(T^{n_k-1}(\chi^*),T(T^{n_k-1}(\chi^*))),\mathcal{H}_s(\chi^*,T(\chi^*)),\\ &\quad\frac{\mathcal{H}_s(\chi^*,T(\chi^*))[1+\mathcal{H}_s(T^{n_k-1}(\chi^*),T(T^{n_k-1}(\chi^*)))]}{1+\mathcal{H}_s(T^{n_k-1}(\chi^*),\chi^*)},\\ &\quad\frac{\mathcal{H}_s(\chi^*,T(T^{n_k-1}(\chi^*)))[1+\mathcal{H}_s(T^{n_k-1}(\chi^*),T(T^{n_k-1}(\chi^*)))]}{1+\mathcal{H}_s(T^{n_k-1}(\chi^*),\chi^*)},\\ &\quad\frac{\mathcal{H}_s(T^{n_k-1}(\chi^*),T(T^{n_k-1}(\chi^*)))[1+\mathcal{H}_s(T^{n_k-1}(\chi^*),T(T^{n_k-1}(\chi^*)))]}{1+\mathcal{H}_s(T^{n_k-1}(\chi^*),\chi^*)}\Bigg\}\\ &=\max\Bigg\{\mathcal{H}_s(T^{n_k-1}(\chi^*),\chi^*),\mathcal{H}_s(T^{n_k-1}(\chi^*),T^{n_k}(\chi^*)),\mathcal{H}_s(\chi^*,T(\chi^*)),\end{aligned}$$

$$\frac{\mathcal{H}_s(\chi^*, T(\chi^*))[1+\mathcal{H}_s(T^{n_k-1}(\chi^*), T^{n_k}(\chi^*))]}{1+\mathcal{H}_s(T^{n_k-1}(\chi^*),\chi^*)},$$
$$\frac{\mathcal{H}_s(\chi^*, T^{n_k}(\chi^*))[1+\mathcal{H}_s(T^{n_k-1}(\chi^*), T(T^{n_k-1}(\chi^*)))]}{1+\mathcal{H}_s(T^{n_k-1}(\chi^*),\chi^*)},$$
$$\left.\frac{\mathcal{H}_s(T^{n_k-1}(\chi^*), T^{n_k}(\chi^*))[1+\mathcal{H}_s(T(T^{n_k-1}(\chi^*)), T^{n_k-1}(\chi^*))]}{1+\mathcal{H}_s(T^{n_k-1}(\chi^*),\chi^*)}\right\}.$$

As $\lim_{n_k\to\infty} M_T(T^{n_k-1}(\chi^*),\chi^*) = \mathcal{H}_s(\chi^*, T(\chi^*))$, and so by using upper limit as $n_k \to \infty$ on both sides of (16.4), implies

$$\mathcal{H}_s(T(\chi^*),\chi^*) \le \phi(\mathcal{H}_s(\chi^*, T(\chi^*))),$$

gives $T(\chi^*) = \chi^*$. Therefore, $\chi^* \in F(T)$. We now show that the fixed point χ^* is unique. Consider opposite case that we have $\chi^*, V^* \in F(T)$ and let $\mathcal{H}_s(\chi^*, V^*) > 0$. Then, relevant case to (1) of Theorem 16.3 implies $\mathcal{H}_s(V^*,\chi^*) \le \phi(\mathcal{H}_s(\chi^*, V^*))$, whence $\mathcal{H}_s(V^*,\chi^*) = 0$, implying $\chi^* = V^*$. We also conclude that $|F(T)| = 1$.

4) Consider $F(T) \neq \emptyset$ is complete. By (1) of Theorem 16.3, $F(T)$ is a singleton. Now, to show the sufficiency case, consider $F(T)$ is a singleton having $\chi \in F(T)$, then by (2) of Theorem 16.3, implies $\chi \in Y_T$, hence $F(T)$ is complete. □

Now, we present an example in a semi-metric space (Y, d_s).

Example 16.2. Let $Y = \{0,1,4\} = V(G)$ and equipped with semi-metric $\mathrm{d}_s : Y \times Y \to \mathbb{R}^+$, where $\mathrm{d}_s(\beta, \alpha) = (\beta - \alpha)^2$. Clearly the semi-metric d_s does not satisfy the triangle inequality. Let the edge set taken by

$$E(G) = \{(0,0),(1,1),(4,4),(0,1),(0,4),(1,4)\}.$$

Note that the sets $\{0\}$, $\{0,1\}$ and $\{0,4\}$ are all bounded and closed. We illustrate that this is so for the set $\{0,4\}$, the others follow a similar argument.

$$y \in \overline{\{0,4\}} \Leftrightarrow d(y,\{0,4\}) = 0 \Leftrightarrow \min\{d(y,0), d(y,4)\} = 0 \Leftrightarrow y \in \{0,4\}.$$

Hence, $\{0,4\}$ is closed with respect to the semi-metric d_s. Boundedness is clear.

We use the notations $\bar{0} = \{0\}$, $\bar{1} = \{0,1\}$ and $\bar{4} = \{0,4\}$, where, $CB(X) = \{\bar{0},\bar{1},\bar{4}\}$. We now employ the PH-metric to calculate the distances between the elements of $CB(Y)$.

$$\mathcal{H}_s(\chi, V) = \begin{cases} 0 & \text{if } \chi = V = \bar{0}, \chi = V = \bar{4} \text{ or } \chi = V = \bar{1}, \\ 1 & \text{if } \chi = \bar{1} \text{ and } V = \bar{0}, \\ 9 & \text{if } \chi = \bar{1} \text{ and } V = \bar{4}, \\ 16 & \text{if } \chi = \bar{0} \text{ and } V = \bar{4}. \end{cases}$$

We now define the mapping $T : CB(Y) \to CB(Y)$ as

$$T(Z) = \begin{cases} \bar{1} & \text{if } Z = \bar{4}, \\ \bar{0} & \text{if } Z \in \{\bar{0}, \bar{1}\}. \end{cases}$$

Since any two elements $V, \chi \in CB(Y)$ that have $(\chi, V) \subseteq E(G)$ implies $(T(\chi), T(V)) \subseteq E(G)$.

Take $\phi : [0, \infty) \to [0, \infty)$ defined as

$$\phi(s) = \begin{cases} \frac{7s}{8}, & \text{if } s \in [0, 40), \\ \frac{2^{n+2}s-5}{2^{2n+1}+1}, & \text{if } s \in [\frac{5(2^{2n+3}+8)}{2^{2n}+1}, \frac{5(2^{2n+5}+8)}{2^{2n+2}+1}],\ n \in \mathbb{N}. \end{cases}$$

It is straightforward to prove that the map ϕ is upper semi-continuous and non-decreasing on $[0, \infty)$ satisfy $\phi(s) < s$ for every $s > 0$.

Now, we show that for all $V, \chi \in CB(Y)$ having $(\chi, V) \subseteq E(G)$, mapping T is satisfies

$$\mathcal{H}_s(T(\chi), T(V)) \le \phi(M_T(\chi, V)),$$

where

$$\begin{aligned} M_T(\chi, V) = \max\Big\{ & \mathcal{H}_s(\chi, V), \mathcal{H}_s(\chi, T(\chi)), \mathcal{H}_s(V, T(V)), \\ & \frac{\mathcal{H}_s(V, T(V))[1 + \mathcal{H}_s(\chi, T(\chi))]}{1 + \mathcal{H}_s(\chi, V)}, \\ & \frac{\mathcal{H}_s(V, T(\chi))[1 + \mathcal{H}_s(\chi, T(\chi))]}{1 + \mathcal{H}_s(\chi, V)}, \\ & \frac{\mathcal{H}_s(\chi, T(\chi))[1 + \mathcal{H}_s(\chi, T(\chi))]}{1 + \mathcal{H}_s(\chi, V)} \Big\}. \end{aligned}$$

Considering the cases:

Case 1. If $V, \chi \in \{\bar{0}, \bar{1}\}$ or $\chi = V = \bar{4}$, then $\mathcal{H}_s(T(\chi), T(V)) = 0$.

Case 2. In case $\chi = \bar{0}$ and $V = \bar{4}$, we have

$$\begin{aligned} \mathcal{H}_s(T(\bar{0}), T(\bar{4})) &= \mathcal{H}_s(\bar{0}, \bar{1}) = 1 \\ &< \frac{63}{8} = \phi(9) = \phi(\mathcal{H}_s(\bar{4}, \bar{1})) \\ &= \phi(\mathcal{H}_s(V, T(V)) \le \phi(M_T(\chi, V)). \end{aligned}$$

Case 3. When $\chi = \bar{1}$ and $V = \bar{4}$, we have

$$\begin{aligned} \mathcal{H}_s(T(\bar{1}), T(\bar{4})) &= \mathcal{H}_s(\bar{0}, \bar{1}) = 1 \\ &< \frac{63}{8} = \phi(9) = \phi(\mathcal{H}_s(\bar{1}, \bar{4})) \\ &= \phi(\mathcal{H}_s(\chi, V)) \le \phi(M_T(\chi, V)). \end{aligned}$$

Clearly, the inequality (16.1) is valid for all cases and we deduce that T is a generalized rational graphic ϕ-contractive mapping. Thus, all the conditions of Theorem 16.3 are satisfied. Moreover, $T(\{0\}) = \{0\}$ making the singleton $\{0\}$, the fixed point for T, from which we infer that $F(T)$ is reduced to the unit set $\{0\}$. Equivalently, the set $F(T)$ is a complete set.

It is evident that the inequality (16.1) holds universally, thereby establishing that T qualifies as a generalized rational graphic ϕ-contractive map. Consequently, all the axioms of Theorem 16.3 hold. Furthermore, since $T(\{0\}) = \{0\}$, the singleton set $\{0\}$ constitutes a fixed point of T. Thus, the fixed-point set $F(T)$ is reduced to $\{0\}$, implying that $F(T)$ forms a complete set.

Corollary 16.1. *Let (Y, d_s) be a d_s-Cauchy complete semi-metric space equipped with directed graph G such that $V(G) = Y$ and $E(G) \supseteq \Delta$. If G is a weakly connected, satisfies $\overset{*}{\mathcal{P}}$ and a map $T : CB(Y) \to CB(Y)$ satisfies the following:*
(1) *An edge (path) between χ and V implies an edge (path) between $T(\chi)$ and $T(V)$, for all $\chi, V \in CB(Y)$.*
(2) *A non-decreasing upper semi-continuous map $\phi : \mathbb{R}_+ \to \mathbb{R}_+$ exists having $\phi(s) < s$ for each $s > 0$ and $\sum_{k=0}^{\infty} \phi^k(s)$ is convergent such that for all $\chi, V \in CB(Y)$ with $(V,\chi) \subseteq E(G)$,*

$$\mathcal{H}_s(T(\chi), T(V)) \leq \phi(H_s(\chi, V)).$$

Then, T attains a fixed point.

Corollary 16.2. *Let (Y, d_s) be an ϵ-chainable space for some $\epsilon > 0$, the mapping $T : CB(Y) \to CB(Y)$ and $\lambda \in [0,1)$ with $\mathcal{H}_s(\chi, V) < \epsilon$. Then T has a fixed point, whenever $\mathcal{H}_s(T(\chi), T(V)) \leq \lambda M_T(\mathcal{H}_s(\chi, V))$ for all $\chi, V \in CB(Y)$.*

Proof. From our assumption and Lemma 16.3, for every $u \in \chi$, we have $v \in V$ such that $d(u, v) < \epsilon$. Since (Y, d_s) is an ϵ-chainable space, we have by Definition 16.6, a sequence $\{y_n\}$, where $d(y_i, y_{i+1}) < \epsilon$ for $i = 0, \ldots, n-1$. Equipping the semi-metric space with a graph, as follows: $V(G) = Y$ and $E(G) = \{(y_i, y_{i+1}) : d(y_i, y_{i+1}) < \epsilon\}$, and we obtain easily that G is weakly connected, hence $(Y_i, Y_{i+1}) \subseteq E(G)$. Then by hypothesis, we have that T is a generalized rational graphic ϕ-contractive map for $\phi(t) = \lambda t$ and Theorem 16.3 implies T has a fixed point. □

As in [5, 17], let Γ denote the set of all non-negative real-valued functions γ with domain $\mathbb{R}_+$ such that
(1) γ is Lebesgue integrable.
(2) γ is a finite integral on every subset of $\mathbb{R}_+$ that is compact.
(3) $\int_0^{\epsilon} \gamma(\eta) d\eta > 0$ for each $\epsilon > 0$.

We obtain as a consequence of Theorem 16.3 fixed-point result satisfying integral-type contraction conditions on a semi-metric space (Y, d_s).

Corollary 16.3. *Let (Y, d_s) be a d_s-Cauchy complete semi-metric space endowed with a directed graph G, where $V(G) = Y$ and $\Delta \subseteq E(G)$. If mapping $T : CB(Y) \to CB(Y)$ satisfies the following conditions:*

(1) *An edge (path) between V and χ implies an edge (path) between $T(V)$ and $T(\chi)$, for all $\chi, V \in CB(X)$.*

(2) *An upper semi-continuous and non-decreasing map $\phi : \mathbb{R}_+ \to \mathbb{R}_+$ exists having $\phi(t) < t$ for each $t > 0$ with $\sum_{k=0}^{\infty} \phi^k(t)$ is convergent such that for $\chi_1, \chi_2 \in CB(Y)$ with $(\chi_1, \chi_2) \subseteq E(G)$, it holds that*

$$\mathcal{H}_s(T(\chi_1), T(\chi_2)) \leq \int_0^{\phi(M_T(\chi_1,\chi_2))} \gamma(t)dt,$$

where

$$\begin{aligned} M_T(\chi_1, \chi_2) = \max\Big\{ & \mathcal{H}_s(\chi_1, \chi_2), \mathcal{H}_s(\chi_2, T(\chi_2)), \mathcal{H}_s(\chi_1, T(\chi_1)), \\ & \frac{\mathcal{H}_s(\chi_2, T(\chi_2))[1 + \mathcal{H}_s(T(\chi_1), \chi_1)]}{1 + \mathcal{H}_s(\chi_1, \chi_2)}, \\ & \frac{\mathcal{H}_s(\chi_2, T(\chi_2))[1 + \mathcal{H}_s(\chi_1, T(\chi_1))]}{1 + \mathcal{H}_s(\chi_1, \chi_2)}, \\ & \frac{\mathcal{H}_s(\chi_1, T(\chi_1))[1 + \mathcal{H}_s(\chi_1, T(\chi_1))]}{1 + \mathcal{H}_s(\chi_1, \chi_2)} \Big\}. \end{aligned}$$

Then the following are true:

(3) *If $F(T) \neq \emptyset$ is complete, then PH-weight assigned to $\chi_1, \chi_2 \in F(T)$ is zero.*

(4) *If $F(T) \neq \emptyset$, then $Y_T \neq \emptyset$.*

(5) *If $Y_T \neq \emptyset$ with the weakly connected graph $\tilde{G}$ satisfies property $\overset{*}{\mathcal{P}}$, then T has a unique fixed point.*

(6) *$F(T)$ is singleton if $F(T)$ is complete and conversely.*

Proof. Define $\vartheta : [0, \infty) \to [0, \infty)$ as $\vartheta(x) = \int_0^x \gamma(t)dt$, then by (2) implies

$$\mathcal{H}_s(T(\chi_1), T(\chi_2)) \leq \vartheta(\phi(M_T(\chi_2, \chi_1))),$$

which implies

$$\mathcal{H}_s(T(\chi_1), T(\chi_2)) \leq \psi(M_T(\chi_1, \chi_2)),$$

having $\psi = \vartheta \circ \phi$. Now, our result follows from applying Theorem 16.3. □

16.5 Conclusions

We established fixed-point theorems for set-valued mappings that satisfy generalized rational graphic ϕ-contractive conditions, developed on the collection of closed as well as bounded subsets of semi-metric space equipped with a graph. To illustrate the applicability of our results, we presented an example that supports the formulations and findings discussed in this chapter. Our results not only unify various related findings from the given literature, but also derived them in multiple directions.

Bibliography

[1] Abbas, M., Alfuraidan, M., Rashed, K., Rahim, A. and Nazir, T., *Fixed point results for set-contractions on metric spaces with a directed graph*, Fixed Point Theory Appl. 1 (2015), 1–9.

[2] Aamri, M., Bassou, A. and Moutawakil, D. E., *Common fixed points for weakly compatible maps in symmetric spaces with application to probabilistic spaces*, Appl. Math. E-Notes 5 (2005), 171–175.

[3] Aamri, M. and Moutawakil, D. E., *Common fixed points under contractive conditions in symmetric spaces*, Appl. Math. E-Notes 3 (2003), 156–162.

[4] Aranelović, I. D. and Kečkić, D. J., *Symmetric spaces approach to some fixed point results*, Nonlinear Anal. 75 (2012), 5157–5168.

[5] Branciari, A., *A fixed point theorem for mappings satisfying a general contractive condition of integral type*, Int. J. Math. Math. Sci. 29(9) (2002), 531–536.

[6] Cho, S. H., Lee, G. Y. and Bae, J. S., *On coincidence and fixed-point theorems in symmetric spaces*, Fixed Point Theory Appl. 2008 (2008), 562130, 9 pp.

[7] Fréchet, M., *Sur quelques points du calcul fonctionnel*, Rend. Circ. Palermo 22 (1906), 1–74.

[8] Hicks, T. L., *Fixed point theorem for multivalued mappings II*, Indian J. Pure Appl. Math. 29 (1998), 133–137.

[9] Imdad, M., Ali, J. and Khan, L., *Coincidence and fixed points in symmetric spaces under strict contractions*, J. Math. Anal. Appl. 320 (2006), 352–360. Corrigendum: J. Math. Appl. 329 (2007).

[10] Jachymski, J., Matkowski, J. and Swiatkowski, T., *Nonlinear contractions on semimetric spaces*, J. Appl. Anal. 1 (1995), 125–134.

[11] Jachymski, J., *The contraction principle for mappings on a metric space with a graph*, Proc. Am. Math. Soc. 136(4) (2008), 1359–1373.

[12] Kutbi, M. A., Latif, A. and Nazir, T., *Generalized rational contractions in semi metric spaces via iterated function system*, RACSAM, Ser. A, Mat. 114 (2020), 187. 1–16.

[13] Latif, A., Nazir, T. and Kutbi, M. A., *Common fixed point results for class of set-contraction mappings endowed with a directed graph*, Rev. R. Acad. Cienc. Exactas Fís. Nat. Ser. A Mat. RACSAM 113 (2019), 3207–3222.

[14] Latif, A., *A fixed point result for multivalued generalized contraction maps*, Filomat 26(5) (2012), 929–933.

[15] Moţ, G. and Petruşel, A., *Fixed point theory for a new type of contractive multivalued operators*, Nonlinear Anal. 70 (2009), 3371–3377.

[16] Nadler, S. B., *Multivalued contraction mappings*, Pac. J. Math. 30 (1969), 475–488.

[17] Radenović, S. and Kadelburg, Z., *Generalized weak contractions in partially ordered metric spaces*, Comput. Math. Appl. 60(6) (2010), 1776–1783.

[18] Ran, A. C. M. and Reurings, M. C. B., *A fixed point theorem in partially ordered sets and some applications to matrix equations*, Proc. Am. Math. Soc. 5 (2004), 1435–1443.

[19] Shahzad, N., Alghamdi, M. A., Alshehri, S. and Aranelović, I., *Semi-metric spaces and fixed points of α-φ-contractive maps*, J. Nonlinear Sci. Appl. 9 (2016), 3147–3156.
[20] Wilson, W. A., *On semi-metric spaces*, Am. J. Math. 53 (1931), 361–373.
[21] Zhu, J., Cho, Y. J. and Kang, M., *Equivalent contractive conditions in symmetric spaces*, Comput. Math. Appl. 50 (2005), 1621–1628.

Kallol Bhuyan and Pradip Debnath

17 A new generalization of Kannan's fixed-point theorem via simple *F*-contraction

Abstract: We present a novel generalization of Kannan's fixed-point theorem by using the recent innovative concept of Cvetković's simple F-contraction [9]. This work addresses a significant gap in the existing literature, as Wardowski's and Kannan's classical theorems are unable to address a specific class of mappings, leaving their fixed point properties unexplored. Our new framework not only bridges this theoretical gap but also provides a powerful tool for establishing the existence of fixed points in previously unstudied scenarios. To underscore the practical utility of our results, we present an example that demonstrates its applicability in the sense that Kannan's and Wardowski's theorem is not applicable, however, our extended theorem provides a positive result. Also, the obtained results have been used to establish a sufficient condition for proving the existence of a solution to a particular class of integral equations. Furthermore, we extend our main theorem to derive a Reich-type generalization, offering new insights into fixed-point theory.

Keywords: Fixed point, simple F-contraction map, complete metric space, Kannan-type contraction, Cauchy sequence, integral equation

MSC 2020: 47H10, 54H25, 54E50

17.1 Introduction and preliminaries

In metric fixed-point theory, as a fundamental theorem the Banach fixed-point theorem assures the existence and uniqueness of fixed points for specific self-maps in complete metric spaces. Kannan [23, 24] later introduced a generalized version providing the inclusion of fixed points for some discontinuous functions as well.

Conflict of interests: The authors declare that they have no known competing financial interests or personal relationships with anyone that could have appeared to influence the work reported in this paper.

Data availability: Not applicable.

Authors' contributions: Author K. B. contributed in Conceptualization, Investigation, Methodology, Validation, Writing the original draft; Author P. D. contributed in Conceptualization, Investigation, Methodology, Validation, Supervision and Editing.

Funding: This research received no external funding.

Kallol Bhuyan, Pradip Debnath, Department of Mathematical Sciences, Tezpur University, Napaam, Assam-784028, India, e-mails: kallol75760@gmail.com, pradip@tezu.ernet.in

https://doi.org/10.1515/9783112215760-017

Theorem 17.1 ([23]). *Suppose $(\mathcal{M}, \delta)$ be a complete MS. The self-mapping $\mathfrak{J} : \mathcal{M} \to \mathcal{M}$ is called a Kannan map if there exists $\lambda \in [0,1)$ such that*

$$\delta(\mathfrak{J}\mathfrak{x}, \mathfrak{J}\mathfrak{y}) \le \frac{\lambda}{2}[\delta(\mathfrak{x}, \mathfrak{J}\mathfrak{x}) + \delta(\mathfrak{y}, \mathfrak{J}\mathfrak{y})]$$

$\forall \mathfrak{x}, \mathfrak{y} \in \mathcal{M}$. Then $\mathfrak{J}$ admits a unique fixed point $\mathfrak{x}^* \in \mathcal{M}$, i. e., $\mathfrak{J}\mathfrak{x}^* = \mathfrak{x}^*$.

Reich [26–28] further extended Banach's contraction principle by introducing a more general contractive criterion involving multiple control coefficients, allowing fixed-point qualities to be satisfied by a larger class of self-maps. Notably, with the right choice of the involved constants, Reich's result simplifies to Kannan's theorem. For more information in this connection, we refer to famous works of Altun et al. [2], Chatterjea [7] and Rus et al. [29].

Theorem 17.2 ([27]). *Consider a complete metric space $(\mathcal{M}, \delta)$. Assume the self-mapping $\mathfrak{J} : \mathcal{M} \to \mathcal{M}$ satisfies*

$$\delta(\mathfrak{J}\mathfrak{x}, \mathfrak{J}\mathfrak{y}) \le \sigma_1 \delta(\mathfrak{x}, \mathfrak{J}\mathfrak{x}) + \sigma_2 \delta(\mathfrak{y}, \mathfrak{J}\mathfrak{y}) + \sigma_3 \delta(\mathfrak{x}, \mathfrak{y}),$$

$\forall \mathfrak{x}, \mathfrak{y} \in \mathcal{M}$, where $\sigma_1, \sigma_2, \sigma_3 \in \mathbb{R}_+$ are such that $\sigma_1 + \sigma_2 + \sigma_3 < 1$. As a result, $\mathfrak{J}$ has exactly one fixed point.

Setting $\sigma_1 = \sigma_2 = 0$ reduces the condition to the Banach fixed-point theorem, while picking $\sigma_1 = \sigma_2$ and $\sigma_3 = 0$ recovers Kannan's one.

In 2012, Wardowski [30] developed the idea of F contraction.

Definition 17.1. Suppose $F : (0, \infty) \to \mathbb{R}$ is a function that satisfies the following requirements:
(F_1) F is strictly increasing, which means $0 < \mathfrak{x} < \mathfrak{y} \implies F(\mathfrak{x}) < F(\mathfrak{y})$.
(F_2) For any given sequence $(\mathfrak{x}_n) \subseteq (0, \infty)$,

$$\lim_{n \longrightarrow \infty} \mathfrak{x}_n = 0 \iff \lim_{n \longrightarrow \infty} F(\mathfrak{x}_n) = -\infty.$$

(F_3) There exists $k \in (0,1)$ such that $\lim_{\mathfrak{x} \longrightarrow 0^+} \mathfrak{x}^k F(\mathfrak{x}) = 0$.

We define $\mathfrak{X}$ as the collection of all functions $F : (0, \infty) \to \mathbb{R}$ fulfilling $(F_1) - (F_3)$. Considering a class $\mathfrak{X}$, the F-contraction is given by the following definition:

Definition 17.2. Let $(\mathcal{M}, \delta)$ be a metric space and $\mathfrak{J} : \mathcal{M} \to \mathcal{M}$ a mapping. If there exists $F \in \mathfrak{X}$ and $\varphi > 0$ such that, for all $\mathfrak{x}, \mathfrak{y} \in \mathcal{M}$,

$$\delta(\mathfrak{J}\mathfrak{x}, \mathfrak{J}\mathfrak{y}) > 0 \implies \varphi + F(\delta(\mathfrak{J}\mathfrak{x}, \mathfrak{J}\mathfrak{y})) \le F(\delta(\mathfrak{x}, \mathfrak{y})),$$

then a mapping $\mathfrak{J}$ is called a F-contraction.

With regard to the idea of F-contraction, Wardowski generalized a altered form of the Banach fixed-point theorem.

Theorem 17.3 ([30]). *Consider a complete metric space $(\mathcal{M}, \delta)$ and suppose $\mathfrak{J} : \mathcal{M} \to \mathcal{M}$ be an F-contraction. Then $\mathfrak{J}$ possesses a unique fixed point $\mathfrak{x}^* \in \mathcal{M}$ and, for every $\mathfrak{x} \in \mathcal{M}$, a sequence $(\mathfrak{J}^n \mathfrak{x})$ converges to $\mathfrak{x}^*$.*

The class $\mathcal{F}^{\nearrow}$, introduced recently by Cvetković [9], consists of functions that satisfy the condition F_1, forming a wider class of F-contractions. Also, $\mathfrak{X} \subseteq \mathcal{F}^{\nearrow}$.

Definition 17.3 ([9]). A set $\mathcal{F}^{\nearrow}$ is the set of all strictly increasing functions $F : (0, \infty) \to \mathbb{R}$.

Definition 17.4 ([9]). Let $(\mathcal{M}, \delta)$ be a metric space and $\mathfrak{J} : \mathcal{M} \to \mathcal{M}$ a mapping. If there exists $F \in \mathcal{F}^{\nearrow}$ and $\varphi > 0$ such that, for all $\mathfrak{x}, \mathfrak{y} \in \mathcal{M}$,

$$\delta(\mathfrak{J}\mathfrak{x}, \mathfrak{J}\mathfrak{y}) > 0 \implies \varphi + F(\delta(\mathfrak{J}\mathfrak{x}, \mathfrak{J}\mathfrak{y})) \leq F(\delta(\mathfrak{x}, \mathfrak{y})),$$

then a mapping $\mathfrak{J}$ is called a simple F-contraction (sF-contraction).

The following primary result focuses on whether a fixed point exists for a simple F-contraction and whether it is unique.

Theorem 17.4 ([9]). *Let $(\mathcal{M}, \delta)$ be a complete metric space and $\mathfrak{J} : \mathcal{M} \to \mathcal{M}$ a simple F-contraction. Then $\mathfrak{J}$ has a unique fixed point $\mathfrak{x}^* \in \mathcal{M}$ and, for every $\mathfrak{x} \in \mathcal{M}$, a sequence $(\mathfrak{J}^n \mathfrak{x})$ converges to $\mathfrak{x}^*$.*

Through this work, we aim to contribute to the evolving field of fixed-point theory by introducing generalized results that build upon and extend the foundational theorems of Kannan and Reich. A key novelty of our approach lies in the application of simple F (sF)-contractions, which provides a more versatile and comprehensive framework for proving fixed points' existence and uniqueness. This addresses a critical gap in the existing literature, as classical results such as Wardowski's and Kannan's theorems are not directly applicable to certain classes of mappings, leaving their fixed point properties under explored. Our work bridges this gap by offering a unified and flexible approach that extends the scope of fixed point analysis.

In this endeavor, we acknowledge the foundational contributions of Aydi et al. [3, 4], Debnath et al. [19, 20], Debnath and Srivastava [21], Górnicki [22], Debnath [15], Bojor [5, 6], and Kundu and Choudhury [8], whose pioneering studies have shaped the field. For more relevant literature, we refer to the recent works of Debnath [10–14, 16–18].

By integrating and extending these works, we not only deepen the theoretical insights into fixed-point theory, in addition, we offer some useful resources for resolving problems in previously unexplored domains. Our results, including a Reich-type extension, offer deeper insights and open new avenues for future research, marking a substantial step forward in the field.

We begin by presenting a suitable advanced version of Kannan's fixed-point theorem for sF-contractions. Also, an illustrative example has been discussed in relation to this generalization. Subsequently, we have proposed a Reich-type generalization of the above.

17.2 Kannan-type simple F-contraction

This section presents a significant generalization of Kannan's theorem through sF-contraction.

Definition 17.5. Let $(\mathcal{M}, \delta)$ be a metric space. The self map $\mathfrak{J} : \mathcal{M} \to \mathcal{M}$ is said to be a generalized Kannan-type sF-contraction (GKSFC) if there exists $\sigma_1, \sigma_2 \in (0,1)$ with $\sigma_1 + \sigma_2 < 1$, $\varphi > 0$ and $F \in \mathcal{F}^{\nearrow}$ such that

$$\varphi + F(\delta(\mathfrak{J}\mathfrak{x}, \mathfrak{J}\mathfrak{y})) \leq \sigma_1 F(\delta(\mathfrak{x}, \mathfrak{J}\mathfrak{x})) + \sigma_2 F(\delta(\mathfrak{y}, \mathfrak{J}\mathfrak{y})) \tag{17.1}$$

$\forall \mathfrak{x}, \mathfrak{y} \in \mathcal{M}$ with $\mathfrak{x}, \mathfrak{y} \notin \text{Fix}(\mathfrak{J})$, where Fix $(\mathfrak{J})$ represents the collection of all fixed points of $\mathfrak{J}$.

Theorem 17.5. *Let $(\mathcal{M}, \delta)$ be a complete MS. A GKSFC, $\mathfrak{J} : \mathcal{M} \to \mathcal{M}$ admits an unique fixed point.*

Proof. Consider a complete metric space $(\mathcal{M}, \delta)$ and let $\mathfrak{J} : \mathcal{M} \to \mathcal{M}$ be a generalized Kannan-type simple F contraction.

Fix $\mathfrak{x}_0 \in \mathcal{M}$.

Let us define a sequence $\{\mathfrak{x}_n\}$ in $(\mathcal{M}, \delta)$ such that

$$\mathfrak{x}_n = \mathfrak{J}(\mathfrak{x}_{n-1}), \quad \text{for all } n \in \mathbb{N} \cup \{0\}. \tag{17.2}$$

Then, $\mathfrak{J}(\mathfrak{x}_0) \in \mathcal{M}$, $\mathfrak{J}^2(\mathfrak{x}_0) \in \mathcal{M}$, $\mathfrak{J}^3(\mathfrak{x}_0) \in \mathcal{M}$ and so on. By using equation (17.2), we can write

$$\begin{aligned} \mathfrak{x}_1 &= \mathfrak{J}(\mathfrak{x}_0) \\ \mathfrak{x}_2 &= \mathfrak{J}^2(\mathfrak{x}_0) \\ \mathfrak{x}_3 &= \mathfrak{J}^3(\mathfrak{x}_0) \end{aligned}$$

and so on. In general, $\mathfrak{x}_n = \mathfrak{J}^n(\mathfrak{x}_0)$, $\forall n \in \mathbb{N} \cup \{0\}$.

If $n_0 \in \mathbb{N}$ satisfies

$$\begin{aligned} \mathfrak{x}_{n_0} &= \mathfrak{x}_{n_0+1}, \\ \text{i. e., } \mathfrak{x}_{n_0} &= \mathfrak{J}\mathfrak{x}_{n_0}, \end{aligned}$$

then $\mathfrak{x}_{n_0}$ turns out to be a fixed point of $\mathfrak{J}$, leaving nothing further to prove. Therefore, assume that $\mathfrak{x}_n \neq \mathfrak{x}_{n+1}$, $\forall n \geq 0$.

Taking $\mathfrak{x} = \mathfrak{x}_n$ and $\mathfrak{y} = \mathfrak{x}_{n-1}$ in (17.1), we derive

$$\begin{aligned}\varphi + F(\delta(\mathfrak{J}\mathfrak{x}_n, \mathfrak{J}\mathfrak{x}_{n-1})) &= \varphi + F(\delta(\mathfrak{x}_{n+1}, \mathfrak{x}_n)) \\ &\leq \sigma_1 F(\delta(\mathfrak{x}_n, \mathfrak{J}\mathfrak{x}_n)) + \sigma_2 F(\delta(\mathfrak{x}_{n-1}, \mathfrak{J}\mathfrak{x}_{n-1})) \\ &< \sigma_1 F(\delta(\mathfrak{x}_n, \mathfrak{x}_{n+1})) + (1-\sigma_1) F(\delta(\mathfrak{x}_{n-1}, \mathfrak{x}_n)). \end{aligned} \tag{17.3}$$

Let, $\delta(\mathfrak{x}_n, \mathfrak{x}_{n-1}) \leq \delta(\mathfrak{x}_{n+1}, \mathfrak{x}_n)$.
Hence, from (17.3), we find that

$$\begin{aligned}&\varphi + F(\delta(\mathfrak{J}\mathfrak{x}_n, \mathfrak{J}\mathfrak{x}_{n-1})) < \sigma_1 F(\delta(\mathfrak{x}_n, \mathfrak{x}_{n+1})) + (1-\sigma_1) F(\delta(\mathfrak{x}_n, \mathfrak{x}_{n+1})) \\ \Longrightarrow\ &\varphi + F(\delta(\mathfrak{x}_{n+1}, \mathfrak{x}_n)) < F(\delta(\mathfrak{x}_n, \mathfrak{x}_{n+1})),\end{aligned}$$

which is not possible.

Therefore, $\delta(\mathfrak{x}_{n+1}, \mathfrak{x}_n) < \delta(\mathfrak{x}_n, \mathfrak{x}_{n-1})$ for all $n \geq 1$.
So from (17.3), we obtain

$$\varphi + F(\delta(\mathfrak{x}_{n+1}, \mathfrak{x}_n)) < F(\delta(\mathfrak{x}_n, \mathfrak{x}_{n-1})).$$

Consequently,

$$\begin{aligned}F(\delta(\mathfrak{x}_{n+1}, \mathfrak{x}_n)) &< F(\delta(\mathfrak{x}_{n-1}, \mathfrak{x}_n)) - \varphi \\ &< F(\delta(\mathfrak{x}_{n-1}, \mathfrak{x}_{n-2})) - 2\varphi \\ &\ \ \vdots \\ &< F(\delta(\mathfrak{x}_0, \mathfrak{x}_1)) - n\varphi. \end{aligned} \tag{17.4}$$

From (17.4), by taking the limit ($n \longrightarrow \infty$), we obtain

$$\lim_{n\longrightarrow\infty} F(\delta(\mathfrak{x}_n, \mathfrak{x}_{n+1})) = -\infty. \tag{17.5}$$

We will consider two cases:

(i) If we suppose that (F_2) holds, then using (17.10), we have

$$\lim_{n\longrightarrow\infty} \delta(\mathfrak{x}_n, \mathfrak{x}_{n+1})) = 0. \tag{17.6}$$

To establish that $(\mathfrak{x}_n) \subseteq \mathcal{M}$ is a Cauchy sequence and hence convergent, assume to the contrary that there exists $\xi > 0$ along with strictly increasing sequences (n_k) and (m_k) in $\mathbb{N}$ satisfying $n_k < m_k$, $k \in \mathbb{N}$.

$$\delta(\mathfrak{x}_{n_k}, \mathfrak{x}_{m_k}) \geq \xi \quad \text{and} \tag{17.7}$$

$$\delta(\mathfrak{x}_{n_k}, \mathfrak{x}_{m_k-1}) < \xi. \tag{17.8}$$

Let k_0 such that $\delta(\mathfrak{x}_{n_k}, \mathfrak{x}_{n_k+1}) < \xi$ for $k \geq k_0$.

Recall that,

$$\lim_{k \longrightarrow \infty} \delta(\mathfrak{x}_{n_k}, \mathfrak{x}_{m_k}) = \xi. \tag{17.9}$$

Also,

$$\delta(\mathfrak{x}_{n_k+1}, \mathfrak{x}_{m_k+1}) \leq \delta(\mathfrak{x}_{n_k}, \mathfrak{x}_{m_k}) - \delta(\mathfrak{x}_{n_k+1}, \mathfrak{x}_{n_k}) - \delta(\mathfrak{x}_{m_k}, \mathfrak{x}_{m_k+1}). \tag{17.10}$$

So,

$$\lim_{k \longrightarrow \infty} \delta(\mathfrak{x}_{n_k+1}, \mathfrak{x}_{m_k+1}) = \xi. \tag{17.11}$$

Therefore,

$$\begin{aligned} F(\xi) &= \lim_{k \longrightarrow \infty} F(\delta(\mathfrak{x}_{n_k+1}, \mathfrak{x}_{m_k+1})) \\ &\leq \sigma_1 \lim_{k \longrightarrow \infty} F(\delta(\mathfrak{x}_{n_k}, \mathfrak{x}_{n_k+1})) + \sigma_2 \lim_{k \longrightarrow \infty} F(\delta(\mathfrak{x}_{m_k}, \mathfrak{x}_{m_k+1})) - \varphi \\ &= -\infty. \end{aligned}$$

That is, $F(\xi) \leq -\infty$, which contradicts our assumption. As a result, $\mathfrak{x}_n \subseteq \mathcal{M}$ forms a Cauchy sequence in a complete metric space, therefore converges. Hence, we have $\lim_{n \longrightarrow \infty} \mathfrak{x}_n = \mathfrak{x}^\circ \in \mathcal{M}$.

Now,

$$\begin{aligned} F(\delta(\mathfrak{x}_{n+1}, \mathfrak{J}\mathfrak{x}^\circ)) &= F(\delta(\mathfrak{J}\mathfrak{x}_n, \mathfrak{J}\mathfrak{x}^\circ)) \\ &\leq \sigma_1 F(\delta(\mathfrak{x}_{n+1}, \mathfrak{x}_n)) + \sigma_2 F(\delta(\mathfrak{J}\mathfrak{x}^\circ, \mathfrak{x}^\circ)). \end{aligned} \tag{17.12}$$

By taking the limit ($n \rightarrow \infty$) in both sides of (17.12) and using (17.5), we get $\lim_{n \longrightarrow \infty} F(\delta(\mathfrak{x}_{n+1}, \mathfrak{J}\mathfrak{x}^\circ)) = -\infty$, thus from the condition (F_2), we get

$$\lim_{n \longrightarrow \infty} \delta(\mathfrak{J}\mathfrak{x}_n, \mathfrak{J}\mathfrak{x}^\circ) = 0. \tag{17.13}$$

Using $\lim_{n \longrightarrow \infty} \mathfrak{x}_n = \mathfrak{x}^\circ$ and (17.13), it follows that $\mathfrak{J}\mathfrak{x}^\circ = \mathfrak{x}^\circ$.

(*ii*) Assume, $\inf F \in \mathbb{R}$.

If for some $\mathfrak{x} \in \mathcal{M}$ and $n \in \mathbb{N}$, it holds that $\mathfrak{J}^n \mathfrak{x} = \mathfrak{J}^{n+1}\mathfrak{x}$, then $\mathfrak{J}^n \mathfrak{x}$ remains unchanged under the mapping $\mathfrak{J}$. Assume that such n does not exist and we denote

$$b = \inf F(\delta(\mathfrak{J}^n \mathfrak{x}, \mathfrak{J}^{n+1}\mathfrak{x})) \in \mathbb{R}. \tag{17.14}$$

Let $n_0 \in \mathbb{N}$ such that $\varphi > 1/n_0$ and observe the sequence $(\mathfrak{x}_m)$ such that

$$F(\delta(\mathfrak{J}^{n_m} \mathfrak{x}_m, \mathfrak{J}^{n_m+1} \mathfrak{x}_m)) < b + 1/m \leq b + 1/n_0, \quad m \geq n_0. \tag{17.15}$$

Moreover,

$$F(\delta(\mathfrak{J}^n \mathfrak{x}_m, \mathfrak{J}^{n+1}\mathfrak{x}_m)) \leq b + 1/n_0, \quad n \geq n_m, m \geq n_0. \tag{17.16}$$

Since $\mathfrak{J}^n \mathfrak{x}_m \neq \mathfrak{J}^{n+1}\mathfrak{x}_m$, $n \geq n_m$, $m \geq n_0$, we have

$$\begin{aligned} b &\leq F(\delta(\mathfrak{J}^n \mathfrak{x}_m, \mathfrak{J}^{n+1}\mathfrak{x}_m)) \\ &\leq \sigma_1 F(\delta(\mathfrak{J}^n \mathfrak{x}_m, \mathfrak{J}^{n-1}\mathfrak{x}_m)) + \sigma_2 F(\delta(\mathfrak{J}^{n+1}\mathfrak{x}_m, \mathfrak{J}^{n}\mathfrak{x}_m)) - \varphi \\ &\leq \sigma_1 F(\delta(\mathfrak{x}_{m+n}, \mathfrak{x}_{m+n-1})) + \sigma_2 F(\delta(\mathfrak{x}_{m+n+1}, \mathfrak{x}_{m+n})) - \varphi. \end{aligned} \tag{17.17}$$

By taking the limit in (17.17) ($n \longrightarrow \infty$) and using (17.5), we have $b \leq -\infty$, which is a contradiction.

As a result, in both cases, we confirm the presence of a fixed point for the mapping $\mathfrak{J}$.

Finally, we prove the uniqueness. Suppose, if possible, we consider two fixed points $\mathfrak{x}^\circ$ and $\mathfrak{y}^\circ$ for which $\mathfrak{x}^\circ \neq \mathfrak{y}^\circ$. Then $\mathfrak{J}\mathfrak{x}^\circ = \mathfrak{x}^\circ \neq \mathfrak{y}^\circ = \mathfrak{J}\mathfrak{y}^\circ$.

Thus, we have $\varphi + \delta(\mathfrak{J}\mathfrak{x}^\circ, \mathfrak{J}\mathfrak{y}^\circ) \leq \sigma_1 F(\delta(\mathfrak{x}^\circ, \mathfrak{J}\mathfrak{x}^\circ)) + \sigma_2 F(\delta(\mathfrak{y}^\circ, \mathfrak{J}\mathfrak{y}^\circ)) = 0$, a contradiction. Hence, $\mathfrak{x}^\circ = \mathfrak{y}^\circ$. □

Next, to establish the non-triviality of our main result, we provide an example to show that there are instances where our extended version of Kannan's theorem is applicable, but neither Kannan's theorem nor that of Wardowski is appropriate.

Example 17.1. Let $\mathcal{M} = \{p, q, r\}$ having the metric

$$\delta(\mathfrak{x}, \mathfrak{y}) = \begin{cases} 0, & \mathfrak{x} = \mathfrak{y}, \\ 4/3, & (\mathfrak{x}, \mathfrak{y}) = (q, r) \text{ or } (\mathfrak{x}, \mathfrak{y}) = (r, q), \\ 1, & \text{otherwise.} \end{cases}$$

Obviously, $(\mathcal{M}, \delta)$ is complete. We introduce $\mathfrak{J} : \mathcal{M} \to \mathcal{M}$ by

$$\mathfrak{J}\mathfrak{x} = \begin{cases} p, & \mathfrak{x} = p, q, \\ q, & \mathfrak{x} = r. \end{cases}$$

Let, $\mathfrak{x} = p$, $\mathfrak{y} = r$. Then $F(\delta(\mathfrak{J}\mathfrak{x}, \mathfrak{J}\mathfrak{y})) = F(\delta(p, q)) = F(1)$ and $F(\delta(\mathfrak{x}, \mathfrak{y})) = F(\delta(p, r)) = F(1)$. Here, it is not possible to find $\varphi > 0$ so that the inequality $\varphi + F(\delta(\mathfrak{J}\mathfrak{x}, \mathfrak{J}\mathfrak{y})) \leq F(\delta(\mathfrak{x}, \mathfrak{y}))$ can be satisfied. Therefore, Wardowski's theorem is not applicable.

Moreover, by setting $\mathfrak{x} = p$, $\mathfrak{y} = r$, if the condition $\delta(\mathfrak{J}\mathfrak{x}, \mathfrak{J}\mathfrak{y}) \leq \lambda\{\delta(\mathfrak{x}, \mathfrak{J}\mathfrak{x}) + \delta(\mathfrak{y}, \mathfrak{J}\mathfrak{y})\}$ holds, this implies that $1 \leq \lambda \cdot 4/3$, which simplifies to $\lambda \geq 3/4 > 1/2$. Consequently, Kannan's theorem does not apply in this case either.

Now assume that $\mathfrak{x}, \mathfrak{y} \in \mathcal{M} \backslash Fix(\mathfrak{J})$ with $\mathfrak{J}\mathfrak{x} \neq \mathfrak{J}\mathfrak{y}$. Then $(\mathfrak{x}, \mathfrak{y}) = (q, r)$ or $(\mathfrak{x}, \mathfrak{y}) = (r, q)$. Without the loss of generality, consider $(\mathfrak{x}, \mathfrak{y}) = (q, r)$ and choose $\sigma_1 = \frac{1}{5}, \sigma_2 = \frac{1}{2}, \varphi = \frac{1}{3} \ln \frac{4}{3}$ and $F(\mathfrak{y}) = \ln \mathfrak{y}$ for $\mathfrak{y} > 0$.

Thus, we have

$$\begin{aligned}\varphi + F(\delta(\mathfrak{J}\mathfrak{x},\mathfrak{J}\mathfrak{y})) &= \frac{1}{3}\ln\frac{4}{3} + F(\delta(\mathfrak{J}q,\mathfrak{J}r))\\ &= \frac{1}{3}\ln\frac{4}{3} + F(\delta(p,q))\\ &= \frac{1}{3}\ln\frac{4}{3} + F(1)\\ &= \frac{1}{3}\ln\frac{4}{3} + \ln(1)\\ &= \frac{1}{3}\ln\frac{4}{3}.\end{aligned}$$

On the other hand,

$$\begin{aligned}\sigma_1 F(\delta(\mathfrak{x},\mathfrak{J}\mathfrak{x})) + \sigma_2 F(\delta(\mathfrak{y},\mathfrak{J}\mathfrak{y})) &= \frac{1}{5}F(\delta(q,\mathfrak{J}q)) + \frac{1}{2}F(\delta(r,\mathfrak{J}r))\\ &= \frac{1}{5}F(\delta(q,p)) + \frac{1}{2}F(\delta(r,q))\\ &= \frac{1}{5}F(1) + \frac{1}{2}F\left(\frac{4}{3}\right)\\ &= \frac{1}{5}\ln(1) + \frac{1}{2}\ln\left(\frac{4}{3}\right)\\ &= \frac{1}{2}\ln\left(\frac{4}{3}\right).\end{aligned}$$

Hence, the condition $\varphi + F(\delta(\mathfrak{J}\mathfrak{x},\mathfrak{J}\mathfrak{y})) \le \sigma_1 F(\delta(\mathfrak{x},\mathfrak{J}\mathfrak{x})) + \sigma_2 F(\delta(\mathfrak{y},\mathfrak{J}\mathfrak{y}))$ is satisfied for all $\mathfrak{x},\mathfrak{y} \in \mathcal{M}\backslash Fix(\mathfrak{J})$ with $\mathfrak{J}\mathfrak{x} \neq \mathfrak{J}\mathfrak{y}$ and using the values that were selected above for σ_1, σ_2, φ and F. It is easy to see that $\mathfrak{J}$ admits a unique fixed point $\mathfrak{x} = p$.

17.3 Reich-type simple F-contraction

This section presents the concept of generalized Reich-type sF-contraction (GRSFC).

Definition 17.6. Let $(\mathcal{M},\delta)$ be a complete metric space. A self map $\mathfrak{J} : \mathcal{M} \to \mathcal{M}$ is said to be a *GRSFC* if there exists $\sigma_1,\sigma_2,\sigma_3 \in (0,1)$ with $\sigma_1+\sigma_2+\sigma_3 < 1$, $\varphi > 0$ and $F \in \mathcal{F}^{\nearrow}$ such that

$$\varphi + F(\delta(\mathfrak{J}\mathfrak{x},\mathfrak{J}\mathfrak{y})) \le \sigma_1 F(\delta(\mathfrak{x},\mathfrak{J}\mathfrak{x})) + \sigma_2 F(\delta(\mathfrak{y},\mathfrak{J}\mathfrak{y})) + \sigma_3 F(\delta(\mathfrak{x},\mathfrak{y})) \tag{17.18}$$

for all $\mathfrak{x},\mathfrak{y} \in \mathcal{M}$ with $\mathfrak{x},\mathfrak{y} \notin$ Fix $(\mathfrak{J})$ with $\mathfrak{J}\mathfrak{x} \neq \mathfrak{J}\mathfrak{y}$.

Theorem 17.6. *Let $(\mathcal{M},\delta)$ be a complete MS. A GRSFC, $\mathfrak{J} : \mathcal{M} \to \mathcal{M}$ admits an unique fixed point.*

Proof. As in proof of Theorem 17.5, we produce a sequence $\{\mathfrak{x}_n\}$.

Putting $\mathfrak{x} = \mathfrak{x}_n$ and $\mathfrak{y} = \mathfrak{x}_{n-1}$ in (17.18), we have

$$\begin{aligned}\varphi + F(\delta(\mathfrak{x}_{n+1}, \mathfrak{x}_n)) &= \varphi + F(\delta(\mathfrak{J}_{\mathfrak{x}_n}, \mathfrak{J}_{\mathfrak{x}_{n-1}}))\\ &\le \sigma_1 F(\delta(\mathfrak{x}_n, \mathfrak{J}\mathfrak{x}_n)) + \sigma_2 F(\delta(\mathfrak{x}_{n-1}, \mathfrak{J}\mathfrak{x}_{n-1})) + \sigma_3 F(\delta(\mathfrak{x}_n, \mathfrak{x}_{n-1}))\\ &= \sigma_1 F(\delta(\mathfrak{x}_n, \mathfrak{x}_{n+1})) + \sigma_2 F(\delta(\mathfrak{x}_{n-1}, \mathfrak{x}_n)) + \sigma_3 F(\delta(\mathfrak{x}_n, \mathfrak{x}_{n-1}))\\ &< \sigma_1 F(\delta(\mathfrak{x}_n, \mathfrak{x}_{n+1})) + \sigma_2 F(\delta(\mathfrak{x}_{n-1}, \mathfrak{x}_n))\\ &\quad + (1 - \sigma_1 - \sigma_2) F(\delta(\mathfrak{x}_n, \mathfrak{x}_{n-1})) \quad (\text{since } \sigma_3 < 1 - \sigma_1 - \sigma_2). \end{aligned} \tag{17.19}$$

Proceeding analogously to the proof of [25, Theorem 2.2], we obtain

$$F(\delta(\mathfrak{x}_n, \mathfrak{x}_{n+1})) \le F(\delta(\mathfrak{x}_{n-1}, \mathfrak{x}_n)) - \varphi \le \cdots \le F(\delta(\mathfrak{x}_0, \mathfrak{x}_1)) - n\varphi. \tag{17.20}$$

And, by taking limit as $n \longrightarrow \infty$ in (17.20), we obtain

$$\lim_{n \longrightarrow \infty} F(\delta(\mathfrak{x}_n, \mathfrak{x}_{n+1})) = -\infty. \tag{17.21}$$

Next, we will divide the proof into two cases, identical to the procedure applied in the proof of Theorem 17.5. By applying the same reasoning to first case, we arrive at the following contradiction:

$$\begin{aligned} F(\xi) &= \lim_{k \longrightarrow \infty} F(\delta(\mathfrak{x}_{n_k+1}, \mathfrak{x}_{m_k+1}))\\ &\le \sigma_1 \lim_{k \longrightarrow \infty} F(\delta(\mathfrak{x}_{n_k}, \mathfrak{x}_{n_k+1})) + \sigma_2 \lim_{k \longrightarrow \infty} F(\delta(\mathfrak{x}_{m_k}, \mathfrak{x}_{m_k+1})) + \sigma_3 \lim_{k \longrightarrow \infty} F(\delta(\mathfrak{x}_{n_k}, \mathfrak{x}_{m_k}))\\ &= -\infty, \end{aligned}$$

which is not possible, followed by establishing that a fixed point exists in a similar way.

Using a similar approach, in the second case, we arrive at a contradiction as follows:

$$\begin{aligned} b &\le F(\delta(\mathfrak{J}^n \mathfrak{x}_m, \mathfrak{J}^{n+1} \mathfrak{x}_m))\\ &\le \sigma_1 F(\delta(\mathfrak{J}^n \mathfrak{x}_m, \mathfrak{J}^{n-1} \mathfrak{x}_m)) + \sigma_2 F(\delta(\mathfrak{J}^{n+1} \mathfrak{x}_m, \mathfrak{J}^n \mathfrak{x}_m)) + \sigma_3 F(\delta(\mathfrak{J}^{n-1} \mathfrak{x}_m, \mathfrak{J}^n \mathfrak{x}_m)) - \varphi\\ &\le \sigma_1 F(\delta(\mathfrak{x}_{m+n}, \mathfrak{x}_{m+n-1})) + \sigma_2 F(\delta(\mathfrak{x}_{m+n+1}, \mathfrak{x}_{m+n})) + \sigma_3 F(\delta(\mathfrak{x}_{m+n-1}, \mathfrak{x}_{m+n})) - \varphi. \end{aligned}$$

And, as $n \to \infty$, we get $b \le -\infty$, which is our contradiction.

Finally, we examine the uniqueness aspect. Assume $\mathfrak{x}^*$ and $\mathfrak{y}^*$ be two fixed points such that $\mathfrak{x}^* = \mathfrak{y}^*$. Then, $\mathfrak{J}\mathfrak{x}^* = \mathfrak{x}^* \neq \mathfrak{y}^* = \mathfrak{J}\mathfrak{y}^*$. Thus, we have

$$\varphi + F(\delta(\mathfrak{J}\mathfrak{x}^*, \mathfrak{J}\mathfrak{y}^*)) \le \sigma_1 F(\delta(\mathfrak{x}^*, \mathfrak{J}\mathfrak{x}^*)) + \sigma_2 F(\delta(\mathfrak{y}^*, \mathfrak{J}\mathfrak{y}^*)) + \sigma_3 F(\delta(\mathfrak{x}^*, \mathfrak{y}^*)) = \sigma_3 F(\delta(\mathfrak{x}^*, \mathfrak{y}^*)),$$

which is a contradiction, since $\varphi > 0$ and $0 < \sigma_3 < 1$. Hence, $\mathfrak{x}^* = \mathfrak{y}^*$. □

17.4 Application in integral equations

In this section, by applying Theorem 17.5, we prove the existence of solutions for integral equations.

Consider $\mathcal{H} = [0, l]$, with $l > 0$, and the function space $\Lambda = C(\mathcal{H}, \mathbb{R})$, which includes all continuous mapping $\mathcal{H}$ into $\mathbb{R}$. The distance between any two functions in Λ is given by the metric:

$$\delta(\mathfrak{g}, \mathfrak{h}) = \sup_{t \in \mathcal{H}} |\mathfrak{g}(t) - \mathfrak{h}(t)| = \|\mathfrak{g} - \mathfrak{h}\|. \tag{17.22}$$

Now, we examine the following integral equation:

$$\mathfrak{g}(\mathfrak{x}) = \mathfrak{q}(\mathfrak{x}) + \xi \int_0^l M(\mathfrak{x}, z) \arctan(\mathfrak{g}(\mathfrak{z}))d\mathfrak{z}, \tag{17.23}$$

where $\mathfrak{x} \in [0, l]$ and $\xi > 0$ is a parameter.

Here,

(i) $\mathfrak{q} : \mathcal{H} \longrightarrow \mathbb{R}$ is continuous and $\arctan : \mathbb{R} \longrightarrow \mathbb{R}$ is Lipschitz continuous with constant $\mathcal{L}$.

(ii) $M : \mathcal{H} \times \mathcal{H} \longrightarrow \mathbb{R}$ is measurable and continuous at $\mathfrak{z} \in \mathcal{H}$ for all $\mathfrak{x} \in \mathcal{H}$.

(iii) $M(\mathfrak{x}, \mathfrak{z}) \geq 0$ for all $\mathfrak{x}, \mathfrak{z} \in \mathcal{H}$.

Theorem 17.7. *Assume that (i) – (iii) hold. Given there exists $\varphi > 0$ and σ_1, σ_2 from $(0, 1)$ with $\sigma_1 + \sigma_2 < 1$, such that:*

$$\int_0^l M(\mathfrak{x}, \mathfrak{z})d\mathfrak{z} \leq C\left(\left\|\mathfrak{g} - \int_0^l M(\mathfrak{x}, \mathfrak{z}) \arctan(\mathfrak{g}(\mathfrak{z}))d\mathfrak{z}\right\|\right)^{\sigma_1} \cdot \left(\left\|\mathfrak{h} - \int_0^l M(\mathfrak{x}, \mathfrak{z}) \arctan(\mathfrak{h}(\mathfrak{z}))d\mathfrak{z}\right\|\right)^{\sigma_2} \tag{17.24}$$

for each $\mathfrak{x} \in \mathcal{H}$ and for all $\mathfrak{g}, \mathfrak{h} \in \Lambda$ such that:

$$C = (\xi \mathcal{L} \|\mathfrak{g} - \mathfrak{h}\| e^{\varphi})^{-1},$$

$$\mathfrak{g}(\mathfrak{x}) \neq \int_0^l M(\mathfrak{x}, \mathfrak{z}) \arctan(\mathfrak{g}(\mathfrak{z}))d\mathfrak{z},$$

$$\mathfrak{h}(\mathfrak{x}) \neq \int_0^l M(\mathfrak{x}, \mathfrak{z}) \arctan(\mathfrak{h}(\mathfrak{z}))d\mathfrak{z},$$

$$\mathfrak{g}(\mathfrak{x}) \neq \mathfrak{h}(\mathfrak{x}) \quad \textit{for all } \mathfrak{x} \in \mathcal{H},$$

then a solution to the integral equation (17.23) exists in Λ.

Proof. Consider a self-map $\mathfrak{J} : \Lambda \longrightarrow \Lambda$ defined by:

$$\mathfrak{J}(\mathfrak{g}(\mathfrak{r})) = \mathfrak{q}(\mathfrak{r}) + \xi \int_0^l M(\mathfrak{r}, \mathfrak{z}) \arctan(\mathfrak{g}(\mathfrak{z}))d\mathfrak{z}, \quad \mathfrak{r} \in [0, l]. \tag{17.25}$$

We obtain for every $\mathfrak{r} \in [0, l]$:

$$\begin{aligned} |\mathfrak{J}(\mathfrak{g}(\mathfrak{r})) - \mathfrak{J}(\mathfrak{h}(\mathfrak{r}))| &= \xi \left| \int_0^l M(\mathfrak{r}, \mathfrak{z})\{\arctan(\mathfrak{g}(\mathfrak{z})) - \arctan(\mathfrak{h}(\mathfrak{z}))\}d\mathfrak{z} \right| \\ &\leq \xi \int_0^l M(\mathfrak{r}, \mathfrak{z})|\arctan(\mathfrak{g}(\mathfrak{z})) - \arctan(\mathfrak{h}(\mathfrak{z}))|d\mathfrak{z}. \end{aligned}$$

Using the Lipschitz condition:

$$\begin{aligned} |\mathfrak{J}(\mathfrak{g}(\mathfrak{r})) - \mathfrak{J}(\mathfrak{h}(\mathfrak{r}))| &\leq \xi\mathcal{L} \int_0^l M(\mathfrak{r}, \mathfrak{z})|\mathfrak{g}(\mathfrak{z}) - (\mathfrak{h}(\mathfrak{z})|d\mathfrak{z} \\ &\leq \xi\mathcal{L}\|\mathfrak{g} - \mathfrak{h}\| \int_0^l M(\mathfrak{r}, \mathfrak{z})d\mathfrak{z} \\ &\leq (\xi\mathcal{L}\|\mathfrak{g} - \mathfrak{h}\|)(\xi\mathcal{L}\|\mathfrak{g} - \mathfrak{h}\|)^{-1} e^{-\varphi}(\|\mathfrak{g} - \mathfrak{J}\mathfrak{g}\|)^{\sigma_1}(\|\mathfrak{h} - \mathfrak{J}\mathfrak{h}\|)^{\sigma_2} \\ &\leq e^{-\varphi}(\|\mathfrak{g} - \mathfrak{J}\mathfrak{g}\|)^{\sigma_1}(\|\mathfrak{h} - \mathfrak{J}\mathfrak{h}\|)^{\sigma_2}. \end{aligned} \tag{17.26}$$

Applying the supremum over $\mathfrak{r} \in \mathcal{H}$ in (17.26), we get:

$$\begin{aligned} \delta(\mathfrak{J}\mathfrak{g}, \mathfrak{J}\mathfrak{h}) = \|\mathfrak{J}\mathfrak{g} - \mathfrak{J}\mathfrak{h}\| &\leq e^{-\varphi}\|\mathfrak{g} - \mathfrak{J}\mathfrak{g}\|^{\sigma_1} \cdot \|\mathfrak{h} - \mathfrak{J}\mathfrak{h}\|^{\sigma_2} \\ &\leq e^{-\varphi}(\delta(\mathfrak{g}, \mathfrak{J}\mathfrak{g}))^{\sigma_1} \cdot (\delta(\mathfrak{h}, \mathfrak{J}\mathfrak{h}))^{\sigma_2}. \end{aligned}$$

Therefore,

$$\begin{aligned} &\frac{\delta(\mathfrak{J}\mathfrak{g}, \mathfrak{J}\mathfrak{h})}{(\delta(\mathfrak{g}, \mathfrak{J}\mathfrak{g}))^{\sigma_1} \cdot (\delta(\mathfrak{h}, \mathfrak{J}\mathfrak{h}))^{\sigma_2}} \leq e^{-\varphi} \\ &\Rightarrow \ln\left[\frac{\delta(\mathfrak{J}\mathfrak{g}, \mathfrak{J}\mathfrak{h})}{(\delta(\mathfrak{g}, \mathfrak{J}\mathfrak{g}))^{\sigma_1} \cdot (\delta(\mathfrak{h}, \mathfrak{J}\mathfrak{h}))^{\sigma_2}}\right] \leq -\varphi \\ &\Rightarrow \ln[\delta(\mathfrak{J}\mathfrak{g}, \mathfrak{J}\mathfrak{h})] - \ln[(\delta(\mathfrak{g}, \mathfrak{J}\mathfrak{g}))^{\sigma_1} \cdot (\delta(\mathfrak{h}, \mathfrak{J}\mathfrak{h}))^{\sigma_2}] \leq -\varphi \\ &\Rightarrow \varphi + \ln[\delta(\mathfrak{J}\mathfrak{g}, \mathfrak{J}\mathfrak{h})] \leq \ln[(\delta(\mathfrak{g}, \mathfrak{J}\mathfrak{g}))^{\sigma_1} \cdot (\delta(\mathfrak{h}, \mathfrak{J}\mathfrak{h}))^{\sigma_2}] \\ &\Rightarrow \varphi + \ln[\delta(\mathfrak{J}\mathfrak{g}, \mathfrak{J}\mathfrak{h})] \leq \sigma_1 \ln[(\delta(\mathfrak{g}, \mathfrak{J}\mathfrak{g}))] + \sigma_2 \ln[(\delta(\mathfrak{h}, \mathfrak{J}\mathfrak{h}))]. \end{aligned} \tag{17.27}$$

Setting $F(\mathfrak{y}) = \ln(\mathfrak{y})$ for $\mathfrak{y} > 0$, we rewrite (17.27) as:

$$\varphi + F(\delta(\mathfrak{J}\mathfrak{g}, \mathfrak{J}\mathfrak{h})) \leq \sigma_1 F(\delta(\mathfrak{g}, \mathfrak{J}\mathfrak{g})) + \sigma_2 F(\delta(\mathfrak{h}, \mathfrak{J}\mathfrak{h})).$$

Since $\mathfrak{g}, \mathfrak{h} \in \Lambda \setminus Fix(\mathfrak{J})$ with $\mathfrak{J}\mathfrak{g} \neq \mathfrak{J}\mathfrak{h}$, it follows by Theorem 17.5 that $\mathfrak{J}$ has a unique fixed point. Consequently, the solution to the integral equation (17.23) is unique. □

17.5 Conclusions and future work

(i) We introduced generalized Kannan-type simple F-contraction (GKSFC), which extends Kannan's fixed-point theorem by incorporating the idea of simple F-contractions, thereby providing a more flexible framework for establishing fixed point results.
(ii) We introduced generalized Reich-type simple F-contraction, where by selecting appropriate values for the coefficients, GRSFC reduces to GKSFC, showing that our formulation connects and generalizes existing results while preserving their fundamental properties.
(iii) To highlight the validity and relevance of the main result, an example has been provided. The proposed example clarifies the cases where neither Kannan's theorem nor Wardowski's theorem is applicable.
(iv) Additionally, we have explored in this study how the GKSFC can be used to solve a certain class of integral equations. This depicts how fixed-point theory is an effective method for solving different kinds of integral problems. Our study on this section is backed by knowledge from earlier studies [15, 1], etc.
(v) Similar to the discussion in Section 4 of [15] and in [1], by using this framework to analyze nonlinear models of real-world issues, including the spread of infectious diseases, economic growth, ecological systems, etc., future research may investigate the improvisation of integral equation-based solution approaches.
(vi) Further investigations can be conducted on the metric completeness properties within this setting.

Bibliography

[1] Alam, K. H., Rohen, Y. and Tomar, A., *On fixed point and its application to the spread of infectious diseases model in Mvb-metric space*, Math. Methods Appl. Sci. 47(7) (2024), 6489–6503.
[2] Altun, I., Minak, G. and Dag, H., *Multivalued F-contractions on complete metric spaces*, J. Nonlinear Convex Anal. 16(4) (2015), 659–666.
[3] Aydi, H., Chen, C. and Karapınar, E., *Interpolative Ćirić–Reich–Rus-type contractions via the Branciari distance*, Mathematics 7(1) (2019), 84.
[4] Aydi, H., Karapinar, E. and Roldán López de Hierro, A. F., *ω-Interpolative Ćirić–Reich–Rus-type contractions*, Mathematics 7(1) (2019), 57.
[5] Bojor, F., *Fixed-point theorems for Reich type contractions on metric spaces with a graph*, Nonlinear Anal., Theory Methods Appl. 75(9) (2012), 3895–3901.
[6] Bojor, F., *Fixed points of Kannan mappings in metric spaces endowed with a graph*, An. Ştiinţ. Univ. 'Ovidius' Constanţa 20(1) (2012), 31–40.

[7] Chatterjea, S. K., *Fixed-point theorems*, C. R. Acad. Bulgare Sci. 25(6) (1972), 727–730.

[8] Choudhury, B. S. and Kundu, A., *A Kannan-like contraction in partially ordered spaces*, Demonstr. Math. 46(2) (2013), 327–334.

[9] Cvetković, M., *Behind the concept of F-contraction*, Fixed Point Theory 25(2) (2024), 519–530.

[10] Debnath, P., *Optimization through best proximity points for multivalued F-contractions*, Miskolc Math. Notes 22(1) (2021), 143–151.

[11] Debnath, P., *Set-valued Meir–Keeler, Geraghty and Edelstein type fixed point results in b-metric spaces*, Rend. Circ. Mat. Palermo (2) 70(3) (2021), 1389–1398.

[12] Debnath, P., *Banach, Kannan, Chatterjea, and Reich-type contractive inequalities for multivalued mappings and their common fixed points*, Math. Methods Appl. Sci. 45(3) (2022), 1587–1596.

[13] Debnath, P., *New common fixed-point theorems for Gornicki-type mappings and enriched contractions*, São Paulo J. Math. Sci. 16(2) (2022), 1401–1408.

[14] Debnath, P., *Common fixed-point and fixed-circle results for a class of discontinuous F-contractive mappings*, Mathematics 10(9) (2022), 1605.

[15] Debnath, P., *A new extension of Kannan's fixed-point theorem via F-contraction with application to integral equations*, Asian-Eur. J. Math. 15(07) (2022), 2250123.

[16] Debnath, P., *Results on discontinuity at fixed point for a new class of contractive mappings*, Sahand Commun. Math. Anal. 20(4) (2023), 21–32.

[17] Debnath, P., *Best proximity points of multivalued Geraghty contractions*, Miskolc Math. Notes 24(1) (2023), 119–127.

[18] Debnath, P., *On some open questions about multivalued enriched interpolative Kannan and Ciric–Reich–Rus type contractions in a Banach space*, Fixed Point Theory 26(2) (2025), 465–476.

[19] Debnath, P., Mitrović, Z. and Radenović, S., *Interpolative Hardy–Rogers and Reich–Rus–Ćirić type contractions in b-metric spaces and rectangular b-metric spaces*, Mat. Vesn. 72 (2020), 368–374.

[20] Debnath, P., Neog, M. and Radenović, S., *Set valued Reich type G-contractions in a complete metric space with graph*, Rend. Circ. Mat. Palermo (2) 69 (2020), 917–924.

[21] Debnath, P. and Srivastava, H. M., *New extensions of Kannan's and Reich's fixed-point theorems for multivalued maps using Wardowski's technique with application to integral equations*, Symmetry 12(7) (2020), 1090.

[22] Górnicki, J., *Fixed-point theorems for Kannan type mappings*, J. Fixed Point Theory Appl. 19 (2017), 1–8.

[23] Kannan, R., *Some results on fixed points*, Bull. Calcutta Math. Soc. 60(1) (1968), 71–77.

[24] Kannan, R., *Some results on fixed points—II*, Am. Math. Mon. 76(4) (1969), 405–408.

[25] Mohammadi, B., Parvaneh, V. and Aydi, H., *On extended interpolative Ćirić–Reich–Rus type F-contractions and an application*, J. Inequal. Appl. 2019 (2019), 290.

[26] Reich, S., *Kannan's fixed-point theorem*, Boll. Unione Mat. Ital. 4(4) (1971), 1–11.

[27] Reich, S., *Some remarks concerning contraction mappings*, Can. Math. Bull. 14(1) (1971), 121–124.

[28] Reich, S., *Fixed points of contractive functions*, Boll. Unione Mat. Ital. 5(4) (1972), 26–42.

[29] Rus, I. A., Petruşel, A. and Petruşel, G., *Fixed Point Theory*, Cluj Univ. Press, 2008.

[30] Wardowski, D., *Fixed points of a new type of contractive mappings in complete metric spaces*, Fixed Point Theory Appl. 2012 (2012), 1–6.

Index

https://doi.org/10.1515/9783112215760-018

FSC
www.fsc.org
MIX
Papier | Fördert gute Waldnutzung
FSC® C083411

Kolibri 360 GmbH
Ferdinand-Jühlke-Straße 7
99095 Erfurt, Deutschland
produktsicherheit@kolibri360.de